Methods in Molecular Biology™

Series Editor
John M. Walker
School of Life Sciences
University of Hertfordshire
Hatfield, Hertfordshire, AL10 9AB, UK

For further volumes:
http://www.springer.com/series/7651

Plant Transposable Elements

Methods and Protocols

Edited by

Thomas Peterson

Department of Agronomy, Department of Genetics, Development and Cell Biology, Iowa State University, Ames, IA, USA

Editor
Thomas Peterson
Department of Agronomy
Department of Genetics,
Development and Cell Biology
Iowa State University
Ames, IA, USA

ISSN 1064-3745 ISSN 1940-6029 (electronic)
ISBN 978-1-62703-567-5 ISBN 978-1-62703-568-2 (eBook)
DOI 10.1007/978-1-62703-568-2
Springer New York Heidelberg Dordrecht London

Library of Congress Control Number: 2013942868

Printed on acid-free paper

Humana Press is a brand of Springer
Springer is part of Springer Science+Business Media (www.springer.com)

Dedication

This volume is dedicated to Tom Gerats, patron and advocate of the petunia *dTph1* system, for his generosity, spirit, and enthusiasm for science and philosophy.

Preface

Methods in Molecular Biology: Plant Transposable Elements

Transposable elements have played a major role in shaping plant genome structure and gene expression. Transposons not only drive sequence expansion, induce mutations, and generate chromosome rearrangements, they also help to shape the epigenetic topology of the eukaryotic genome. This volume contains a remarkable collection of chapters authored by recognized authorities in transposon biology. The subjects covered here reflect the inherent diversity of transposable elements, and include reviews of major transposon systems, with a focus on the structures and activities of the major Class II (DNA) elements. Included are insights on how to distinguish transposon-based variegation from other causes, how to use transposons for clonal analysis of development, and what genetic markers are favored for use as markers of transposon activity. As transposons can be extremely useful as experimental tools, additional chapters provide detailed descriptions and methods for using transposable elements for gene isolation and activation tagging. Finally, several chapters describe how computational methods can be applied to study the ancient transposon remnants that comprise the bulk of plant genomes, as well as how to identify recent and ongoing transposition events. Overall, this volume contains the background information and specific experimental protocols needed by students and scientists interested in understanding and applying the unique properties of transposable elements in their own research.

Ames, IA, USA *Thomas Peterson*

Acknowledgements

I would like to thank Melissa Lang for editorial assistance. I am grateful to the many friends and colleagues who have contributed herein their personal insights and experience for the benefit of readers interested in understanding and working with transposable elements.

Contents

Contributors

WAYNE T. AVIGNE • *Plant Molecular and Cellular Biology Program, Horticultural Sciences Department, University of Florida, Gainesville, FL, USA*
PHILIP W. BECRAFT • *Department of Genetics, Development and Cell Biology, Iowa State University, Ames, IA, USA; Department of Agronomy, Iowa State University, Ames, IA, USA*
CRISTIAN CHAPARRO • *Laboratoire Génome et développement des plantes, Université de Perpignan Via Domitia, Perpignan, Cedex, France*
JOSEPH COLLINS • *Plant Molecular and Cellular Biology, Horticultural Sciences Department, University of Florida, Gainesville, FL, USA*
ANNE DIÉVART • *CIRAD, UMR AGAP, Montpellier cedex 5, France*
HUGO K. DOONER • *Waksman Institute, Rutgers University, Piscataway, NJ, USA*
GAËTAN DROC • *CIRAD, UMR AGAP, Montpellier cedex 5, France*
MY-LINH DOLL • *Institute of Biology/Applied Genetics, Freie Universitat Berlin, Berlin, Germany*
MOAINE ELBAIDOURI • *Laboratoire Génome et développement des plantes, Université de Perpignan Via Domitia, Perpignan, Cedex, France*
MOO YOUNG EUN • *Genomics Division, Gene Function Lab, Department of Agricultural Biotechnology , National Academy of Agricultural Science, RDA, Suwon, Republic of Korea*
NINA V. FEDOROFF • *Huck Institutes of the Life Sciences, Penn State University, University Park, PA, USA; King Abdullah University of Science and Technology (KAUST), Thuwal, Saudi Arabia*
TOM GERATS • *Department of Plant Genetics/IWWR/Radboud University, Nijmegen, The Netherlands*
EMMANUEL GUIDERDONI • *CIRAD, UMR AGAP, Montpellier cedex 5, France*
CHANG-DEOK HAN • *Division of Applied Life Science, Department of Biochemistry, Plant Molecular Biology and Biotechnology Research Center (PMBBRC), Gyeongsang National University, Jinju, Korea*
AMAL HARB • *Department of Biological Sciences, Faculty of Science, Yarmouk University, Irbid, Jordan*
AN-PING HSIA • *Iowa State University, Ames, IA, USA*
JIN HUANG • *Division of Applied Life Science, Plant Molecular Biology and Biotechnology Research Center, Gyeongsang National University, Jinju, Korea*
CHARLES HUNTER • *Plant Molecular and Cellular Biology Program, Horticultural Sciences Department, University of Florida, Gainesville, FL, USA*
NING JIANG • *Department of Horticulture, Michigan State University, East Lansing, MI, USA*
TAE HO KIM • *Genomics Division, Gene Function Lab, Department of Agricultural Biotechnology, National Academy of Agricultural Science, RDA, Suwon, Korea*

Karen E. Koch • *Plant Molecular and Cellular Biology, Horticultural Sciences Department, University of Florida, Gainesville, FL, USA*

Lakshminarasimhan Krishnaswamy • *Department of Biology, Singapore University of Technology and Design (SUTD), Singapore, Singapore*

Brent A. Kronmiller • *Bioinformatics and Computational Biology Graduate Program, Department of Plant Pathology and Microbiology, Iowa State University, Ames, IA, USA*

Reinhard Kunze • *Dahlem Centre of Plant Sciences–Angewandte Genetik, Albrecht-Thaer-Weg, Freie Universitat Berlin, Berlin, Germany*

Nadège Lanau • *CIRAD, UMR AGAP, Montpellier cedex 5, France*

Katina Lazarow • *Leibniz-Institute for Molecular Pharmacology (FMP), Berlin, Germany*

Gang-Seob Lee • *Genomics Division, Gene Function Lab, Department of Agricultural Biotechnology, National Academy of Agricultural Science, RDA, Suwon, Korea*

Yubin Li • *Waksman Institute, Rutgers University, Piscataway, NJ, USA*

Damon Lisch • *Department of Plant and Microbial Biology, University of California, Berkeley, CA, USA*

Sanzhen Liu • *Iowa State University, Ames, IA, USA*

Donald R. McCarty • *Horticultural Sciences Department, Plant Molecular and Cellular Biology Program, University of Florida, Gainesville, FL, USA*

Delphine Mieulet • *CIRAD, UMR AGAP, Montpellier cedex 5, France*

Olivier Panaud • *Laboratoire Génome et développement des plantes, Université de Perpignan Via Domitia, Perpignan, Cedex, France*

Dong-Soo Park • *Department of Functional Crop Science, National Institute of Crop Sciences Rural Development Administration, Milyang, Republic of Korea*

Soo Kwon Park • *Department of Functional Crop Science, National Institute of Crop Sciences Rural Development Administration, Milyang, Republic of Korea*

Andy Pereira • *Department of Crop, Soil, and Environmental Sciences, University of Arkansas, Fayetteville, AR, USA*

Peter A. Peterson • *Department of Agronomy, Iowa State University, Ames, IA, USA*

Thomas Peterson • *Department of Agronomy, Department of Genetics, Development and Cell Biology, Iowa State University, Ames, IA, USA*

Julia Qüesta • *John Innes Centre, Norwich Research Park, Norwich, UK; Formerly located at Centro de Estudios Fotosintéticos y Bioquímicos (CEFOBI), Facultad de Ciencias Bioquímicas y Farmacéuticas, Universidad Nacional de Rosario, Rosario, Argentina*

Patrick S. Schnable • *Center for Plant Genomics, Roy J Carver Co-Lab, Iowa State University, Ames, IA, USA*

Gregorio Segal • *Waksman Institute, Rutgers University, Piscataway, NJ, USA*

Masaharu Suzuki • *Plant Molecular and Cellular Biology Program, Horticultural Sciences Department, University of Florida, Gainesville, FL, USA*

Michiel Vandenbussche • *UMR 5667 CNRS-INRA-ENS Lyon-Unversité Lyon I, RDP Laboratory, ENS Lyon, Lyon, Cedex, France*

Virginia Walbot • *Department of Biology, Stanford University, Stanford, CA, USA*

Dafang Wang • *Iowa State University, Ames, IA, USA*

Qinghua Wang • *Waksman Institute, Rutgers University, Piscataway, NJ, USA*

ROGER P. WISE • *Department of Plant Pathology and Microbiology, Corn Insects and Crop Genetics Research Unit, USDA-ARS, Iowa State University, Ames, IA, USA*
YUAN HU XUAN • *Division of Applied Life Science (BK21 Program), Plant Molecular Biology and Biotechnology Research Center (PMBBRC), Gyeongsang National University, Jinju, Korea*
GIHWAN YI • *College of Agriculture & Life Science Department of Agro-industry & Farm management, Kyungpook National University, Daegu, Milyang, Republic of Korea*
DOH WON YUN • *Genomics Division, Gene Function Lab, Department of Agricultural Biotechnology , National Academy of Agricultural Science, RDA, Suwon, Korea*
JAN ZETHOF • *Plant Genetics, IWWR, Radboud University Nijmegen, Nijmegen, The Netherlands*

Chapter 1

Historical Overview of Transposable Element Research

Peter A. Peterson

Abstract

Research on transposable elements began nearly 100 years ago with classical genetic experiments. Remarkably, many of the activities of transposable elements, such as the ability to transpose, to induce chromosome rearrangements, to undergo cycles of activity and inactivity, and to affect expression of neighboring genes, were described by geneticists long before transposons were molecularly isolated. This chapter traces the historical roots of transposable element research, describing the scientists, their observations, and interpretations as they sought to understand the enigma of transposable elements.

Key words Variegation, Transposition, Chromosome breakage, *Ac/Ds*, *En/Spm*

1 Genetic Era

1.1 Variegation

Variegation in both ornamental and crop plants has long been observed. Most prominently, the early Native Americans were impressed with the colors and variegation in the ears of maize that they harvested (*see* Fig. 1). These Native Americans very likely saved these various colored ears and in some instances made the exceptional ears part of the identity of their cultural group.

Many of these exceptional types found their way, eventually, to the early American colonists, where they soon became part of academic pursuits. One of the first to be attracted to these variegated maize ears was Emerson, who in 1914 [1] started the study of kernel pericarp variegation at the University of Nebraska. This variegation affected the color of the ear causing red stripes on pericarp of individual kernels, and in some instances covering major portions of the ear (*see* Fig. 2). Emerson interpreted the results in terms of two variables, *S* and *V*. He considered the variable factor, *S*, to be self-color (i.e., red kernel pericarp), which is inhibited by an associated factor, *V*. However, the inhibition of *S* by *V* is temporary; as Emerson wrote, "sooner or later, *V* loses its power to inhibit and changed the self-color gene to *S*." Emerson continued his study of variegated pericarp over a period of 10 years, and

Thomas Peterson (ed.), *Plant Transposable Elements: Methods and Protocols*, Methods in Molecular Biology, vol. 1057, DOI 10.1007/978-1-62703-568-2_1, © Springer Science+Business Media New York 2013

Fig. 1 Variegated ears that the Native Americans grew

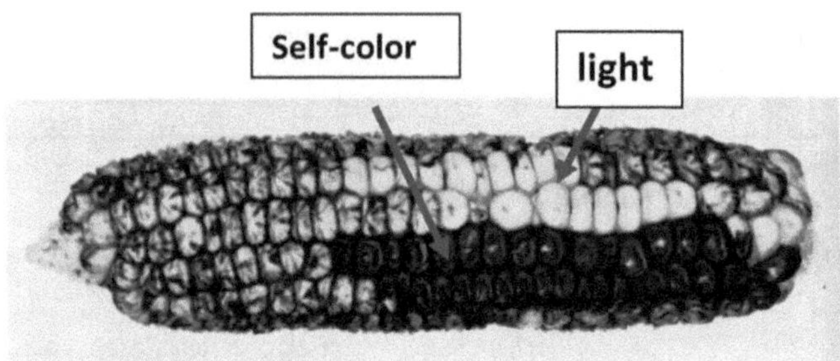

Fig. 2 P^{vv}: Twin sector of P^{vv} showing the change from P^{vv} to *S (self-color)* and light variegated [5]

published a number of seminal papers which are models of careful genetic research [2, 3]. These and other studies showed that red pericarp and cob pigmentation are controlled by a single gene termed *P-rr* (red pericarp, red cob), which was derived from and allelic with the variegated form *P-vv* (variegated pericarp, variegated cob) (*see* Fig. 2). However, it was only many years later that Brink and his students [4, 5] identified Emerson's "temporary" factor *V* as a transposable element, *Modulator of pericarp*, (*Mp*), which was subsequently shown to be identical to McClintock's *Activator* (*Ac*) element (see below).

1.2 Movement of Genetic Elements in the Genome: The Ac/Ds System

Emerson moved from Nebraska to Cornell University where he continued his study of *P-vv* and the genetics of pericarp and cob variegation. At Cornell he and his students identified a new phenotype of light variegated pericarp. Interestingly, the light variegated pattern required the presence of another factor which segregated

Fig. 3 McClintock in her lab at the Carnegie Institution of Washington at Cold Spring Harbor, NY. Here she is examining her crosses to identify the choromosome breakage events

independently of the *P-vv* allele on chromosome one (*see* Fig. 2, *Light pattern as part of twin sector*). It would be some years before this independent controlling genetic element was determined to be a copy of *Modulator* which had transposed from the original *P-vv* [4]. Further studies verified that there was a simultaneous loss of the original factor from *P-vv*, and a reversion to a fully red *P-rr* allele (*see* Fig. 2, *red sector*). Coincident with the change from *P-vv* to *P-rr* was the relocation of the element (*Modulator*) to a new site on the chromosome where it induced light variegation of a *P-vv* allele (*P-vv* + *Modulator* = light variegated sector). Thus, Emerson's original interest in the kernel pericarp variegation ultimately yielded one of the most compelling demonstrations of transposed genetic material in the genome. Although transposition *per se* was not recognized or stated as such by Emerson, the significance of his work on the genetics of variegation was certainly not lost on Emerson's protege Barbara McClintock.

Following her studies at Cornell in the 1930s, McClintock was invited by Lewis Stadler to become an assistant professor at the University of Missouri at Columbia. At the time, Stadler was conducting X-irradiation to increase variation and change genes in maize, which McClintock also took an interest in. As the X-rays physically broke the chromosomes, McClintock discovered the fascinating behavior related to the broken ends (*see* Fig. 3). Then, in the late 1930s, among her stocks of X-rayed plants, McClintock discovered plants whose chromosomes broke spontaneously

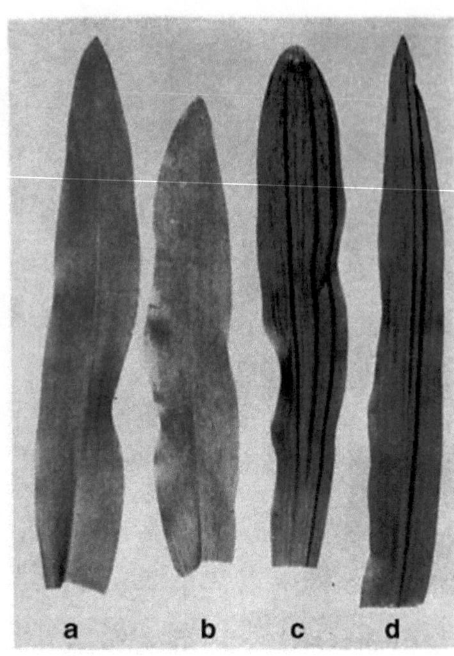

Fig. 4 *pg*^{m}: (**a**) pale stable; (**b**) late, fine stripes; (**c**) early, wide stripes; (**d**) early- and late-occurring striping. Peterson [11]

without further irradiation [6]. This was the beginning of her work with *Ac/Ds*; in a series of elegant experiments McClintock showed that chromosome breakage was induced by a factor she termed *Dissociation* (*Ds*), and that chromosome breakage at *Ds* was controlled by a second independent factor, *Activator* (*Ac*). Moreover, *Ac* could induce the transposition of both itself and *Ds* within the genome [7, 8].

1.3 Discovery of the En and Spm Systems

In 1948, following a year studying Drosophila genetics in the Demerec lab at the Carnegie labs in Cold Spring Harbor, NY, Peter Peterson joined the Botany Department at the University of Illinois to study with Professor M.M. Rhoades, who had worked previously at Columbia University. The Botany Department provided Peterson with a part-time teaching assistantship; this proved to be an enlightening experience which was both enjoyable and rewarding. In addition, Professor Rhoades provided a research problem that Peterson unknowingly at the time would be following for the next 60 years. The research involved a pale-green seedling mutable gene (*pg*^{m}) (*see* Fig. 4). This *pg*^{m} mutability arose from the progeny of maize seed exposed to the Crossroads atomic bomb test at Bikini Atoll in the South Pacific in 1946 [9, 10].

As the *pg*^{m} mutability was enhanced by another factor controlling mutability of this gene, this factor was called *Enhancer* (*En*) [11]. Soon a number of alleles were isolated that included, in addition to

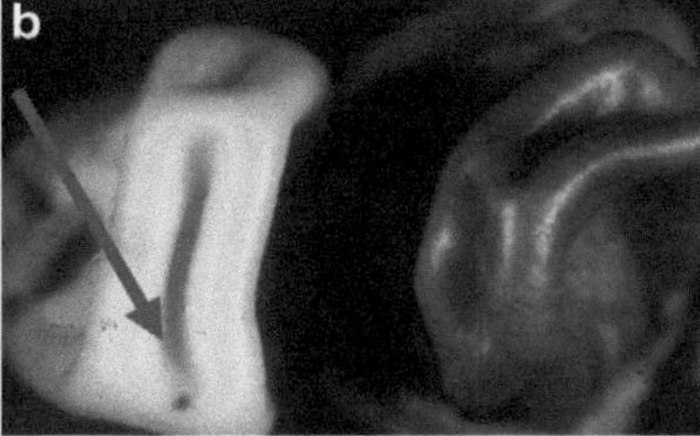

Fig. 5 (**a**) $a1^{m-1}$: This pale allele is receptive to *Spm* and *En*; each of the transposons suppresses the pale coloration causing mutations (**b**) $a1^{m-1}$ *sh2*: With (*left*) and without (*right*) *En—low excision rate* (*arrow*)

the original pg^m, some with different patterns of mutability (timing of mutability) (*see* Fig. 4b, c). Among the exceptions of pg^m, a stable pg^s allele containing an *I* (*Inhibitor*) element arose that responded to signals from a segregating factor that was identified as *En*. This two-factor (*En/I*) genetic interaction was analogous to the *Ac/Ds* system previously described by McClintock [12].

While working with the *Ac/Ds* system, McClintock [12–14] uncovered another unstable phenotype (*see* Fig. 5) which she experimentally confirmed to be unrelated to *Ac/Ds*. Because this new phenotypic expression could be turned off (*pale with no colored sectors*, *see* Fig. 5a) and turned on (*colorless with colored sectors in crosses*, *see* Fig. 5b) she termed this new system *Spm* (*Suppression-mutation*) [12]. When *Spm* was absent the kernels were colored (*see* Fig. 5a). However, when *Spm* was introduced via genetic crosses, the color was abolished and mutations appeared on the colorless kernels (*see* Fig.5b). This genetic interaction was unusual and unlike any expression that she had observed with *Ac/Ds*. Surprisingly, it seemed as though *Spm* controlled both loss of function (purple to colorless) and gain of function (appearance of colored spots). This was more than simple mutability, and appeared to be more like the control of gene expression.

1.4 Relationships Among Controlling Element Systems

During the 1950s, several labs other than McClintock were studying variegation and controlling element systems in maize. These included Rhoades' *Dt-adt*, Emerson and Brink's *P-vv*, and Peterson's pg^m [15–18]. As might be expected, these researchers recognized the similarities of their respective systems and began to devise genetic tests to detect functional interactions among them. One of the first overlaps was reported by Barclay and Brink [17], who tested whether the *Modulator* system at *P-vv* was functionally related with McClintock's *Ac/Ds*. The test was positive, indicating that *Modulator*, the factor controlling *P-vv* variegation, was homologous with *Ac*. The genetic equivalence of *Modulator* and *Ac* was the basis for the eventual cloning of the *P-vv* allele using *Ac* as a molecular probe [19, 20].

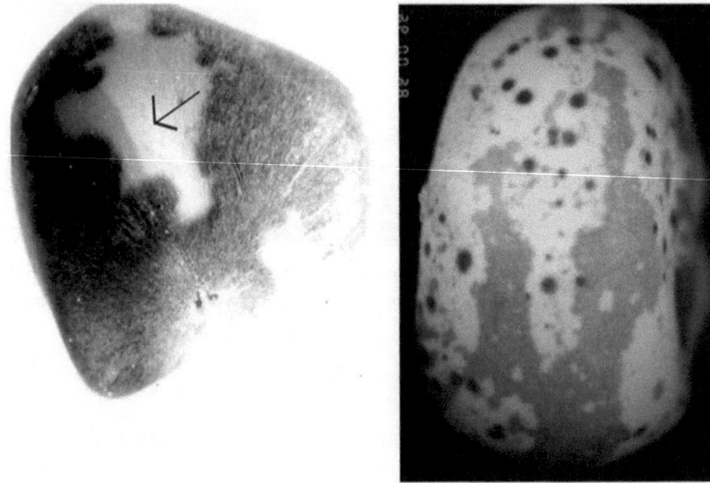

Fig. 6 Two *En* $a1^m$ alleles: *Left*: $a1^{(Au)}$ with frequent colorless sectors (*arrow*). *Right*: $a1^{m(papu)}$ gives rise to somatic sectors and germinal derivatives that are pale, colorless, and purple. Both are located in the second exon of the *a1* gene [40]

Similarly, it had been suggested that Peterson's pg^m, which conditioned mutable green stripes on a pale green background, might be related to McClintock's mutable *Ac/Ds*. This possibility was tested using the reporter allele $pg^{(mr)}$, which elicits a stable pale green seedling phenotype that responds to the presence of *En*. When crossed together, the $pg^{(mr)}$ allele did not respond to *Ac*, indicating their functional independence. The next test was designed to detect whether *En* and *Spm* are related. This test used McClintock's $a1^{m-1}$ (*see* Fig. 5a) in combination with Peterson's *En* (*see* Fig. 6). This test was positive, but with the caveat that *Spm* and *En* might represent independent systems which were present together in the same lines. In later tests with other responding alleles, it was confirmed that *Spm* and *En* trigger all of the same reporter alleles [18]. Subsequently, phase variation of regulatory elements was identified and tested [21].

1.5 The Controversy Surrounding the Controlling Element Hypothesis

McClintock's $a1^m$ allele is colored in the absence of *Spm* (*see* Fig. 5a); when *Spm* is introduced to $a1^m$, the ground color is abolished and darker spots appear (*see* Fig. 5b). This ability of *Spm* to control the expression of a gene (*A1* in this case) appealed to McClintock and was a central theme in her later writings [22]. Certainly, gene expression was turned off by *Spm* and colored spots did appear. Yet, this was in the mid 1950s, 6 years before Jacob et al. introduced the operon model for control of gene expression [23]; the generality and significance of control of gene expression by transposable elements was unclear. In 1968–1969 while on sabbatical at the Karolinska Institute in Stockholm, Sweden, researching bacteria with Professor Joe Bertani, Peterson wrote two papers [24, 25] questioning the controlling aspect of these mobile elements in the genome. It would be 7 years before a molecular description of eukaryotic gene regulation emerged [26] and 13 years before the cloning of a maize transposable element [27]. Only with the advent of modern molecular biology could the role of transposable elements in controlling gene expression be more fully investigated.

2 Molecular Era

2.1 Molecular Isolation of the Transposons *Ac/Ds* and *En*

As transposable element theory gained wider acceptance and molecular biology approaches became available, a number of researchers initiated efforts to capture a transposon in the lab. One of these was the Fedoroff group in Baltimore, who sought the *Ac/Ds* elements inserted in the maize *Waxy* gene. This group identified and isolated *Wx* mRNA and corresponding polypeptides, and used these as probes to eventually fish out genomic clones from the maize *wx* locus [28]. Using these clones they confirmed the homology to McClintock's well-characterized insertion (the *wx-m6* allele) mutation at the *Wx* locus. This allele was advantageous because of the availability of the progenitor *Wx* allele and several newly derived *Wx* revertants. Using these materials, Fedoroff and coworkers were able to demonstrate a convincing correlation between structural changes in the maize *wx* DNA, and the genetic effects of *Ac/Ds* insertions in *wx* [27]. A second team comprised the group of Professor Heinz Saedler at the Max Planck Institute, Cologne, Germany, in collaboration with the P. Peterson lab in Ames, IA. The Max Planck group independently isolated a maize *Wx* gene clone [29]. It fell to the Peterson lab to find an *En* element insertion in the *Wx* gene. Up to this time in 1980, the Peterson lab did not have an *En* inserted into the *waxy* gene; thus, an effort was started to find such an event, namely, *wx-En*.

Because a waxy-mutable is difficult to recognize in a colored kernel, and all the *En*-containing stocks were in colored kernel lines, the lines were converted by crossing to the color inhibiter *C1-I* allele to generate the homozygous genotype *C1-I C1-I*, *En En*, *Wx Wx*. Plants of this genotype were grown in an isolation plot and their ears fertilized with pollen from plants of genotype *c1 c1*, *sh sh*, *wx wx*. Approximately 1 acre (roughly 20,000 plants) was crossed in this way by growing four rows of detassled ear parent plants (*C1-I En Wx*) alternating with two rows of pollen parent plants (*c1 sh wx*). Following harvest, ears were individually examined for the presence of waxy-mutable kernels. The screen was designed to detect kernels in which *En* had transposed into the *Waxy* gene yielding *wx-En*. A single waxy-mutable was recovered from 1.9×10^{-6} kernels screened [30]. The rescued allele, *wx-84-4*, was sent to Cologne where *En* was isolated and sequenced [31, 32].

2.2 Transposon Sequences as Probes to Isolate Tagged Genes

Once the transposons *Ac/Ds* and *En* were molecularly cloned, they could be used as hybridization probes to detect and isolate other genes containing the inserted elements. Using this gene tagging approach, genes could be isolated based solely on the availability of a TE-tagged allele. In this way a relatively large number of maize genes were rapidly isolated and characterized; perhaps most impressive was the isolation of many of the genes involved in anthocyanin biosynthesis, including *A1*, *A2*, *Bz1*, *C1*, *C2*, and *Whp* [33–39]. These successes spurred further attempts to transposon tag and clone genes involved in a variety of metabolic and developmental pathways, and employing not only *Ac/Ds* and *En/Spm* but also other transposons such as the maize *Mutator* element. Moreover, the development of plant transformation technologies ultimately enabled the maize transposable elements to be employed as insertional mutagens in a variety of other species. Overall, the impact of transposable elements on modern genetic research is hard to overstate. In the current era where high-throughput sequencing and computational analyses are the norm, it is important to remember that the most fundamental aspects of transposon biology were elucidated by McClintock and colleagues employing only classical genetics and cytogenetics approaches. Today's researchers would do well to heed the historical lesson that careful observation, well-designed experiments, and maintaining one's connections with the biological realm have their rewards.

References

1. Emerson RA (1914) The inheritance of a recurring somatic variation in variegated ears of maize. Am Nat 48:87–115
2. Emerson RA (1917) Genetical studies of variegated pericarp in maize. Genetics 2:1–35
3. Anderson EG, Emerson RA (1923) Pericarp studies in maize. 1. The inheritance of pericarp color. Genetics 8:466–476
4. Van Schaik NW, Brink RA (1959) Transpositions of modulator, a component of the variegated pericarp allele in maize. Genetics 44:725–738
5. Greenblatt IM, Brink RA (1962) Twin mutations in medium variegated pericarp maize. Genetics 47:489–501
6. McClintock B (1941) Spontaneous alterations in chromosome size and form in *Zea mays*. Genes Chromosome Quant Biol IX:72–80
7. McClintock B (1950) The origin and behavior of mutable loci in maize. Proc Natl Acad Sci 36:344–355
8. McClintock B (1951) Chromosome organization and genic expression. In: Warren B (ed) Genes and mutations, vol XVI. Cold Spring Harbor symposia on quantitative biology, The Biological Laboratory, Cold Spring Harbor, Long Island, NY, pp 13–47
9. Anderson EG (1948) On the frequency and transmitted chromosome alterations and gene mutations induced by atomic bomb radiations in maize. Proc Natl Acad Sci USA 34: 387–390
10. Anderson EG et al (1949) Hereditary effects produced in maize by radiations from the bikini atomic bomb I. Studies on seedlings and pollen of the exposed generation. Genetics 34:639–646
11. Peterson PA (1959) The pale green mutable system in maize. Genetics 45:115–133
12. McClintock B (1956) Intranuclear systems controlling gene action and mutation. Brookhaven Symp Biol 8:58–74
13. McClintock B (1956) Controlling elements and the gene. Cold Spring Harb Symp Quant Biol 221:197–216
14. McClintock B (1954) Mutations in maize and chromosomal aberrations in *Neurospora*. Carnegie Inst Wash Yr Bk 53:254–260
15. Rhoades MM (1936) The effect of varying gene dosage on aleurone colour in maize. J Genet 33:347–354
16. Rhoades MM (1941) The genetic control of mutability in maize. Cold Spring Harb Symp Quant Biol 9:138–144

17. Barclay PC, Brink RA (1954) The relation between modulator and activator in maize. Proc Natl Acad Sci U S A 40:1118–1126
18. Peterson PA (1965) A relationship between the *Spm* and *En* control systems in maize. Am Nat 99:391–398
19. Chen J, Greenblatt IM, Dellaporta SL (1987) Transposition of Ac from the P locus of maize into unreplicated chromosomal sites. Genetics 117:109–116
20. Lechelt C, Peterson T, Laird A, Chen J, Dellaporta SL, Dennis E, Peacock WJ, Starlinger P (1989) Isolation and molecular analysis of the maize P locus. Mol Gen Genet 219:225–234
21. Peterson PA (1966) Phase variation of regulatory elements in maize. Genetics 54:249–266
22. McClintock B (1957) Genetic and cytological studies of maize: types of *Spm* elements. A modifier element within the Spm system. The relation between *a1m-1* and *a1m-2*. Aberrant behavior of a fragment chromosome. Carnegie Inst Wash Yr Bk 56:393–401
23. Jacob F, Monod J (1961) Genetic regulatory mechanisms in the synthesis of proteins. J Mol Biol 3:318–356
24. Peterson PA (1970) Controlling elements and mutable loci in maize: their relationship to bacterial episomes. Genetica 41:33–56
25. Peterson PA (1970) The *En* mutable system in maize. III. Transposition associated with mutational events. Theor Appl Genet 40:367–377
26. Jeffreys AJ, Flavell RA (1977) A physical map of the DNA regions flanking the rabbit β-globin gene. Cell 12:429–439
27. Fedoroff N, Wessler S, Shure M (1983) Isolation of the transposable maize controlling elements Ac and Ds. Cell 35:235–242
28. Shure M, Wessler S, Fedoroff N (1983) Molecular identification and isolation of the waxy locus in maize. Cell 35:225–233
29. Schwarz-Sommer ZS et al (1984) The *Spm* (*En*) transposable element controls the excision of a 2-kb DNA insert at the *wxm-8* allele of *Zea mays*. EMBO J 3:1021–1028
30. Peterson PA (1985) The isolation of *En1* in the *wx-84-4* allele. Maize Genet Coop News Lett 59:3
31. Pereira A et al (1985) Genetic and molecular analysis of the enhancer (*En*) transposable element system of *Zea mays*. EMBO J 4: 17–23
32. Pereira A et al (1986) Molecular analysis of the En/Spm transposable element system of Zea mays. EMBO J 5:835–841
33. O'Reilly C et al (1985) Molecular cloning of the *a1* locus of *Zea mays* using the transposable elements *En* and *Mu1*. EMBO J 4:877–882
34. Menssen A (1990) The *En/Spm* transposable element of *Zea mays* contains splice sites at the termini generating a novel intron from a dSpm element in the *A2* gene. EMBO J 9:3051–3057
35. Fedoroff NV, Furtek DB, Nelson OE (1984) Cloning of the bronze locus in maize by a simple and generalizable procedure using the transposable controlling element activator (Ac). Proc Natl Acad Sci USA 81(12):3825–3829
36. Cone KC, Burr FA, Burr B (1986) Molecular analysis of the maize anthocyanin regulatory locus C1. Proc Natl Acad Sci USA 83(24): 9631–9635
37. Paz-Ares J et al (1986) Molecular cloning of the *C1* locus of *Zea mays*: a locus regulating the anthocyanin pathway. EMBO J 5:829–833
38. Wienand U et al (1986) Molecular cloning of the *c2* locus of *Zea mays*, the gene coding for chalcone synthase. Mol Gen Genet 203:202–207
39. Franken P et al (1991) The duplicated chalcone synthase genes *C2* and *Whp* (white pollen) of Zea mays are independently regulated; evidence for translational control of Whp expression by the anthocyanin intensifying genein. EMBO J 10:2605–2612
40. Peterson PA (1961) Mutable *a(1)* of the En system in maize. Genetics 46:759–771

Chapter 2

Distinguishing Variable Phenotypes from Variegation Caused by Transposon Activities

Virginia Walbot

Abstract

Variable phenotypes are common in nature and in laboratory materials. Guidelines and illustrations are presented to help distinguish developmental, environmental, disease, and somatic recombination-generated variation from the phenotypes caused by transposable elements.

Key words Disease symptoms, Color pattern, Anthocyanin, Physiological regulation, Clonal analysis

1 Introduction

Those working with defined transposable elements (TE) typically rely on visible phenotypes to monitor transposon activity. Phenotypes such as restoration of anthocyanin synthesis after element excision (*see* Fig. 1a) have been used to define the timing and frequency of TE movement in maize and many other plants. In the example shown, a *Mu* transposon is inserted into the reading frame of a maize *bronze2* allele (*bz2::mu4-MuDR* and the other allele is a recessive *bz2-reference*), and as expected for this element family, there is a high frequency of sectoring represented by small spots of dark color on a beige background [1]. *MuDR/Mu* elements are unusual in that somatic excision and heritable insertions occur very late in the life cycle, generating only small clonal sectors [1]. Spontaneous silencing of the element system results in a nearly completely beige kernel with rare spots or a completely beige kernel. Other transposable element systems such as *Ac/Ds* and *Spm/dSpm* exhibit excision throughout aleurone development in maize endosperm. There are sectors up to 1/2 kernel representing very early excision events down to single-cell sectors representing excision after the last division in the tissue (*see* Fig. 1b) [2].

Thomas Peterson (ed.), *Plant Transposable Elements: Methods and Protocols*, Methods in Molecular Biology, vol. 1057, DOI 10.1007/978-1-62703-568-2_2, © Springer Science+Business Media New York 2013

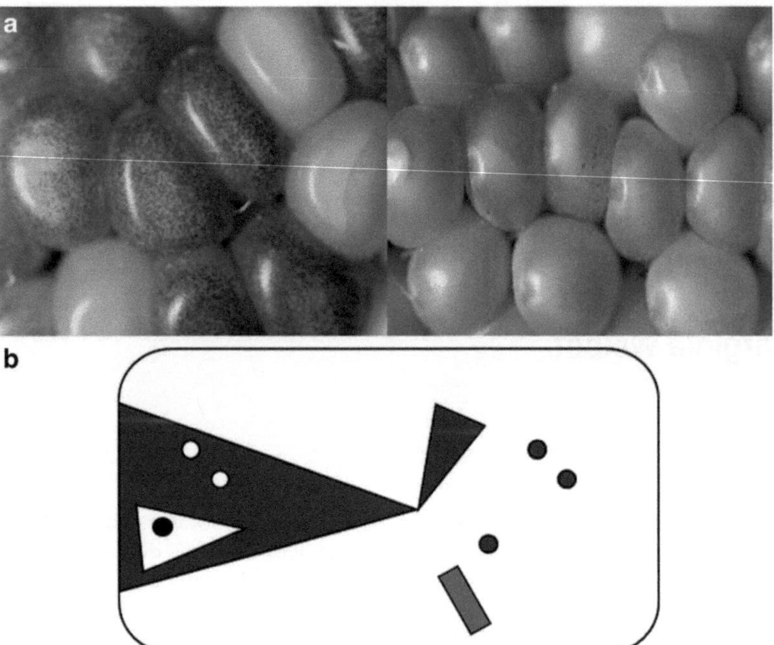

Fig. 1 Transposable element phenotypes and plant development. (**a**) Somatic excision of a *Mu* transposable element monitored with a mutable *bz2* allele in active (*left*) and inactive (*right*) *Mutator* individuals. In the inactive individual, a few kernels exhibit a few excision sectors. The kernels are heterozygous with a single copy of the mutable allele in the triploid aleurone *bz2/bz2/bz2-mu*. (**b**) Drawing of the top (crown) of a maize kernel illustrating a range of excision sector sizes, as expected with *Ac/Ds*-mediated variegation, with a single-copy anthocyanin reporter allele heterozygous with the tester allele (i.e., *al/al/ a1-mutable*). A range of color intensities is illustrated. Anthocyanin is found in the aleurone of maize endosperm, and this tissue arises from fertilization of the central cell by a sperm followed by a period of syncitial development. For a few cell nuclear cycles, the nuclei remain central, but then migrate to the plasma membrane and undergo further divisions up to about 250–500 nuclei before cellularization starts. Because plant cells do not move during development, sectors should remain coherent once cells form. In this diagram, a revertant event corresponding to ~1/8 of the aleurone is shown along with later events. Early events are generally triangular with the narrow end pointed towards the top of the kernel. Later events (after cellularization) are generally square or rectangular; the simple explanation based on the logic of clonal analysis is that aleurone cells alternate their anticlinal division planes to generate squares that make rectangles that make squares. The final events appear to be *dots*, but when observed under a microscope are small squares and rectangles. Within the largest sector, the phenotype of element reinsertion into the reporter gene is illustrated as *dots* of pigment loss within a purple sector

In other laboratory stocks or in nature, variable phenotypes are often observed, some of which are caused by transposons. The purpose of this chapter is to illustrate how to determine whether a phenotype is likely to be caused by a TE or not. Of course, for most cases no genetic proof is possible; however, with the cost of DNA sequencing plummeting and knowledge of the anthocyanin, chlorophyll, carotenoid, and other pigment biosynthetic pathways, it should be possible in the near future to test whether visible variation in color results from TE activities. Prior to that, careful observation of environmental conditions and of a plant's development can offer clues for distinguishing TE-mediated events from other sources of variable phenotypes. There are also stereotyped aspects of TE behavior that will favor the conclusion that you are observing TE-mediated events.

2 Materials

The materials presented here were grown and photographed by the author or obtained from experts, as listed in the caption of Fig. 2. For more information on dahlias please consult http://www.stanford.edu/group/dahlia_genetics/dahlia_database_cultivars.htm.

3 Methods

3.1 Observing the Phenotype

In most cases, TE insertion into or near a gene disrupts expression and generates a recessive allele. Provided the other allele is also recessive, a mutant phenotype will be observed (i.e., a white rather than a purple flower). In this simple case, TE excision during somatic growth can restore gene expression, provided DNA repair following excision restores the reading frame or other required features of the gene, resulting in a purple sector (*see* Fig. 1). Each revertant sector is an independent event, and given the diversity of DNA repair products resulting from deletion and insertion of bases at the double-strand break, you should expect variation in the intensity of purple pigment ranging from very pale coloration to intensely dark. For the *bronze2* locus, we suspect that a very low level of protein function suffices; consequently most revertant sectors are dark, but some pale and very pale sectors are routinely observed among the thousands of spots on each kernel.

The wide range of phenotypes is even more obvious with TEs that transpose throughout development (*see* Fig. 1b). Some large sectors will be dark, and some will be pale. In the case of *Ac/Ds* elements, which exhibit mainly local transposition [3], reinsertion of the excised element can occur, resulting in colorless sectors within a larger revertant purple area. In a sector with 30,000 cells

Fig. 2 Variable phenotypes. Developmentally programmed (**a–d**). (**a**) This collarette dahlia variety has a regular appearance in which the outer petal tips are *yellow* and the inner petalloid organs are completely *yellow*. The extent of outer petal *yellow* marking is variable ranging from a small *dot* of color as seen in the uppermost flower to a crescent of *yellow* in the lower *right*, but the feature of tip marking is regular. (**b**) This lacinated dahlia variety has twisted petals with the upper (adaxial) surface *bright red* and the lower (abaxial) surface *white*. The twisted petals make the flower as a whole appear highly variegated, but the pattern is actually simple. (**c**) Venosa patterning in snapdragon; photo is from Kevin Davies. Anthocyanin pigmentation concentrated around the veins can result from environmental conditions (e.g., metal nutrient deficiency) or, as shown here, be a developmental pattern. (**d**) Epidermal peel of C. Peel and photo by Kathy Schwinn Kathy.Schwinn@plantandfood.co.nz. Variation caused by environmental factors (**e**, **f**). (**e**) In maize and other grasses defects in

(about 12 % of the aleurone) as shown here and with TE reinsertion possible at any time in development, multiple white sectors would be expected within the large dark area. Occasionally one of these reinsertion mutations will generate a quite large sector, and within it will be smaller revertant purple sectors. The number of excision and reinsertion events can be quantified and normalized for the number of cells to infer timing, permitting calculation of the frequency of excision and reinsertion over developmental time.

Although TE events are stochastic, most TEs exhibit certain patterns of transposition that can be observed in the repetitive plant organs. TE systems such as *MuDR/Mu* always produce small sectors wherever they are scored, while other TE systems produce a wider but characteristic range of sector sizes. Each TE pattern can be affected by epigenetic mechanisms, element copy number, or other controls within a particular individual. Observing multiple organs for regularity—not uniformity—is a good way to separate possible TE-mediated events from developmentally programmed patterns.

Fig. 2 (continued) chlorophyll synthesis or stabilization are common during cool nights (low-temperature-sensitive defect) or in particularly bright light (intense-light-sensitive defect). Over the course of 24 h, ranks of cells in the leaf blade mature and proplastids differentiate as chloroplasts. If either the day/light or night/cool temperature condition is deleterious to chlorophyll acquisition, an alternative "zebra" striping pattern can result with alternate dark *green* and *yellow–green* chlorotic bands. One pair of stripes represents a single day. The defective zones are not repaired to full *green*, and the striping persists as a record of the environmental conditions that alternately favored or inhibited leaf chlorophyll content. (**f**) In this water lily variety of dahlia, all of the petal tips are bright *pink*, while petal bases and the centers of flowers are *white*. This phenotype results from exposure of the outer petal tips to sunlight during bud break; the petal bases and inner petals are shielded from sunlight by the outer petals. There must be a short phenocritical period (time during which petals respond to sunlight by synthesizing anthocyanin) to result in the well-demarcated pattern observed. Likely transposon activities. (**g**) Snapdragon variety with many small *dots* of color. (**h**) Single flower of the Rio Fiesta dahlia with striping matching the expectation of TE activity. The base color of *yellow* typically results from a block early in the anthocyanin pathway yielding accumulation of chalcone or a chalcone derivative. (**i**) Dahlia with striping in the pattern expected for TE excision. The background color of *beige–pink* is similar to what is observed when a terminal step of anthocyanin is blocked such that pigment accumulates in the cytoplasm where it is oxidized to a brownish pigment. (**j**) Dahlia with pale petal sectors. (**k**) Blackberry Ripple dahlia with half flower revertant sectors of intense *purple* pigmentation. Biotic agents. (**l**) Apple (*Malus domestica*, Ellstar variety) leaves are host to *Phyllonorycter blancardella* (LEPIDOPETRA: GRACILLARIDAE). The leaves are uniformly *green* when females lay their eggs; later when the leaves start to senesce from natural cues they turn *yellow* but areas with larvae stay *green* and photosynthetically active. Photo from Mélanie Body and David Giron, Institut de Recherche sur la Biologie de l'Insecte, Faculté des Sciences et Techniques, Parc Grandmont, 37,200 Tours, France. (**m**) Fungal damage in poplar of (*Populus trichocarpa*) yield *dots* of cell death (hypersensitive response) surrounded by *yellow* islands of chlorosis on an otherwise healthy *green* seedling leaf. (**n**) The fungal pathogen *Leptosphaeria maculans* causes *green* islands to form on infected leaves of *Brassica napus*. Photograph supplied by Dr. Yongju Huang, University of Hertfordshire, UK

3.2 Observing the Plant Over Time

Regular patterns are expected for TE-mediated events, although exceptions are numerous: for example, slightly larger sectors with *Mu* excision, and zones of leaves, flowers, and/or entire organs where the element system has shut down, resulting in a uniform mutant phenotype. Organ development is much more uniform than TE-mediated variation. For example, in Fig. 2a, a collarette dahlia is shown in which outer petals are red and inner petalloid organs are yellow. These are defining characteristics of this dahlia variety: all flowers will exhibit these organ-specific color differences. Figure 2b illustrates another common developmental pattern of differential coloration within an organ; here petal upper (adaxial) surfaces are deeply pigmented while the lower (abaxial) surface is pale. In this dahlia variety, petals are twisted and curled such that both ab- and adaxial surfaces are visible, giving each flower a variegated appearance from a distance. Closer inspection shows, however, that there is a highly uniform underlying pattern to each petal, rather than transposon action through ontogeny of individual flowers giving rise to sectors of red and pale petals that appear to be random when the flower is viewed from a distance.

The *Venosa* gene phenotype of snapdragon illustrates another type of variable pattern (*see* Fig. 2c, d). Here anthocyanin pigmentation is restricted to tissue surrounding the veins. In this case the pattern is developmentally programmed [4], but a similar pattern could also result if there were a gradient from the vein to the petal field of a signal or molecule required to sustain high anthocyanin synthesis. Anthocyanin is stabilized in the plant vacuole after binding divalent metal ions, and a deficiency in metal can diminish pigmentation significantly (http://www.stanford.edu/group/dahlia_genetics/meta_reports/lab_4/lab_4_meta.htm). For either developmental or physiological patterns highlighting the veins, what is important is the uniformity of the phenotype. The observed variation of red and *white* petals is not from TE-mediated events.

3.3 Variegated Phenotypes Caused by Light or Temperature

Plants develop in a fluctuating environment, and organ phenotypes can be modulated by environmental conditions. This plasticity of form and other phenotypes is a key strategy of plant acclimation to the environment, and the range of such phenotypes is highly adaptive [5, 6]. Both biosynthesis and stabilization of chlorophyll pigment can be affected by prevailing light and temperature in maize. Adult maize leaves develop over approximately a 10-day period. During this period, each day bands of cells at the base of the blade become competent to differentiate as chlorophyll-containing photosynthetic cells. When chlorophyll content is sensitive to low or high temperature or to excess light (or perhaps other factors), transverse bands of dark and light tissue termed zebra striping (*see* Fig. 2e) are observed. Each dark plus light band represents 1 day; chlorophyll development occurred under permissive conditions (for example, warm temperature) in the dark area, and under restrictive conditions (cool temperature) in the paler stripe. In grasses, poor chlorophyll deposition is not corrected or corrected very slowly, and consequently zebra bands persist for a season or for a substantial period of time. It is easy to imagine that a short interval of restrictive conditions would generate a portion of a leaf with a transverse zebra banding pattern that might be mistaken for TE-induced variegation. But knowledge of how maize leaves grow would rule out TE activities, because the clonal sectors within the leaf run from the leaf tip to the base. Consequently, developmental sectors—as would be produced by a transposon excision or insertion event in a single cell that generated a large clone—would be long and narrow, not transverse.

The *iojap* mutant of maize illustrates the complexity of phenotype that is possible [7]. The background color is a highly variegated pale green, light green, and white, not a uniform color. The defining allele harbors a *Ds* element, and TE excision results in long or short narrow stripes of dark green pigmentation [8]. The background color is variable, because the nuclear mutation results in failure to maintain plastid ribosomes. During the phenocritical period of rapid plastid division during leaf greening some plastids lose all of their ribosomes, resulting in white cells; in heteroplastidic cells, some plastids are normal and others are albino resulting in a wide range of color phenotypes because there are dozens of plastids per cell.

Figure 2f illustrates a picotee phenotype in which petal outer tips and edges are darkly pigmented compared to the rest of the petal. This might have a developmental genetic basis, but it is more commonly the result of light-dependent anthocyanin pigmentation. When buds first open, only the petal tips receive light. If pigmentation requires light but the phenocritical period for responding to light is brief, then only the initially exposed tissue will accumulate anthocyanin. Similarly, stems in the shade are often green compared to purple-tinged stems in full sun. A simple experimental protocol to determine if light is setting the pattern of pigmentation is to cover a small section of an expanding organ with aluminum foil for a few days, then remove the foil, and compare the covered area to neighboring tissue (http://www.stanford.edu/group/dahlia_genetics/meta_reports/lab_1/lab_1_meta.htm).

Figure 2g is a snapdragon with fine-spotting on the lower petal overlaying within-organ variation in color. Do the spots represent a late excision TE such as *MuDR/Mu*, or is this a developmental or a physiological pattern? Looking at just one flower a TE seems likely as the spots are within zones producing either an orange or a white background color. However, the presence of spots is characteristic of this snapdragon variety—presumably an evolved feature to attract pollinators, or a feature selected by horticulturists—making it less likely to be caused by TEs. With TEs, some large sectors without spots would be expected from element silencing, as well as some variability in spot size and intensity. However, a large planting of this variety produced significant variation in color intensity, with the plants in shade having lower pigmentation than plants in continuous full sun. Based on looking at a large suite of individuals, it appears that what variation exists in this variety is conditioned by the environment.

3.4 Likely TE-Mediated Variation

Figure 2h–j shows dahlias with variation that I propose is caused by transposons. Like the examples just discussed, there is variation based on environmental conditions—petals in full sun are more intense than those in shade—but the pattern of the variegation is similar to the TE expectation rather than strictly physiological or developmental control. Dahlia petals are longer than they are wide. Early (larger) developmental sectors run from base to tip and comprise about 1/8 of petal area. Smaller sectors are also elongated but are restricted to the tip, middle, or base of the petal. The darker areas in Fig. 2h, i shows flowers that conform to this clonal expectation. Additionally, the diversity of sector sizes—from 1/8 petal down to very tiny sectors—is similar to Fig. 1b, and dissimilar to typical developmental patterns in which the uniformity of the phenotype from organ to organ is the expectation. In the dahlias of Fig. 2h the frequency of pigment restoration is quite high suggesting a highly active system. Observing successive flowers on these plants, I noted that some flowers had far fewer sectors, a possible indication that the TE system could be silencing. Some flowers had very large sectors, encompassing the floral center; in these cases the revertant phenotype might be heritable if the lineages leading to pollen or egg are derived from a revertant sector within the flower.

Figure 2j meets the criteria for sector shape, albeit with a low frequency of events visualized as discrete pale stripes from petal tip to petal base (see the lower left of the flower). Closer inspection shows that the outer edges of all petals are pale. Petal and leaf edges have relatively greater contributions of epidermal cells than the central parts of these organs. If anthocyanin is accumulating to lower levels in the epidermis compared to the subepidermal cells, the petal centers will be dark and the petal edges will be pale. In chimeric plants, cells of different genotype are juxtaposed (i.e., an epidermis genetically capable of making anthocyanin overlaying an internal layer mutant for this phenotype) [9]. In such an organ there will be uniform moderate anthocyanin pigmentation. Routine developmental errors in cell division can introduce an epidermal cell into the subepidermal layer, and this event will produce a sector of darker pigmentation; reciprocally a subepidermal cell can become part of the epidermis, and this will result in a white sector. Chimerism could explain the flower in Fig. 2j; dahlias are typically propagated vegetatively from tubers, and over time, somatic mutations could accumulate that ultimately would result in a chimeric apical meristem with a genotypic difference between the epidermal and subepidermal layers. Yet another possible explanation is somatic recombination during petal growth in a plant heterozygous for functional and nonfunctional alleles at one step in anthocyanin biosynthesis. In this situation heterozygosity results in medium color. Somatic recombination can generate sectors that are homozygous mutant in one layer of the meristem. As the petal grows there would be pink epidermis overlying white subepidermal cells or vice versa; in either case we see a pale sector. Microscopy could confirm or refute this explanation and might provide evidence of darker colored sectors (somatic recombination product now homozygous for the functional allele) that are not easily observed by eye. Somatic recombination can produce twin spots: adjacent sectors of white and darker cells. If this is observed, then somatic recombination is the most likely explanation for the phenotype.

Figure 2k is the Blackberry Ripple dahlia variety which is a profusely branching tall variety. The large number of branches favors observation of very large sectors in flowers, because the buds produced in the axil of each leaf are a small sample of cells from the original apical meristem. When a TE excises, a cell capable of accumulating anthocyanin can result, establishing a purple clone. When this initial bud produces a branch, the buds in the axils of each leaf on the branch are an even smaller sample of the original apex. Consequently, an event in the apical meristem of hundreds of cells would be a small sector on the terminal flower, but in the first rank of branches, this sector might have expanded tenfold to encompass several petals and parts of petals. In the second rank of branches, the sector either disappears (the bud meristem did not include the marked purple cell) or expands. The sector would now be of striking size, about 10–25 % of a flower, and one can predict that the next rank of branches will have half-purple flowers as shown here. These are two flowers from buds on opposite sides of the stem, and each has a good sample of purple cells in the floral meristem. Buds from leaves on the branches subtending these flowers will predictably produce some all-purple flowers.

3.5 Biotic Factors That Can Mimic TE

Green islands on senescing leaves can be caused by cytokinin; hence the apple leaf in Fig. 2l might represent a TE mutation in a cytokinin synthesis or response pathway with somatic excision restoring a greener late-season phenotype. The sector conforms to the boundaries expected for a dicot leaf [10], because it runs along the major veins of the leaf and does not cross the midrib. This sectoring is not caused by a TE, however, but by developing insect larvae that "trick" the plant into extending the useful life of this part of the leaf by manipulating cytokinin despite normal senescence of the rest of the leaf [11]. Figure 2m shows a poplar leaf with dots of chlorosis and some necrosis, all caused by fungal infection (N. Zimmerman, pers. comm.). Particularly the chlorotic spots could be mistaken for a sector that results in excess carotenoid relative to chlorophyll. The sectors are variable in size and without further analysis could readily be mistaken for TE activity. Figure 2n illustrates the green island phenomenon of another fungal pathogen on *Brassica napus* [12]. For these pathogen cases, different extents of infection from year to year, or between parts of a plant, or between individuals during a growing season might be the best clue that something other than TEs cause the variegated phenotype.

In closing, I would like to recommend a short field guide that provides genetic explanations for variation observed in nature [13]. This compact book is focused on the Pacific Northwest, but the illustrations and explanations provide a good introduction to the types of variation to expect in plants and lucid explanations of the underlying developmental genetics when known.

Acknowledgments

I thank Professor Zheng-Hui He for explaining green islands on white orchid flowers and pointing me to the extensive literature on this phenomenon in plant pathology. I am very grateful to those who supplied pictures as cited in the caption of Fig. 2. Research on plant development in my laboratory is supported by a grant from the National Science Foundation (PGRP 07-01880).

References

1. Walbot V, Rudenko GN (2002) *MuDR/Mu* transposons of maize. In: Craig NL, Craigie R, Gellert M, Lambowitz A (eds) Mobile DNA II. Amer. Soc. Microbiology, Washington, DC, pp 533–564
2. Levy AA, Walbot V (1990) Regulation of the timing of transposable element excision during maize development. Science 248: 1534–1537
3. Athma P, Grotewold E, Peterson T (1992) Insertional mutagenesis of the maize *P* gene by intragenic transposition of *Ac*. Genetics 131: 199–209
4. Shang YJ et al (2011) The molecular basis for venation patterning of pigmentation and its effect on pollinator attraction in flowers of Antirrhinum. New Phytol 189: 602–615
5. Walbot V (1985) On the life strategies of plants and animals. Trends Genet 1:165–169
6. Walbot V (1996) Sources and consequences of phenotypic and genotypic plasticity in flowering plants. Trends Plant Sci 1:27–32
7. Coe EH Jr, Thompson D, Walbot V (1988) Phenotypes mediated by the *iojap* genotype in maize. Am J Bot 75:634–644
8. Han CD, Coe EH Jr, Martienssen RA (1992) Molecular cloning and characterization of *iojap* (*ij*), a pattern striping gene of maize. EMBO J 11:4037–4046
9. Szymkowiak EJ, Sussex IM (1996) What chimeras can tell us about plant development. Annu Rev Plant Physiol Plant Mol Biol 47:351–376
10. Poethig RS (1997) Leaf morphogenesis in flowering plants. Plant Cell 9:1077–1087
11. Giron D, Kaiser W, Imbault N, Casas J (2007) Cytokinin-mediated leaf manipulation by a leafminer caterpillar. Biol Lett 3:340–343
12. Huang Y-J et al (2006) Temperature and leaf wetness duration affect phenotypic expression of *Rlm6*-mediated resistance to *Leptosphaeria maculans* in *Brassica napus*. New Phytol 170:129–141
13. Griffiths AJF, Ganders FR (1983) Wildflower genetics: a field guide for British Columbia and the Pacific Northwest. Flight Press, Vancouver, BC

Chapter 3

Using Transposons for Genetic Mosaic Analysis of Plant Development

Philip W. Becraft

Abstract

Genetic mosaics, or chimeras, are individual organisms composed of cells or tissues of two or more distinct genotypes. They are experimentally useful for addressing several key biological questions. These include fate mapping through analysis of marked clonal lineages, analyzing cell or tissue interactions such as the induction of developmental events, and analyzing whether a gene acts cell autonomously. Genetic mosaics can arise in many ways, including through the action of transposable elements. Naturally occurring transposons can generate genetic mosaics by somatically inserting into a gene to cause a mutant sector, somatically excising from a mutant gene to create a revertant wild-type sector, or causing chromosomal breaks or rearrangements leading to loss of a gene or genes. Transposons have also been cleverly engineered to allow the generation of marked somatic sectors, sometimes in controlled ways. Here we review ways in which transposon-induced genetic mosaics have been used experimentally, the various methods that have been used, and general considerations for designing genetic mosaic studies using transposon methods.

Key words Chimera, Sector, Clonal analysis, Lineage analysis, Gene action, Cell autonomy, Cell interaction

1 Introduction

Genetic mosaics, or chimeras, are individual organisms containing cells with two or more distinct genotypes. Genetic mosaics can arise through a variety of mechanisms including several involving transposable elements (TEs or transposons). Transposable elements can produce genetic mosaics by virtue of genetic reversion of a transposon-induced mutant allele, through chromosome breakage caused by aberrant transposition of particular TEs, through excision of TEs from engineered constructs introduced as transgenes, or via epigenetic modification of TE activity. Events occur where the genotypes of individual progenitor cells are altered during the development of the organism. A clone of genetically distinct cells

Thomas Peterson (ed.), *Plant Transposable Elements: Methods and Protocols*, Methods in Molecular Biology, vol. 1057, DOI 10.1007/978-1-62703-568-2_3,

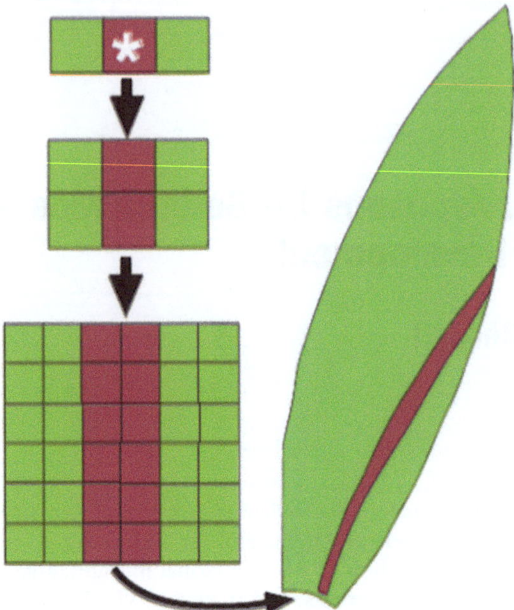

Fig. 1 Formation of a genetic mosaic sector. A transposition event in a precursor cell (*asterisk*) creates a genotypic difference (*purple color*) from the background genotype. Cell divisions create a clone of daughter cells derived from the original progenitor, forming a genetically distinct sector

arises from such an event and, due to the absence of cell migration in plants, forms a somatic sector (*see* Fig. 1).

One of the advantages of transposon-based methods is that, depending on the transposon used, transposition events may occur at any time during development. This is in contrast to methods that rely on ionizing radiation to induce chromosome breakage for generating sectors. With most radiation sources, it is convenient to treat seeds but may be difficult to treat growing plants. Thus, such analyses are typically restricted to post-embryonic development because it is problematic to treat late stages of plant development or early embryonic development, which occurs while seeds are still developing within floral organs of the mother plant.

Genetic mosaics afford a powerful tool for addressing several aspects of developmental biology and genetics, including lineage analysis, studying gene action, studying cell and tissue interactions, and examining the cellular phenotypes of embryo-lethal mutants. General aspects of plant genetic mosaic analysis have been extensively reviewed [1–4]. Here, examples of each of these types of analyses will first be considered from a biological perspective. This will be followed by a technical treatment considering experimental design and methods by which transposons can be used to generate genetic mosaics.

2 Materials

2.1 Lineage Analyses

2.1.1 Endosperm

As mentioned, transposon-mediated genetic mosaic plants involve somatic sectors that each derive from a single cell in which a genetic event occurred (*see* Fig. 1). When such sectors are marked with an identifiable phenotype, it serves as a powerful means to ascertain ontological cell lineage relationships and patterns. The sizes and shapes of sectors can be used to infer the patterns of cellular growth and division during development. Barbara McClintock recognized the clonal basis of "mutable" sectors even as she was performing her historic research discovering transposable elements [5]. Subsequently, she demonstrated the correlation between cell division patterns and the appearance of sectors in maize endosperm using starch and anthocyanin markers [6]. *Waxy1* (*wx1*) encodes a granule-bound starch synthase required for amylose synthesis [7]. *Wx1* sectors can be detected in internal starchy endosperm tissues by assaying amylose content with iodine stain. In particular genotypes, the surface aleurone tissue of the endosperm accumulates anthocyanin and many of the genes required for anthocyanin accumulation are affected by TEs.

Early-occurring TE-induced changes in *wx1* gene expression produced conical sectors radiating from the central regions of the endosperm to the surface. Later events produced smaller cones that emanated from more peripheral regions out toward the surface. These were consistent with observed patterns of cell division that become localized to the peripheral regions of the endosperm during kernel growth [8, 9]. Most strikingly, very early sectors occupied consistent regions of the endosperm. Half-kernel sectors of either *wx1* or anthocyanin markers consistently divided the kernel into left and right halves. Quarter-kernel sectors divided the kernels into consistent quadrants and so on. Such sectors indicated that the orientations of early nuclear divisions during the coenocytic stage of endosperm development were stereotypical [6].

Histological descriptions of endosperm development described cell divisions in the aleurone layer as restricted to the anticlinal plane [9], suggesting that the aleurone cell lineage was distinct from the rest of the endosperm [10]. Very-late-occurring sectors were studied to examine questions of cell lineage of the aleurone in relation to the internal starchy endosperm. A chromosome-breaking *Ds* element was used to simultaneously uncover *wx1*, to mark starchy endosperm cells, and *C1*, to confer anthocyanin to otherwise unpigmented aleurone cells [11]. Late-occurring *wx1* sectors were observed to be "capped" by purple aleurone indicating that the starchy endosperm and aleurone lineages were not separate, contrary to what had previously been posited.

2.1.2 Shoot Apical Meristems and Leaves

Lineage analysis was used to investigate the establishment and stability of apical initial cells in the maize embryonic shoot apical meristem (SAM) [12]. Maize typically produces five embryonic leaves during seed development. *Gl1* encodes a component of cuticular wax biosynthetic machinery and *gl1* mutants are recognized by their lack of epicuticular wax on juvenile leaves [13]. The *gl1-m8* allele is unstable due to insertion of an autonomous *En/Spm* transposon, and genetic reversion produces sectors that are visible by the restoration of surface wax on juvenile leaves. A prevalence of early events occupied lateral halves of the plant, from the leaf midvein on one side of the plant to the midvein on the opposite leaf. This implied that the SAM was often derived from two initial cells that were oriented side by side with respect to the bilateral symmetry of the plant. Furthermore, the relative portions of the leaves occupied by such sectors often changed (increased or decreased) going up the plant, demonstrating that the apical initials are dynamic rather than a permanent population of cells [12].

The concept of a developmental "compartment" was introduced in the study of clonal mosaics in *Drosophila* wings, where somatic sectors obeyed morphological boundaries despite having varying proliferative capacities [14]. Additional analysis of revertant *gl1-m8* sectors in maize suggested that developmental compartments may also occur in plant development. Revertant leaf sectors occupied longitudinal stripes whose boundaries typically corresponded to lateral veins and were often bounded by successive lateral veins [15]. The area between successive lateral veins was considered a developmental module and the lateral veins represent compartment boundaries. Sector analysis further suggested that a module descended from a group of four founder cells during leaf primordium initiation, and that these four cells were ultimately derived from a single progenitor cell.

2.1.3 Leaf Epidermis

Leaf epidermises contain a variety of cell types that show characteristic spacing patterns. There are several mechanisms by which patterns could theoretically be established during development and knowing the lineage relationships of cells can help distinguish among these mechanisms. In grasses and other monocots, stomata tend to occur in "chains," files of cells that average 22 cells long in juvenile maize leaves [16]. A discrete clonal relationship among cells in a chain was proposed to be an element of specifying cell fate [17]. Several predictions derive from this hypothesis, for example that genetically marking a "chain progenitor" cell should produce a clonal sector that corresponds in length and width to the stomatal chain, or that those stomatal chains should not cross sector boundaries. Lineage analysis was performed in maize with two different marker systems, one using epidermal wax characteristics and the other using anthocyanin pigments [16]. The maize *glossy15* (*gl15*) gene encodes an AP2-like transcription factor that controls the deposition of epicuticular waxes on juvenile leaves [18]. The *gl15-m1* mutant allele contains a nonautonomous *dSpm* transposon insertion which, in the presence of an autonomous *En/Spm* element, can excise and restore *gl15* gene function. The *bronze1-m* allele contains a *Ds* transposon which can similarly excise in the presence of *Ac* [19]. Lineage relationships were inferred by studying revertant wax-bearing sectors on juvenile leaves, or anthocyanin-pigmented cells in adult leaves. Contrary to explicit predictions of the model, the sizes, shapes, and boundaries of sectors did not conform to stomatal chains [16]. Thus, another mechanism of spatial patterning must be responsible for establishing the fate of stomatal chains.

Similar studies have been conducted to explore the patterning of epidermal cell types on *Arabidopsis* leaves. In *Arabidopsis*, stomata very rarely form adjacent to each other, and most stomatal complexes consist of a pair of guard cells surrounded by three subsidiary cells. Two mechanisms proposed to explain stomatal patterning in dicots include a signaling model involving lateral inhibition, and a cell lineage model by which a stereotypical sequence of cell divisions produce regular spacing. Lineages were marked with a *35S::Ac::GUS* transgene, where excision of the maize *Ac* transposon restored beta-glucuronidase (GUS) gene activity, which is detectable by histochemical staining [20]. The vast majority of sectors in the two-to-five cell size range, that included a stomate, contained the two guard cells and zero to three subsidiary cells. Thus, all the cells of most stomatal complexes are clonally related, supporting the lineage-based patterning model.

Trichomes also occur as part of multicellular complexes in *Arabidopsis* leaf epidermis and complexes rarely occur in adjacent positions. The *35S::Ac::GUS* system was again used to perform a similar lineage analysis [21]. Trichomes that formed on sector boundaries were examined and it was found that complexes could contain clonally unrelated cells. Furthermore, there was no consistent pattern to the portions of complexes that were bisected by sector borders. Therefore, a lineage mechanism does not appear to determine the spacing pattern of trichomes.

2.1.4 Flowers

Maize anther development was studied by analyzing sectors of cells marked with anthocyanin pigments caused by genetic reversion of *Ds*-induced mutant alleles of the *a1* or *bz1* genes [19, 22]. Only half of reversion events that caused pigmented anthers were heritable and it was found that events in either the L1 or the L2 meristem layers could result in anther pigmentation but only the L2 events were heritable. Thus, the pollen lineage was shared with the inner cell layer of the bilayered anther wall, but not the outer layer.

An anthocyanin marker was also used to mark clonal sectors to study growth dynamics of the complexly shaped flower petals of snapdragon, *Antirrhinum majus* [23, 24]. The *pallida* (*pal*) gene is required for anthocyanin pigmentation in the epidermis of flower organs and the *pal^rec^-2* allele has disrupted function due to insertion of a *Tam3* transposon in the gene promoter. A novel feature of the *Tam3* transposon system is that transposition activity is cold inducible; a shift from 25 to 15 °C induces red sector formation in *pal^rec^-2* mutants. By inducing sectors at different developmental time points, the growth of different regions of petals could be inferred. This allowed the development of sophisticated spatiotemporal models describing the dynamic cellular growth behaviors of different regions of petals throughout flower development [23, 24]. It was determined that petal asymmetry could be attributed to differential directions of anisotropic growth (change in length ≠ change in length) but that differential growth rates or differential degrees of anisotropy were not important.

2.1.5 Roots

Cell lineage analysis was conducted in *Arabidopsis* to construct an embryonic fate map of root progenitor cells. The aforementioned *35S::Ac::GUS* system was employed and seedling roots were stained for GUS activity. It was found that the hypophyseal cell begets columella cells and the central cell of the root apical meristem (RAM). In contrast, sectors within the root or hypocotyls showed variable endpoints, indicating that embryonic cells make variable contributions to the seedling root [25].

The *Arabidopsis* RAM consists of four central cells, which are surrounded by initials that undergo stereotypical cell divisions to generate the tissues of the root. The *35S::Ac::GUS* marker system was used to analyze the generation of root epidermal tissues after germination [26]. Three cell lineages were found, one including the columella root cap, one for the epidermis and lateral root cap, and one for the cortex and endodermis. Sectors for each lineage were shown to trace back to an initial cell in the RAM. Thus, genetically marked lineages confirmed histological observations on cell division patterns. The later study was followed up using a temperature-inducible transposon system, conceptually similar to that described for the *Antirrhinum Tam3* transposon, but this one was engineered. A *35S::Ds::GUS* reporter expresses GUS whenever the nonautonomous *Ds* transposes to restore a functional transgene. The transposase source was a second transgene containing the *Ac* transposase coding region controlled by a soybean heat-shock promoter [27]. Seedlings were heat shock-induced 3 days after germination, and sectors were analyzed 2 days later. Results confirmed the earlier study and allowed the determination of the number of meristematic cells present in each tissue lineage. Furthermore, it was found that (a) central cells were mitotically active, (b) initial cells were not permanent and could be replaced by daughters of central cell divisions, and (c) cell layer invasions were common from one cell file to another demonstrating the importance of cell interactions in specifying cell fate [28].

Another *Ds*-containing reporter was developed with a translational fusion between histone 2B and yellow fluorescent protein (YFP) driven by the 35S promoter. Expression of the reporter, upon excision of the *Ds*, results in the accumulation of the fusion protein in nuclei causing them to fluoresce [29]. This allows live cell imaging, enabling individual sectors to be analyzed at multiple time points during development. Three cell files within the pericycle contribute initial cells to the formation of a lateral root primordium. However, it was found that only derivatives of the central file contributed to the entire distal region of the lateral root primordium while derivatives of the flanking initials were restricted to the base of the primordium.

2.2 Gene Action

A primary question in the study of gene action is whether a gene functions cell autonomously or non-autonomously. That is, are the phenotypic consequences of a particular cellular genotype restricted to just the cells of that genotype, or can the phenotypes of neighboring cells of a different genotype be influenced? In cases of non-autonomy, the implication is that the gene either encodes a mobile gene product or regulates, perhaps indirectly, the production of a mobile substance.

There are several examples where the phenotypes of revertant sectors derived from transposon-induced mutants have been examined at the cellular level. The maize *gl15* gene, which encodes an AP2-like transcriptional regulator that controls epicuticular wax formation and other epidermal cell characteristics on juvenile leaves [18, 30], was already mentioned in reference to a lineage analysis [16]. It was originally concluded that the *gl15* gene functioned cell autonomously by microscopically examining revertant sectors from an *En/Spm*-induced mutant allele [18]. The production of juvenile versus adult cell characteristics strictly adhered to cellular boundaries; individual cells either showed the mutant phenotype or a normal phenotype. This strongly suggested that the gene functioned cell autonomously—any epidermal cell carrying the *Gl15+* revertant allele showed a normal phenotype while adjacent homozygous mutant cells showed the mutant phenotype.

Similar conclusions were reached by studying reversion of a *Mutator* (*Mu*) transposon-induced allele of *dek1* in maize [11]. *Dek1* encodes a membrane-localized calpain protease that is required for the specification of an aleurone layer in the endosperm [31]. The occurrence of late, single-celled revertant *dek1* sectors that differentiated as aleurone demonstrated that *Dek1* likely functioned cell autonomously in the endosperm. Furthermore, peripheral endosperm cell fate remained plastic, and the positional cues necessary for aleurone cell fate specification were present throughout development [11].

The same principle was used to study the action of the *rolC* gene of *Agrobacterium rhizogenes* [32]. The *rolC* gene is carried on the T-DNA and induces root formation in transformed plant tissues. It was hypothesized to mediate the synthesis of growth regulators and as such, it was of interest to determine whether the gene functioned cell autonomously or whether transformed cells could induce root formation in neighboring untransformed cells. When the *rolC* gene was expressed in leaf tissues, under the control of a 35S promoter, chlorophyll content was decreased. An *Ac* transposon-inactivated version of the 35S:rolC construct was generated and transformed into tobacco. Upon excision of the *Ac*, the function of *rolC* was restored, resulting in pale, chlorophyll-deficient sectors in otherwise green leaves. The boundaries of these sectors were sharply defined suggesting that the action of *rolC* was cell autonomous in leaf tissues [32].

In all the above examples, transposition restored gene function resulting in sectors of function in otherwise nonfunctional backgrounds. Except for the *rolC* example, this produces wild-type sectors in otherwise mutant tissues. It is also possible to conduct such experiments in the opposite direction, generating loss-of-function sectors in an otherwise functional background. This can be accomplished using chromosome-breaking TEs, described below. In individuals heterozygous for a recessive mutant, and carrying a breaker TE proximal to the dominant wild-type allele, chromosome breakage results in loss of the acentric fragment carrying the wild-type allele, resulting in a clonal sector hemizygous for the mutation (*see* Figs. 2, 3, and 4). Thus, mutant sectors are formed in a background of otherwise normal tissue.

Such studies were first conducted in maize by McClintock who observed that in the presence of *Ac*, a *Ds* element on chromosome 9 caused breakage events that generated sectors of kernel markers [5, 33]. These included the *C1* and *Bz1* genes involved in anthocyanin production of the aleurone layer, and *Wx1* involved in amylose biosynthesis. She presented a striking example of cell-nonautonomy and gene interactions in anthocyanin production. Kernels were generated with the following genotype: *C1-I*, *Bz1*, *Ds/C1*, *bz1*, – Tissue of this genotype lacks anthocyanin pigment because *C1-I* is a dominant-negative allele of the *C1* gene required for anthocyanin synthesis. *Bz1* encodes UDPG-flavonol 3-0-glucosyl transferase enzyme, which catalyzes a late step in anthocyanin biosynthesis. *Ds*-induced chromosome breakage results in loss of the chromosome arm carrying the *C1-I* and *Bz1* alleles, uncovering wild-type *C1* function and the *bz1* mutant allele. Sectors of this genotype are bronze colored due to accumulation of an anthocyanin intermediate compound. Interestingly, the sector boundaries were bordered by anthocyanin pigmentation several cells wide. The interpretation was that *C1-I* blocks production of a precursor for the BZ1-catalyzed reaction. However, this precursor was produced in cells of the *C1*, *bz1* sectors and could diffuse a short distance into the tissue

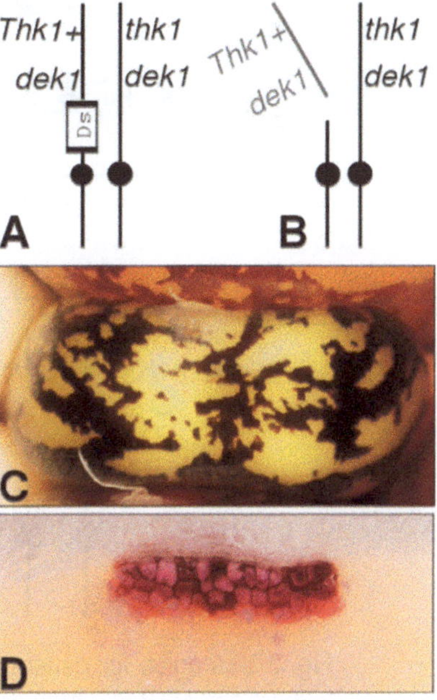

Fig. 2 Example of using genetic mosaics to study gene interactions: Epistasis of *thk1* over *dek1* [37]. (**a**) The starting genotype was heterozygous for *Thk1+/thk1* and homozygous mutant *dek1*. This genotype produces endosperm with no aleurone layer. The *Thk1+* allele is located distal to a chromosome-breaking *Ds* transposon and an *Ac* element is present in the genome. (**b**) Breakage of the chromosome at *Ds* results in loss of an acentric chromosome fragment carrying the wild-type *Thk1+* allele, uncovering the recessive *thk1* mutant allele on the homologous chromosome. (**c**) The regions of the kernel lacking anthocyanin show the *dek1* mutant phenotype lacking aleurone. Where the *Thk1+* allele is lost, the *thk1* mutant phenotype is revealed, visible as anthocyanin-pigmented sectors of aleurone. (**d**) Section showing the multi-layered and pigmented thick aleurone phenotype in a *thk1* sector (double mutant for *dek1*) in a background of *dek1* (single mutant) tissue lacking aleurone

containing the functional *Bz1* allele and thus be converted to anthocyanin.

Chromosome-breaking TEs have subsequently been used to study the action of a number of genes including *vp5* [34]. Mutants of *vp5* are carotenoid deficient, resulting in albino leaves that cause seedling lethality. In the presence of *Ac*, a chromosome-breaking *Ds* element on chromosome 1S caused loss of the wild-type *Vp5+* allele, generating cell-autonomous albino sectors throughout the photosynthetic tissues of the plant.

Rigorous testing of cell autonomy of gene action requires that sector borders be confirmed independently of the gene or the phenotype of interest. This can be accomplished by simultaneously uncovering a cell-autonomous marker along with the gene of interest (GOI). Chromosome-breaking TEs have been used for

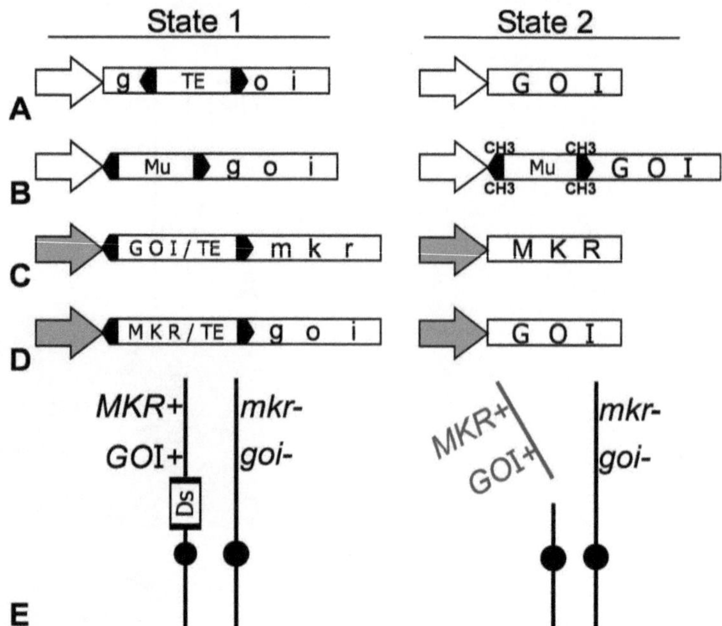

Fig. 3 Strategies for generating genetic mosaics with TEs. GOI is the gene of interest and MKR is a marker gene. *Capital letters* represent the functional state and *small letters* are the nonfunctional state for each. *Arrows* designate promoters, which in (**a**, **b**) are the endogenous gene promoters, and in (**c**, **d**) can be whatever promoter that is desired for the engineered construct (see text). State 1 is the starting condition, which represents the background in which mosaic sectors occur. State 2 represents the condition associated with the genetic change found within somatic sectors. (**a**) A mutable allele. A TE insertion in a GOI disrupts the gene function (State 1). This allele should be homozygous or heterozygous with a stable recessive mutant allele. Transposition of the TE out of the gene restores GOI function (State 2), thus creating sectors of cells with a functional GOI in a nonfunctional background. (**b**) *Mu*-suppressible allele. Insertion of a *Mutator* element in the 5′ region of a gene disrupts GOI function when *Mu* is in the active state (State 1). This allele should be homozygous or heterozygous with a stable recessive mutant allele. Epigenetic silencing of *Mu* is associated with cytosine methylation in the TIRs, which allows expression and restores function of the GOI (State 2). (**c**) A generalized diagram of an engineered construct that was used to study Arabidopsis flower development [43]. The GOI is situated within the TIRs of a TE and expressed under the control of an appropriate promoter. Thus the GOI is functional while the marker gene is not expressed in State 1. The construct is used in a homozygous mutant background for the GOI. Transposition removes the dominant wild-type GOI allele and activates the marker, generating marked sectors of mutant cells in tissues that are otherwise wild type. (**d**) Another version of an engineered construct [43], also used in a homozygous mutant background for the GOI. The marker is contained within the TE and expressed under the control of the promoter in State 1, while the GOI is not expressed. Transposition removes the marker and activates expression of the GOI, generating wild-type, unmarked cells in tissues that are otherwise mutant and marked. (**e**) Chromosome-breaking *Ds* elements. Dominant wild-type alleles of the GOI and a marker are located on a chromosome arm, distal to a chromosome-breaking *Ds*. Recessive mutant alleles of the GOI and marker are located on the homologous chromosome and since both are heterozygous, the initial phenotype in State 1 is wild type and unmarked. With an active *Ac* in the genome, the *Ds* undergoes aberrant transposition resulting in loss of the acentric distal fragment containing the wild-type alleles

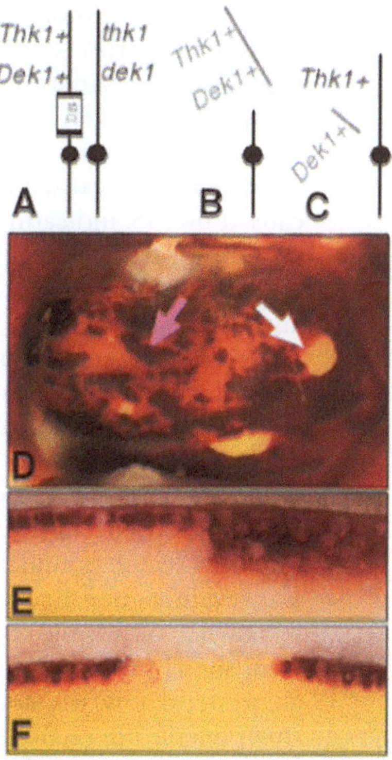

Fig. 4 Evidence of interstitial loss from a chromosome-breaking *Ds* element *Ds1S4* on maize chromosome 1. (**a**) The initial genotype is heterozygous *Thk1+*, *Dek1+/thk1*, *dek1*, which produces a normal phenotype with a single aleurone layer. (**b**) A standard breakage event causing simultaneous loss of *Thk1+* and *Dek1+*. (**c**) An interstitial deletion that causes loss of *Dek1+* but retains *Thk1+*. (**d**) A kernel with the starting genotype shown in (**a**). Simultaneous loss of *Thk1+* and *Dek1+* as shown in (**b**) (or loss of *Thk1+* alone) produces thick aleurone sectors visible as *dark patches* (*shaded arrow*) [37]. Interstitial breaks causing loss of *Dek1+* alone cause unpigmented sectors devoid of aleurone (*white arrow*). (**e**) Section through the border between normal tissue (single layer of pigmented aleurone on the left) and a mutant sector showing the thick aleurone phenotype of *thk1*. (**f**) Section through an unpigmented sector showing the aleuroneless phenotype of *dek1*

several examples of this. The *dek1* mutant is embryo-lethal and shows pleiotropic defects in the endosperm, including floury texture, and a lack of aleurone. Sectors of *dek1* mutant tissue appeared cell autonomous in the aleurone [11, 35, 36] but did not show sharp boundaries of the endosperm texture phenotype, suggesting that the gene functioned non-cell autonomously for this trait [36]. To test this further, the cell-autonomous *vp5* marker was recombined with *dek1* and the doubly mutant chromosome crossed to the *Ds1S4* line which contains a chromosome-breaking *Ds* proximal to *dek1* [35]. In leaves, the boundaries of abnormal epidermal cell morphology generally coincided with boundaries of albinism suggesting that in leaf tissues, *Dek1* gene action was cell autonomous.

Similar conclusions were reached with yet another embryo-lethal mutant on chromosome 1S. The *thick aleurone1* (*thk1*) mutant causes multiple aleurone layers in the endosperm; aleurone sectors appear cell autonomous, at least in the lateral dimension [37]. The *thk1* mutant allele was also marked with *vp5* and again, aberrant cell phenotypes caused by *thk1* generally corresponded to albinism caused by *vp5*, indicating that *Thk1* is likely cell autonomous. One interesting point is that isolated aleurone cells were not observed internal from the surface suggesting that *Thk1* might not act cell autonomously between cell layers of the endosperm. Directional differences in gene action also occur in other tissues as described in the following section on tissue interactions.

A powerful use of genetic mosaics that has not been extensively utilized is in the study of gene interactions. For example, tissues of different genotypes can produce a biochemical complementation resulting in the production of visible anthocyanin pigments at sector boundaries, as described above. Another example was recently published [37]. A double-mutant analysis suggested a somewhat surprising result: the maize *thk1* mutant with multiple aleurone layers was epistatic to *dek1*, which normally is devoid of aleurone; that is, the double mutant had a *thk1* phenotype. To test this genetic interaction, a mosaic experiment was performed to generate *thk1; dek1* double-mutant sectors in kernels that were otherwise *dek1* single mutant (*see* Fig. 2). As predicted by the double-mutant analysis, the epistatic relationship was evident from the formation of multi-aleurone sectors in kernels that otherwise lacked aleurone.

2.3 Tissue Interactions and Cell Fate

The development of multicellular organisms requires extensive coordination among cells and tissues. Developmental events in one tissue often require instructive signaling from a different tissue. Genetic mosaics can be employed to identify and study such inductive events, as well as the action of specific genes involved in the process. For example, flower development is a complex process requiring the differentiation of specialized cell types and the coordinated development of multiple tissues and organs. *Antirrhinum* plants produce spike inflorescences bearing approximately 20 flowers. Mutants in the *floricaula* gene transform flower meristems to inflorescences resulting in a sterile structure where indeterminate inflorescence branches replace flowers. The *flo-613* allele is caused by a *Tam3* transposon insertion and excision can restore gene function. Revertant sectors generate flowers ranging from near-wild-type morphology to variably aberrant [38]. A periclinal sector in any of the meristematic cell layers is sufficient for flower development, with reversion in the L1 able to confer near-wild-type flower phenotype, including sporogenous tissues in the L2 [39]. Expressions of the downstream homeotic genes, *DEFICIENS* (*DEF*) and *PLENNA*, were induced in all three layers even though

FLO was functional in just the L1 or the L3. Subsequent analysis of the *Arabidopsis* ortholog *LEAFY* showed that plants expressing *LFY* RNA only in the L1 layer contained LFY protein throughout the inflorescence meristem, indicating that the protein trafficked between cells [40]. Thus, any cell layer in the flower meristem is capable of inducing flower development throughout.

Flower organ identity is controlled by homeotic genes according to the "ABC" model (reviewed in ref. 41). B class floral homeotic genes encode MADS box transcriptional regulators required for the petal and stamen organ identity. In *Antirrhinum* and *Arabidopsis*, mutations in B function genes cause the second whorl organs to develop as sepals instead of petals and the third whorl to assume carpel identity instead of stamen. Despite evolutionarily conserved sequences and functions, the behavior of B function genes in genetic mosaics differed between these two species. Revertant sectors of TE-induced alleles of *def* and *globosa* (*glo*) showed that wild-type internal layers could confer normal development to the L1 but not vice versa [42]. When RNA was produced internally, protein was observed in all layers including the L1, but when RNA was confined to the L1, so was the protein. The revertant phenotypes correspondingly included all cell layers or were confined to the L1, respectively. Thus DEF and GLO proteins appear to show polar trafficking between layers. In contrast, *AP3*, the *Arabidopsis* ortholog of *DEF*, does not appear itself to be trafficked but rather to exert nonautonomous effects through downstream genes [43]. Mosaics were generated by an innovative engineered transposon system, described below, that results in sectors being marked either by gain or loss of a GUS marker. In both the second and third whorls, AP3 protein was restricted to cells that contained a functional copy of the *AP3* gene, as determined by the GUS marker. In the second whorl (petals), *AP3* function appeared largely cell layer autonomous except that organ shape was determined by the epidermal (L1) genotype. Interestingly, different results were obtained for stamens where *AP3* function in the L2 + L3 was able to confer stamenoid identity to epidermal cells. There were also layer nonautonomous effects on maintaining AP3 protein levels. Thus, the mode of action for this B-class gene varies in different developmental contexts and appears to have diverged during the evolution of *Arabidopsis* and *Antirrhinum*. Also the functional relationships among meristem layers differ among organs.

Mosaic analysis can also be applied to study the modes of action for dominant mutants. The dominant *Lg3* mutant causes ectopic expression of the *Lg3* knox homeobox gene, which results in abnormal formation of the ligular region of maize leaves [44–46]. Genetic mosaics were generated using coordinate epigenetic suppression of *Mu* transposons, described in detail below. Analysis of mutant sectors showed that *Lg3* generally acted cell autonomously in the lateral dimension, but not in the transverse dimension [44, 45]. A mutant genotype in the epidermal layer had no influence on the leaf phenotype, even though many of the characteristics altered in the mutant are epidermal, including the ligule. However, the action of the mutant allele in mesophyll cells did affect epidermal cell fate [44]. The timing of sector formation was also important; earlier-formed sectors caused more proximal cell fate transformations, while progressively later-formed sectors resulted in a gradation of less proximal cell fates [45].

2.4 Cellular Phenotypes of Embryo-Lethal Mutants

Genes often function throughout development or at multiple times. However, if a mutant causes early lethality, the possible function of the gene at later stages of development cannot be ascertained. Genetic mosaics offer the opportunity to observe the phenotype of adult tissues that are mutant for embryo- or seedling-lethal genes. For example, the *thk1* mutant described above causes multiple aleurone layers [37]. Many aleurone mutants show corresponding phenotypes in the sporophyte epidermis [47] but this relationship could not be directly addressed because the *thk1* mutant is embryo-lethal. A chromosome-breaking *Ds* element was used to uncover the *thk1* mutant allele and leaf sectors were examined [37]. Mutant epidermal cells were enlarged and irregularly shaped but were generally only a single cell layer thick. Therefore, it was concluded that the developmental function of *Thk1* is different in leaf epidermis than it is in the aleurone.

Such use of chromosome-breaking TEs was a rationale for an effort to distribute breaker *Ds* elements around the genome of maize [36]. In addition to *thk1*, this system has been used to uncover adult leaf albino sectors of the seedling-lethal, carotenoid-deficient *vp5* mutation [34]. It was also used to show the aberrant morphology of leaf epidermal cells mutant for an embryo-lethal *dek1* allele [35]. Furthermore, this analysis revealed the autonomy of *Dek1* gene action in leaf tissues, whereas autonomy in endosperm tissues had been inconclusive due to an ineffective marker [35].

3 Methods

There are two basic types of genetic mosaics that can be generated with transposons; either a gene can be rendered functional (often wild type) in a background of cells containing a nonfunctional (mutant) gene or a gene can be rendered nonfunctional in a background of cells containing a functional gene. Three general systems that can produce these effects are considered here and summarized in Fig. 3. Also, see notes 1-6 regarding general considerations in designing TE-based genetic mosaic experiments.

3.1 Reversion of "Mutable" Alleles

TE-induced mutants are famous for their propensity to revert to functional states upon excision of the TE, and it is this genetic instability that leads to the term "mutable." This is a well-known means of producing genetic mosaics consisting of revertant sectors of wild-type cells in a background of mutant tissue (*see* Fig. 3a). This is the simplest method of using TEs for mosaic analysis and also the most limited. The major limitation is that a mutable TE-induced allele for a particular GOI must be available; otherwise it is a significant undertaking to tag it. A second major limitation is that sectors are not independently marked—thus one has no basis for determining sector boundaries other than by the phenotype conferred by the mutable gene per se. Although this is adequate for conducting lineage analysis, for examination of gene action or tissue interactions it is highly desirable to have sectors independently marked with a known cell-autonomous marker. Several studies based on such systems were discussed above [11, 12, 18, 39].

3.2 Epigenetic Silencing of "Mutable" Alleles

Several TEs are subject to epigenetic regulation, which can affect the expression of TE-induced mutations. In the case of the Mutator (*Mu*) transposon family, epigenetic silencing of a *Mu* element involves methylation of the terminal inverted repeats and sometimes results in suppression of a mutant phenotype [48]. This generally occurs in alleles where the *Mu* insertion is in the 5′ region of a gene. There is a cryptic promoter in the TIRs of some *Mu* elements but when *Mu* is in the active state, transcription from this promoter is blocked. However, when the *Mu* system becomes silenced, the cryptic promoter element becomes active which can allow expression of the affected gene and restoration of the wild-type phenotype (*see* Fig. 3b) [49].

Suppression can occur and be stably inherited in somatic cells, resulting in sectors of phenotypically normal tissue in an otherwise mutant background. When epigenetic silencing occurs, *Mu* elements throughout the genome are affected, and multiple mutant loci can simultaneously be suppressed [44, 50]. Thus if two suppressible *Mu*-induced mutants are present in an individual (and they need not be on the same chromosome), they both become suppressed in the same cells. Hence, a suppressible allele of a GOI can be combined with a suppressible marker, such as the *a1-mum2* mutant, which affects anthocyanin accumulation. When *Mu* is active, the *a1* anthocyanin-deficient mutant phenotype expresses; when *Mu* becomes silenced, the *a1-mum2* allele is suppressed, allowing anthocyanin production in somatic sectors. This marker was used in conjunction with the *Mu*-suppressible *Lg3-Or211* mutant [44]. Since *Lg3-Or211* is dominant, suppression results in a wild-type phenotype. Simultaneous silencing of *Lg3-Or211* and *a1-mum2* created sectors of *Lg3+* wild-type tissue pigmented red with anthocyanin, in a mutant background that was unpigmented.

In principle, similar analyses can be done using the epigenetic cycling of other TE systems, including *Ac* or *En/Spm*. However, these systems do not provide cellular resolution, nor simultaneous activation as seen with the *Mu* system because markers switch between mutant and mutable [6, 51]. In other words, sectors occur as stable mutant versus unstable mutant (producing revertant sectors) rather than mutant vs. suppressed (wild type) as seen with the *Mu* system.

3.3 Chromosome-Breaking Elements

When more than two TIRs are in close proximity, aberrant transpositions can result in loss of chromosomal integrity. Such configurations have been studied for *Ac/Ds* elements [52–57] and indeed appear to be the basis for the chromosome dissociation events that are the namesake of *Ds* and led to its discovery by Barbara McClintock [5, 33, 57]. As shown in Figs. 2–4, such "breaker" TEs can be used to create genetic mosaics by uncovering one or more recessive alleles on the homologous chromosome opposite the breaker [5, 33, 34, 36]. This can allow simultaneous uncovering of a GOI with a marker [11, 35, 37], or of two genes of interest for studying genetic interactions [37]. The breaker line is simply crossed to the recessive mutant and in the heterozygous F1, sectors of mutant cells will be produced in a wild-type background. Chromosome breakage results in the production of an acentric fragment containing the chromosome arm distal to the site of the breaker TE. The acentric fragment is lost in subsequent cell divisions, allowing expression of the hemizygous recessive allele(s).

Neuffer undertook a systematic effort to distribute chromosome-breaking *Ds* elements around the maize genome and identified breaker lines for 12 of the 20 chromosome arms [36]. Thus, a large portion of the maize genome is accessible to mosaic analysis with this system. The main limitation is that it is highly preferable that target genes should occupy chromosomal positions distal to the breaker sites. Because breaker TEs can also generate rearrangements (*see* Fig. 4), it is also possible to uncover proximal loci. Varying sector sizes and frequencies have been interpreted as resulting from proximal versus distal position of the marker relative to the TE [36], although these interpretations have not been verified by mapping.

3.4 Engineered Elements

Engineered genetic constructs offer abundant opportunity for flexibility and creativity in tailoring genetic mosaics to almost any situation. Analyses based on engineered constructs have been described above [25, 27–29, 32, 43]. There are two major components to engineered systems, the transposase source and the target construct containing the GOI and reporter, each considered below.

3.4.1 Target Constructs

There are two basic configurations for target constructs, summarized in Fig. 3. A TE can be incorporated into the construct such that it disrupts the function of a GOI, restoring the GOI upon transposition. Conversely, the target can be configured such that it is initially functional and then rendered nonfunctional upon transposition. Likewise, it is possible for transposition to render the marker functional or nonfunctional. In either case, the construct is placed in a genetic background mutant for the GOI so as to make the phenotype contingent on whether or not the construct is active. Figure 3c, d depicts a generalized version for constructs that were used for mosaic analysis described for *AP3* in Arabidopsis [43]. As depicted, loss of the GOI is marked by activation of the marker (*see* Fig. 3c) or vice versa the activation of the GOI is marked by loss of the marker.

It is also possible to devise schemes where the GOI and marker are simultaneously activated. One means to accomplish this is to have GOI expressed as a translational fusion with a marker such as GFP. It must first be demonstrated that the fusion protein is functional. Another potential approach is to utilize a two-component system such as the GAL4/UAS [58, 59]. Expression of the exotic *GAL4* transcription factor would be made contingent on (or disrupted by) TE transposition. The GOI and a suitable marker would be placed under the control of a GAL4-responsive UAS promoter element and whatever cells expressed GAL4 would also express the GOI and marker.

One final aspect to the target construct is the promoter that drives expression of the target gene(s). This promoter could be constitutive to reveal sectors in all tissues, or could be specific to reveal sectors only in specific tissues or within the expression domain of a particular gene. For example, in an analysis previously described, the target construct was driven by the *AP3* promoter, so when TE transposition conferred expression, it was within the normal expression context of the GOI [43].

3.4.2 Transposase Constructs

The major consideration with transposase constructs is the promoter used to control transposase expression. There are three basic options: the first uses a constitutive promoter in which case transposition and target gene activation would occur stochastically throughout plant development. The second option uses a gene- or tissue-specific promoter so that transposition and activation of the target construct would be limited to specific cells or tissues. The third option is to use an inducible promoter to allow temporal and spatial control of the transposase source. The soybean heat-shock promoter has been used in this context [27–29]. Particularly exciting was a demonstration that a microscopic laser could be used for targeted activation of the heat shock system in specific cells [29].

4 Concluding Remarks

Genetic mosaic analyses have greatly advanced our understanding of developmental biology. Transposon systems provide many approaches to generating genetic mosaics in plants. Some, based on endogenous elements, are limited to a particular species such as maize or snapdragon. The construction of transgenes designed to restore or abrogate gene function upon TE transposition now allows a genetic mosaic analysis of any recessive mutant in a species that is transformable. Such constructs also provide extensive opportunities for creativity in experimental design.

5 Notes

1. Experimental design centers around the question to be addressed, available tools, and appropriate controls. One important consideration is transposon biology. Different TE families display different behaviors which affect the size, frequency, and distribution of sectors. The objectives of the study should be compatible with the behavior of the corresponding TE.
2. Another major consideration is marking sectors. For lineage analysis, marking cell lineages is the sole objective, whereas for all other types of mosaic analyses, marking sectors independently of the trait of interest is highly desirable. When studying gene action or tissue interactions, the boundary between cellular genotypes is the focus of analysis. As such, precise identification of those boundaries is necessary and since the action (cell autonomy or nonautonomy) of the gene under investigation is presumably unknown, the exact sector boundary cannot be unequivocally determined on the basis of the experimental phenotype alone. It is not always possible or practical to mark sectors, for example when examining revertant sectors of a mutable TE allele. Valuable information can still be gleaned from unmarked sectors but they must be interpreted with the clear caveat that the true sector boundary is in fact unknown.
3. The major points regarding the marker are that it should be easily assayed in the tissue of interest, should be cell autonomous, and should not itself influence development. Several markers, including anthocyanin pigment, albinism, GUS, and fluorescent proteins like GFP, have been used extensively and can safely be assumed to meet most of these criteria. However, several of these markers cannot be used in all tissues. For example, anthocyanin accumulates predominantly in epidermal cells and is often light dependent. Albinism is only appropriate for chlorophyll-containing tissues. GFP can be difficult to detect in chlorophyllous tissues due to autofluorescence.

4. GFP and related fluorescent proteins, anthocyanin and chlorophyll markers, are all extremely convenient since fresh tissues can be examined, requiring minimal processing. For the most part, these methods suffer from the drawback that samples must be analyzed within a short time, before the tissues degrade. It is possible to analyze GFP in fixed tissues, and even embedded sections [60], but samples might be difficult to examine more than once due to photobleaching. However a powerful advantage of GFP variants is that they can be examined in living tissues, allowing the possibility of examining sector progression at multiple timepoints during development [29]. GUS-marked tissues enjoy longevity. A bit of processing is generally undertaken to fix and sometimes clear tissues but samples can be stored nearly indefinitely in 70 % ethanol, and can be embedded for sectioning. A major drawback of GUS-marked sectors is that they are invisible until histochemical staining is performed.
5. In any genetic mosaic analysis, it is always advisable to compare stable wild-type, and, if available, stable mutant to verify that the phenotypes observed are in fact due to the activity of the TE in generating genetic sectors. If examining revertant sectors of an unstable TE-induced mutant allele, these are the only controls for comparison. For marked sectors, major assumptions in mosaic analyses are that markers and the methods of uncovering them do not affect development and that the markers are visible in the tissues of interest. For commonly used markers, it is safe to assume that the marker per se does not influence development, but for untried markers, it is prudent to generate mosaics with the marker alone. The marked sectors should show normal development and, if in doubt, should be compared to another marker to verify similar patterns of development.
6. The method of generating sectors is a particular concern when using transgenic constructs or chromosome-breaking TEs. Transgenes have the possibility of creating unintended mutations upon integration into the genome and may be subject to position effects which could influence their behavior. Therefore it is advisable to examine several independent transgenic events for consistency in the phenotypes of sectors. When chromosome-breaking TEs are used to generate sectors, it is essential to generate control sectors carrying only the marker because it cannot be assumed that monosomy will be developmentally neutral. Furthermore, many (perhaps most) "chromosome-breaking" TEs also produce chromosome rearrangements [37, 61]. As such, one must bear in mind that some sectors may carry only the marker and not the GOI and vice versa (Fig. 4). Therefore, it is essential to generate control

sectors with chromosomes carrying only the marker allele and compare to the array of phenotypes generated with the experimental chromosome to help decipher which sector phenotypes are associated with the GOI.

References

1. Poethig RS (1987) Clonal analysis of cell lineage patterns in plant development. Am J Bot 74:581–594
2. Poethig RS (1989) Genetic mosaics and cell lineage analysis in plants. Trends Genet 5:273–277
3. Szymkowiak EJ, Sussex IM (1989) Chimeric analysis of cell layer interactions during development of the flower pedicel abscission zone. In: Osborne DJ, Jackson MB (eds) Cell separation in plants. Springer, Berlin, pp 363–368
4. Szymkowiak EJ, Sussex IM (1996) What chimeras can tell us about plant development. Annu Rev Plant Physiol Plant Mol Biol 47:351–376
5. McClintock B (1951) Chromosome organization and genic expression. Cold Spring Harb Symp Quant Biol 16:13–47
6. McClintock B (1978) Development of the maize endosperm as revealed by clones. In: Subtelny S, Sussex IM (eds) The clonal basis of development. Academic Press, Inc., New York, pp 217–237
7. Shure M, Wessler S, Fedoroff N (1983) Molecular identification and isolation of the waxy locus in maize. Cell 35:225–233
8. Kiesselbach TA (1949) The structure and reproduction of corn. Nebraska Agric Exp Stn Res Bull 161:1–96
9. Randolph LF (1936) Developmental morphology of the caryopsis in maize. J Agric Res 53:881–916
10. Walbot V (1994) Overview of key steps in aleurone development. In: Freeling M, Walbot V (eds) The maize handbook. Springer, New York, pp 78–80
11. Becraft PW, Asuncion-Crabb YT (2000) Positional cues specify and maintain aleurone cell fate in maize endosperm development. Development 127:4039–4048
12. Bossinger G et al (1992) Formation and cell lineage patterns of the shoot apex of maize. Plant J 2:311–320
13. Sturaro M et al (2005) Cloning and characterization of GLOSSY1, a maize gene involved in cuticle membrane and wax production. Plant Physiol 138:478–489
14. Garcia-Bellido A, Ripoll P, Morata G (1976) Developmental compartmentalization in the dorsal mesothoracic disc of Drosophila. Dev Biol 48:132–147
15. Cerioli S et al (1994) Early event in maize leaf epidermis formation as revealed by cell lineage studies. Development 120:2113–2120
16. Hernandez ML, Passas HJ, Smith LG (1999) Clonal analysis of epidermal patterning during maize leaf development. Dev Biol 216:646–658
17. Charlton WA (1990) Differentiation in leaf epidermis of *Chlorophytum comosum* Baker. Ann Bot 66:567–578
18. Moose SP, Sisco PH (1994) Glossy15 controls the epidermal juvenile-to-adult phase transition in maize. Plant Cell 6:1343–1355
19. Dawe RK, Freeling M (1990) Clonal analysis of the cell lineages in the male flower of maize. Dev Biol 142:233–245
20. Serna L, Torres-Contreras J, Fenoll C (2002) Clonal analysis of stomatal development and patterning in Arabidopsis leaves. Dev Biol 241:24–33
21. Larkin JC et al (1996) The control of trichome spacing and number in Arabidopsis. Development 122:997–1005
22. Dawe RK, Freeling M (1992) The role of initial cells in maize anther morphogenesis. Development 116:1077–1085
23. Rolland-Lagan A-G, Bangham JA, Coen E (2003) Growth dynamics underlying petal shape and asymmetry. Nature 422:161–163
24. Vincent CA, Carpenter R, Coen ES (1995) Cell lineage patterns and homeotic gene activity during Antirrhinum flower development. Curr Biol 5:1449–1458
25. Scheres B et al (1994) Embryonic origin of the arabidopsis primary root and root-meristem initials. Development 120:2475–2487
26. Dolan L et al (1994) Clonal relationships and cell patterning in the root epidermis of Arabidopsis. Development 120:2465–2474
27. Balcells L, Sundberg E, Coupland G (1994) A heat-shock promoter fusion to the Ac transposase gene drives inducible transposition of a Ds element during Arabidopsis embryo development. Plant J 5:755–764
28. Kidner C et al (2000) Clonal analysis of the *Arabidopsis* root confirms that position, not lineage, determines cell fate. Planta 211:191–199
29. Kurup S et al (2005) Marking cell lineages in living tissues. Plant J 42:444–453
30. Moose SP, Sisco PH (1996) Glossy15, an APETALA2-like gene from maize that regulates

leaf epidermal cell identity. Genes Dev 10: 3018–3027

31. Lid SE et al (2002) The *defective kernel 1* (*dek1*) gene required for aleurone cell development in the endosperm of maize grains encodes a membrane protein of the calpain gene superfamily. Proc Natl Acad Sci USA 99: 5460–5465
32. Spena A, Aalen RB, Schulze SC (1989) Cell-autonomous behavior of the rolC gene of Agrobacterium rhizogenes during leaf development: a visual assay for transposon excision in transgenic plants. Plant Cell 1: 1157–1164
33. McClintock B (1950) The origin and behavior of mutable loci in maize. Proc Natl Acad Sci U S A 36:344–355
34. Wurtzel ET (1992) Use of a *Ds* chromosome-breaking element to examine maize *Vp5* expression. J Hered 83:109–113
35. Becraft PW et al (2002) The maize *dek1* gene functions in embryonic pattern formation and in cell fate specification. Development 129:5217–5225
36. Neuffer MG (1995) Chromosome breaking sites for genetic analysis in maize. Maydica 40:99–116
37. Yi G et al (2011) The *thick aleurone1* mutant defines a negative regulation of maize aleurone cell fate that functions downstream of *dek1*. Plant Physiol 156(4):1826–1836
38. Carpenter R, Coen ES (1995) Transposon induced chimeras show that floricaula, a meristem identity gene, acts non-autonomously between cell layers. Development 121:19–26
39. Hantke SS, Carpenter R, Coen ES (1995) Expression of floricaula in single cell layers of periclinal chimeras activates downstream homeotic genes in all layers of floral meristems. Development 121:27–35
40. Sessions A, Yanofsky MF, Weigel D (2000) Cell-cell signaling and movement by the floral transcription factors LEAFY and APETALA1. Science 289:779–782
41. Causier B, Schwarz-Sommer Z, Davies B (2010) Floral organ identity: 20 years of ABCs. Semin Cell Dev Biol 21:73–79
42. Perbal MC et al (1996) Non-cell-autonomous function of the Antirrhinum floral homeotic proteins DEFICIENS and GLOBOSA is exerted by their polar cell-to-cell trafficking. Development 122:3433–3441
43. Jenik PD, Irish VF (2001) The Arabidopsis floral homeotic gene APETALA3 differentially regulates intercellular signaling required for petal and stamen development. Development 128:13–23
44. Fowler JE, Muehlbauer GJ, Freeling M (1996) Mosaic analysis of the liguleless3 mutant phenotype in maize by coordinate suppression of mutator-insertion alleles. Genetics 143:489–503
45. Muehlbauer GJ, Fowler JE, Freeling M (1997) Sectors expressing the homeobox gene liguleless3 implicate a time-dependent mechanism for cell fate acquisition along the proximal-distal axis of the maize leaf. Development 124:5097–5106
46. Muehlbauer GJ et al (1999) Ectopic expression of the maize homeobox gene liguleless3 alters cell fates in the leaf. Plant Physiol 119:651–662
47. Becraft PW, Yi G (2011) Regulation of aleurone development in cereal grains. J Exp Bot 62:1669–1675
48. Martienssen R et al (1990) Somatically heritable switches in the DNA modification of Mu transposable elements monitored with a suppressible mutant in maize. Genes Dev 4:331–343
49. Barkan A, Martienssen RA (1991) Inactivation of maize transposon *Mu* suppresses a mutant phenotype by activating an outward-reading promoter near the end of *Mu1*. Proc Natl Acad Sci U S A 88:3502–3506
50. Martienssen R, Baron A (1994) Coordinate suppression of mutations caused by Robertson's mutator transposons in maize. Genetics 136:1157–1170
51. Fedoroff N, Schläppi M, Raina R (1995) Epigenetic regulation of the maize Spm transposon. Bioessays 17:291–297
52. English JJ, Harrison K, Jones J (1995) Aberrant transpositions of maize double Ds-like elements usually involve Ds ends on sister chromatids. Plant Cell 7:1235–1247
53. Yu C, Zhang J, Peterson T (2011) Genome rearrangements in maize induced by alternative transposition of reversed *Ac/Ds* termini. Genetics 188:59–67
54. Zhang J, Peterson T (1999) Genome rearrangements by nonlinear transposons in maize. Genetics 153:1403–1410
55. Zhang J, Peterson T (2004) Transposition of reversed *Ac* element ends generates chromosome rearrangements in maize. Genetics 167:1929–1937
56. Zhang J et al (2009) Alternative Ac/Ds transposition induces major chromosomal rearrangements in maize. Genes Dev 23:755–765
57. Martinez-Ferez IM, Dooner HK (1997) Sesqui-Ds, the chromosome-breaking insertion at bz-m1, links double Ds to the original Ds element. Mol Gen Genet 255:580–586
58. Brand AH, Perrimon N (1993) Targeted gene expression as a means of altering cell fates and generating dominant phenotypes. Development 118:401–415

59. Laplaze L et al (2005) GAL4-GFP enhancer trap lines for genetic manipulation of lateral root development in Arabidopsis thaliana. J Exp Bot 56:2433–2442
60. Luby-Phelps K et al (2003) Visualization of identified GFP-expressing cells by light and electron microscopy. J Histochem Cytochem 51:271–274
61. Zhang J et al (2011) Transposable elements as catalysts for chromosome rearrangements. In: Birchler JA (ed) Plant chromosome engineering. Humana Press, New York, pp 315–326

Chapter 4

Survey of Natural and Transgenic Gene Markers Used to Monitor Transposon Activity

Lakshminarasimhan Krishnaswamy and Thomas Peterson

Abstract

Marker genes have played a critical role in the discovery of plant transposable elements, our understanding of transposon biology, and the utility of transposable elements as tools in functional genomics. Marker traits in model plants have been useful to detect transposable elements and to study the dynamics of transposition. Transposon-induced changes in the sequence of marker genes and consequently their expression have contributed to our understanding of molecular mechanisms of transposition and associated genome rearrangements. Further, marker genes that have been cloned and are compatible in heterologous systems have found versatile utility in the design of DNA constructs that have enabled us to understand the finer details of transposition mechanisms, and also allowed the use of transposon-based tools for functional genomics. This chapter traces the role of marker traits and marker genes (endogenous and transgenic) in various plant systems, and their contributions to the advancement of transposon biology over the past several decades.

Key words Transposon, Transgenic marker genes, Pigmentation genes, Chromosome rearrangement, Maize

1 History

Gregor Mendel used seven marker traits in the garden pea plant to formulate the fundamental laws of genetics. Ever since, visible phenotypes in experimental organisms have played a lynchpin role in every major discovery in the field. McClintock's discovery of the phenomenon of transposition of genetic elements in maize is no exception. With the rediscovery of Mendel's laws, most scientists recognized the significance and practical utility of genetics in agriculture. While scientists in Europe studied a wide range of plants, in the United States researchers tended to focus on agricultural plants. The number of established experimental field stations and land grant universities contributed significantly to this development.

Thomas Peterson (ed.), *Plant Transposable Elements: Methods and Protocols*, Methods in Molecular Biology, vol. 1057,
DOI 10.1007/978-1-62703-568-2_4, © Springer Science+Business Media New York 2013

During the early part of the twentieth century fruit fly and maize evolved to be the model systems of choice for genetics.

Apart from its importance as a food crop, maize presents several traits desirable for genetic studies: it is easy to carry out self- or cross-pollination; each plant produces large number of seeds and the seeds are viable for several years; further, the large chromosomes facilitate cytological examination. In addition, one outstanding feature of maize as a model system for genetics is the wide variety of phenotypically distinct characters that exhibit qualitative (and not quantitative) variation. Maize presents a wide range of variation in several innocuous traits (traits which do not affect the viability or the survival of the plant). These include pigmentation pattern on the kernel, cob and vegetative tissues, and texture of endosperm determined by the amount and nature of starch stored. A wide range of these variations in maize contributed to many discoveries in plant genetics including the discovery of transposable elements by Barbara McClintock.

2 McClintock and the Role of Marker Traits in the Discovery of *Ac/Ds* Transposable Elements

Barbara McClintock, who had previously characterized the ten maize chromosomes during her graduate work at Cornell, had turned her interests to understanding the biological role of the ends of the chromosomes (telomeres). At the University of Missouri, Columbia, she studied the behavior of chromosomes broken by X-rays. While irradiation is indeed efficient in causing chromosome breaks, the site of break is unpredictable; it could occur in any of the chromosomes. McClintock designed an ingenious genetic technique to break a specific chromosome in a predictable manner and to map the position of breakage relative to certain marker loci (reviewed in ref. 1). The short arm of chromosome 9 (9s) in maize has several genes controlling visible kernel traits, including (*C1 (Colorless kernel1)*, *Shrunken1 (Sh1)*, *Bronze1 (Bz1)*, and *Waxy (Wx)*). McClintock used a maize line that carried an inverse duplication segment in 9s; when crossover occurred between the inverse duplicated regions, it formed a dicentric chromosome at anaphase which broke the chromosome at a random site along the chromosome arm as the two centromeres were pulled towards the opposite poles. McClintock used the loss of marker traits to position the chromosome breaks. In one exceptional plant, McClintock observed that chromosome breaks occurred at the same site near to the *waxy* locus. She called this site "*Dissociation*" *(Ds)*; through further genetic analysis she established that chromosome breakage at this site depended on a *trans* factor at a different locus which she called "*Activator*" *(Ac)*. The presence of several kernel phenotype markers all located on 9s served like Aladdin's lamp in McClintock's hands. She used these markers to more precisely map the position of the *Ds* locus. To her surprise, in one rare kernel she observed that the position of *Ds*-induced chromosome breakage had shifted from its original position proximal to the *waxy* locus to a new position at the *C1* locus, simultaneously giving rise to a new allele c^{m-1} resulting in variegated pigmentation in the kernel. This marked the discovery of transposition of the *Ds* element. McClintock further tested the ability of *Ds* to mobilize in an *Ac*-dependent manner and generated mutable alleles of *Wx* (wx^{m-1}), and *Bz* (bz^{m-1} and bz^{m-2}). Over the decades that followed, the capacity of transposable elements to insert into genes has been exploited extensively to generate mutations and to tag genes across the genome in various species [2–11].

3 Variegation Patterns in Marker Traits Reveal More About Transposons

The variegation patterns observed in marker phenotypes shed light not only on the ability of *Ac/Ds* elements to transpose but also on the frequency of transposition, timing of transposition in relation to developmental stage, and the effect of dosage of *Ac* on these two aspects [12]. Transposition events occurring earlier in development result in larger variegation sectors, while later transposition events produce smaller sectors. The frequency of transposition is reflected by the number of nonoverlapping sectors. McClintock observed that when *Ac* copy number was increased from 1 to 2 to 3 (in the triploid endosperm of the kernel), transposition (or chromosome breakage) occurred later during development [12]. Thus, simply by observing the sizes of colorless or colored sectors, she could gather insight into the effects of *Ac* dosage and its relation to the developmental stage of the kernel.

Working independently, Greenblatt and Brink [13, 14] had studied *Ac* element transposition from the *p1-vv* allele of the maize *p1* locus. Analysis of twinned sectors formed as a result of *Ac* transposition led them to deduce that transposition often occurred during DNA replication, and that *Ac* often transposed to linked sites.

4 Marker Traits as Transposon Traps

Working independently, Marcus Rhoades observed variegation in anthocyanin pigmentation in the aleurone of the maize kernel endosperm. He showed that this was associated with a particular allele of the *a1* locus [15], which he called *Dotted.* Soon after McClintock's discovery of the *Ac/Ds* elements, Neuffer determined that *Dotted* (*Dt*) was a new mobile element [16]. Similarly, Emerson had previously studied a variegation allele at the *P1* locus (*P-vv*), which Brink and Nilan determined to be caused by the activity of the transposon *Mp* [17]. Later on *Mp* was determined to be identical to the *Ac* element [18]. Peterson identified the *Enhancer-Inhibitor* (*En/I*) element at the *pale green mutable* (*pg-m*) locus through his studies on the variegation effect of this marker gene in maize leaf, kernel, anther, and husk [19], while McClintock identified the *Suppressor-mutator* (*Spm/dSpm*) element by the variegation it induced in expression of the *a1* gene [20]. *En/I* and *Spm/dSpm* elements were subsequently determined to be identical [21]. More recently, the *Mx/rMx* element system has been identified and characterized at the maize *Bronze* locus [22].

In the late 1970s Robertson discovered a system of transposable elements in maize that caused a roughly 30-fold enhanced forward mutation rate [23]: this system was aptly termed "Robertson's Mutator element" system. Robertson identified new mutations by screening F2 seedling progeny for pigmentation (albinos, luteus, yellow–green, pale-green, virescent, etc.) and other phenotypes (e.g., dwarfs, glossies, blue fluorescence, necrotics). However, the biochemical and molecular basis of these traits were not known at that time. *Mu* elements were later cloned by using different known markers as traps/tags; 13 different *Mu* elements have been identified so far [24].

These examples illustrate how the discovery and early characterization of transposable elements were made possible by the power of genetics afforded by the range of marker genes available in maize. Transposon discovery in other plants lagged behind, and consequently for several years transposable elements were considered to be an anomalous phenomenon unique to maize. However, in the late 1960s and early 1970s transposable elements were identified and characterized in bacteria (*see* refs. 25, 26), yeast (*see* ref. 27), and fruit fly [28]. Soon it became evident that transposable elements are ubiquitous. The emergence of molecular biology and wide application of molecular techniques rejuvenated the field of plant transposon biology. A significant milestone was the molecular isolation of *Ac/Ds* [29] and *En/Spm* [30–32] elements. Members of both systems were cloned using the same approach of trapping the elements in the maize *waxy* gene. Maize kernels were screened for insertion of the transposon into the *waxy* gene which controls kernel starch composition, and the *waxy* gene sequence was used as a probe to identify and clone the inserted element. A similar strategy has been applied to the molecular isolation of several transposon elements in other plant systems as well.

5 Other Visible Marker Loci in Maize

In addition to the genes mentioned above, several other loci have been extensively used by McClintock and others in their quest to elucidate transposon biology, including *A1*; *A2*; *Bz*; *I* and *C*; *P*; *Pr*;

Sh1; *Sh2*; *Y*; and *Wx* [33]. McClintock's cytological studies revealed that transposition at chromosome-breaking *Ds* (*CBDs*) elements often resulted in chromosomal rearrangements such as deletion, duplication, inversion, and translocation [12, 34–39]. More recent molecular studies of chromosome breakage by *Ac/Ds* elements at the *waxy* [40], *bronze* [41–44], and *P1* [45–51] loci have further advanced our understanding of the structures of chromosome-breaking elements, as well as the types of genome rearrangements associated with chromosome breakage and their utility as tools for chromosome engineering.

A relatively large number of phenotypic marker genes are conveniently distributed along the ten chromosomes in maize; insertion of a transposon directly into one of these marker loci usually results in immediate and observable phenotypic effects. Interestingly, Neuffer devised a strategy to utilize markers which are not tightly linked to the transposon element as testers for transposition-mediated chromosome breakage [52, 53]. In an effort to try and use a *CBDs* element as a tool to cause loss of specific regions of the maize genome (which will enable uncovering of recessive alleles in a chimera), he started with a line that carried a chromosome-breaking *doubleDs* element on chromosome 10. From this, he derived stocks in which the *doubleDs* element had transposed to 16 of the 20 maize chromosome arms. These chromosome arms carry specific visible markers that can serve as reporters for transposition-mediated chromosome breakage and subsequent loss of the chromosome arm carrying the marker gene.

6 Endogenous Marker Genes in Other Plants

Most of the original transposon work was conducted in maize, partly for historical reasons, but also because of the large number of visible markers, and ease of conducting genetic/cytogenetic analyses with maize. Nevertheless, genes involved in pigment synthesis and genes that influence certain key metabolic or developmental pathways have been useful as transposon traps and markers for transposition in various systems.

In *Antirrhinum majus* (snapdragon), the *nivea* locus harbors a chalcone synthase gene that catalyzes an early step in anthocyanin biosynthesis; alleles of *nivea* have contributed to the discovery of three transposons: *tam1*, *tam2*, and *tam3* [54–57]. In addition, the Antirrhinum *pallida* locus, which encodes dihydroflavonol-4-reductase that catalyzes a late step in the anthocyanin pathway, has also been useful for characterization of the effects of transposon insertions [58]. Regulatory genes and structural genes involved in pigment synthesis have also been useful in identifying new transposons in several other plants: some examples include the *candystripe* transposon identified in the *y1* gene in *Sorghum bicolor* [59], the *PsI* insertion in *Hf1* gene of Petunia [60], and *dTph1* element in petunia [61] and *Tpn1* in Japanese morning glory [62] both identified as insertions in DFR (dihydroflavonol-4-reductase) genes involved in anthocyanin biosynthesis.

In plants, genes involved in nitrate uptake and nitrate reduction offer a powerful system to screen for endogenous transposons [63]. Nitrate is taken up by the plant and reduced by nitrate reductase to nitrite. In the same way, the herbicide chlorate can be taken up by the plant and reduced by nitrate reductase to chlorite, a toxic compound. Plants selected for resistance to the herbicide chlorate are usually defective in chlorate uptake or nitrate reduction. The Arabidopsis endogenous transposon *Tag1* was identified as an insertion in the *CHL1* gene following a screen for chlorate resistance [63]. Similar screens led to the identification of the tobacco retrotransposon *Tnt1* by its insertion into the nitrate reductase gene [64], and the transposable element *Jordan* in the algae Volvox, by insertion in the nitrate reductase locus (*nitA*) [65]. Although the chlorate selection can be powerful, it is unlikely to be effective in plants with duplication or redundancy of gene functions involved in nitrate uptake and nitrate reduction.

7 Transgenic Markers Used in Transposon Research

The development of plant transformation techniques beginning in the early 1980s has extended the benefit of transgenic marker and reporter genes to transposon research. The discovery of newer marker genes, development of experimental procedures to include more plant species amenable to transformation, availability of codon optimization tools to enhance the efficacy of marker genes in different plants, and development of sensitive tools (e.g., florescence) to screen for marker gene expression are some of the significant advances made in transgenic capabilities over the past two decades. Naturally, this progress has enhanced the applications of marker gene approaches for transposon research. A list of marker genes used in transposon research is presented in Table 1.

Transgenic constructs containing transposons together with suitable marker genes have enabled researchers to address a number of fundamental questions regarding transposon biology. Although different transposon systems have been characterized to different extents, some of the main aspects addressed include the following: (1) to determine the minimal DNA components required for efficient transposition of a transposable element; (2) to study the frequency of excision and insertion of transposon elements; (3) to understand the molecular mechanisms of DNA repair following excision of transposon elements; and (4) to investigate the various kinds of genome rearrangements caused by abnormal transposition reactions. Further, marker genes have enabled

Table 1
Various marker genes so far used in transposon research and their utility in specific applications

Selection	Gene	Application (refer section in text)
Resistance selection markers	*spt*: Streptomycin phosphotransferase confers resistance to antibiotic streptomycin *nptII*: Neomycin phosphotransferase confers resistance to antibiotic kanamycin *bar*: Phosphinothricin acetyl transferase confers resistance to herbicide phosphinothricin *hyg*: Hygromycin phosphotransferase confers resistance to antibiotic hygromycin *dhfr*: Dihydrofolate reductase confers resistance to methotrexate	• To select transgenic plants with transposon constructs (7.1) • To screen for segregation of transposase gene locus and transposed element (7.1) • To determine the nature of genome rearrangement at transgene constructs with noncanonical transposon ends (7.4)
Negative selection markers	*iaaH*: Confers susceptibility to naphthalene acetamide (NAM) and analogues *codA*: Cytosine deaminase confers susceptibility to 5-flurocytosine	• To screen for segregation of transposase gene locus and transposed element (7.1) • To monitor transposition event at noncanonical transposon ends resulting in genome rearrangement (7.4)
Histochemical staining marker	*uidA*: GUS (beta-glucuronidase) is a versatile marker which enables histochemical staining of plant tissue	• To detect transposon excision events or determine frequency of excision, by cloning the transposon element between the GUS gene and its promoter (7.1) • Used as marker gene in enhancer trap and promoter trap constructs (7.2) • To determine frequency of transposition-induced intrachromosomal recombination (7.3)
Fluorescent protein markers	*GFP*: Green fluorescent protein *DsRed*: Red fluorescent protein	• Used as marker gene in enhancer trap and promoter trap constructs (7.2) • To screen for segregation of transposase gene locus and transposed element (7.1) • To determine the nature of genome rearrangement at transgene constructs with noncanonical transposon ends (7.4)

The numbers in parenthesis indicate relevant sections in the text engineering of transposable elements as tools for functional genomics research.

Recent reviews [66, 67] present a comprehensive discussion on the various markers used for selection of transformed plants in several species. Here we will focus on the design of transgenic constructs used for transposon research, especially aspects of the positioning of selection marker genes and reporter genes relative to transposon sequences within the constructs that have enabled researchers to unravel molecular details of transposition, and also to exploit transposon-based tools for genetics and functional genomics.

7.1 First-Generation Transposon Constructs

Initially, marker genes in transposon constructs were used to monitor or select for transposition events. For example, excision could be monitored by the use of a construct in which a transposon inserted into the untranslated leader sequence blocks the expression of a downstream antibiotic- or herbicide-resistance gene; excision of the transposon allowed expression of the resistance gene, enabling selection of plants carrying germinal transposition events [68, 69]. An improvement over this was the development of two-element systems [3, 70], comprising a stable source of transposase and a nonautonomous element carrying the selection marker gene; this enabled selection for transposon insertion events (not just excision). Because these constructs included two marker genes (one cloned within the mobile element and the other anchored at the transgene locus), this strategy provided a genetic screen to determine the frequency of linked versus unlinked insertion of the transposon element. Moreover, this approach is especially useful for gene tagging by insertional mutagenesis because it allows screening for cosegregation of the selection marker gene (in the transposon) with the observed mutant phenotype. A further advancement was the use of three marker genes: one to monitor element excision, a second to select for transposon reinsertion, and a third to select against the source of transposase after the initial transposition event, and thereby stabilize the transposed element [71]. The choice of marker genes here depends on their effectiveness in the plant species of interest and the tools available. When multiple selection marker genes are effective in a plant, a combination of positive and negative selection can be applied [72, 73]. Otherwise a selectable resistance marker can be used together with screenable fluorescent reporter genes [74].

7.2 Reporter Genes in Enhancer Traps and Promoter Traps

Due to their inherent mobility, transposons have been especially useful in the development of enhancer trap and gene trap constructs (reviewed in refs. 75–77). In enhancer traps a reporter gene and a minimal promoter are inserted inside a transposon; the reporter gene is not expressed except when the transposon inserts in the vicinity of an endogenous enhancer. In gene trap constructs, the reporter gene is preceded by one or more splice acceptor sequences. When the gene trap element inserts within the transcribed region of a gene, the reporter gene exons may be spliced to the host gene exons. The expectation is that the expression pattern of the reporter gene reflects the expression pattern of the endogenous gene into which the enhancer or the gene trap element is inserted. Although T-DNA vectors have been successfully used as vectors for enhancer traps and gene traps, they cannot be mobilized from the site of T-DNA insertion. In contrast, transposons are ideal for enhancer and gene trap approaches because they can be easily mobilized across the genome to saturation. It is important that the enhancer trap and gene trap constructs use reporter genes that can be easily visualized and have good sensitivity of detection. Therefore marker genes that confer resistance to antibiotics or herbicide are not the best candidates for these constructs. In plants, glucuronidase (GUS) and green fluorescent protein (GFP) have been widely used as reporter genes for enhancer and promoter trap systems [77].

7.3 Reporter Genes to Study DNA Repair Following Transposon Excision

The GUS reporter gene has also been useful in probing the effect of transposon excision on homologous recombination between directly repeated sequences. Athma and Peterson [78] had observed that excision of *Ac* element from the *P* locus in maize induced homologous recombination between directly repeated sequences flanking the *p1* gene. In order to study this phenomenon in detail in transgenic Arabidopsis, a *Ds* element was cloned between two partially overlapping, nonfunctional segments of the β-glucuronidase gene; i.e., the 5′ part of the *gus* gene flanked one side of the *Ds* element and the 3′ part of *gus* flanked the other side [79]. The flanking regions had overlap of about 600 bp, which formed a direct repeat on either side of the *Ds* element. Recombination of the *gus* fragments would restore a functional *gus* gene, resulting in blue sectors detectable by histochemical staining. In this way it was shown that the presence of a functional *Ac* transposase induced a 1,000-fold increase in frequency of recombination between the flanking homologous sequence.

7.4 Utility of Marker Genes in Analysis of Transposition-Induced Chromosome Rearrangements and Chromosome Engineering

McClintock had observed cytological evidence that *Ac/Ds* elements sometimes caused chromosomal rearrangements such as deletions, duplications, inversions, and translocations [34–39]. She described two states of the *Ds* element: state II elements transposed from one locus to another, whereas state I elements caused chromosome breaks more often, and were associated with a higher frequency of chromosomal rearrangements. Molecular analysis of naturally occurring state I elements revealed that the chromosome-breaking *Ac/Ds* loci carried multiple transposon elements in proximity [80–84]. In a standard transposition reaction, the transposase recognizes the pair of terminal inverted repeat sequences (at the 5′

and 3′ ends of the transposon element), cuts the element from the donor site, and inserts at a new site. (The 5′ and 3′ "terminal inverted repeat" sequences in a standard transposon are reverse complement of each other, and are critical for transposase recognition.) When multiple transposon elements are present in proximity, transposase may sometimes recognize a pair of terminal sequence from two different elements, the two termini being "direct repeats" with respect to each other, or their relative orientation being "reversed" compared to a standard transposon. Transposition involving such noncanonical pairs of termini could result in a wide variety of rearrangements, and these rearrangements could affect the expression of neighboring marker genes.

Naturally occurring chromosome-breaking *Ac/Ds* alleles with "direct repeat" [40–46] and "inverse repeat" [47, 48, 85] of the terminal ends at the *p1* [44–48, 85] and *bz* [40–43] loci have contributed immensely to our understanding of the wide variety of chromosome rearrangements caused by such transposition (reviewed in ref. 86). The *p1* and *bz* loci are convenient markers because rearrangements that affect these markers affect the kernel pigmentation phenotype. Kernels with altered phenotype of the marker can be selected and used for precise molecular characterization of the rearrangement. English et al. [87] tested this phenomenon in tobacco by cloning a *double-Ds* element adjacent to two reporter genes, *SPT* and *GUS*. (The *double-Ds* element has one *Ds* element inserted within another *Ds* element in the opposite orientation; this places the terminal sequences of the two elements in "direct repeat" configuration with respect to each other.) The use of two marker genes, a selection marker (*SPT*) and a reporter gene (*GUS*), in a transgenic system enabled them to confirm that transposition at the *double-Ds* element frequently caused loss of both marker genes, and the marker gene loss was associated with DNA rearrangements. To study this phenomenon in greater detail, synthetic constructs designed to carry a pair of *Ds* ends as direct or reverse repeats have been used in maize [86], Arabidopsis [72, 86], and rice [86, 88]. These synthetic constructs enabled the study of transposition-induced chromosome rearrangement at various loci in the genome. The use of three marker genes positioned at various sites with respect to the *Ds* element facilitated the phenotypic classification of different types of rearrangement even before molecular characterization. For instance, in the reverse-repeat construct used in Arabidopsis, the *iaaH* gene is cloned between a pair of reversed *Ds* ends; two other selection markers, *bar* and *nptII*, are placed on either side of the reversed-end *Ds* element. A chromosome rearrangement at this pair of noncanonical transposon ends will result in loss of the negative selection marker *iaaH* and render the plants resistant to naphthalene acetamide selection. Deletion events occurring on one side of the reversed-end *Ds* element can be detected by loss of the *bar* gene, while deletions to the other side can be identified by loss of the *nptII* gene. Inversion events will result in plants resistant to all three treatments. Following the initial selection, putative rearrangements can be verified by molecular characterization. Similar constructs have been tested in maize and rice, using a different set of markers. The maize constructs [86] include two genes (*p1* and *c1*) involved in kernel pigmentation and a selection marker (*bar*); the rice construct [86, 88] has two fluorescent protein markers (*GFP* and *RFP*) and a selection gene (*hpt*). The ingenious design of these constructs has widened the scope for chromosome engineering in plants. (Please refer ref. 86 for detailed description of these constructs and their utility.)

8 Notes

1. When frequencies of transposition events are determined based on the number and size of pigmented (or non-pigmented) spots in the maize kernel (or similar systems including transgenic systems with GUS marker), it must be noted that a large spot could actually be two or more overlapping spots. Therefore, the number of distinct spots indicates only the minimum number of events.
2. In maize sometimes it could be challenging to distinguish transposition-induced variegation in pigmentation from irregular pigment distribution caused by other phenomena (e.g., *R*-mottling, insect bite marks). However analysis of the kernel under a simple microscope can distinguish a genuine transposition event; in case of a transposon-induced variegation the spots usually have smooth edges and distinct boundaries, corresponding to cell lineage clones.
3. Transposition events that result in variegation of kernel pigmentation are easy to score. However, variegation of kernel starch phenotype is relatively more difficult especially when the transposition occurs later during development and the variegated sectors are numerous and small [89].
4. The *p1* locus in maize determines pigmentation of pericarp and cob tissues and this system facilitates studying somatic mutation events. One significant advantage of this system is that the pericarp covering a single kernel and the underlying germ cell share a common lineage, except that the germ cell has also undergone one meiotic division; the somatic mutation detected in the pericarp is likely to be shared by the underlying germinal cells, and therefore can be easily recovered in the following generation for genetic and molecular studies [90].

5. A number of regulatory and structural genes influence kernel color in maize. The *p1* gene conditions flavonoid pigmentation in the pericarp, while *Pr1*, *R*, and *C1* genes determine flavonoid pigmentation of aleurone, and *Y* gene determines carotenoid pigmentation of the endosperm. The aleurone layer surrounds the endosperm and the pericarp is external to the aleurone. Therefore, it is important to choose a genetic background such that the phenotype of interest is not masked by pigmentation of the other layers. For instance, a colorless pericarp and aleurone background would be ideal for studying endosperm pigmentation.
6. Exons of *c1* and *p1* genes have been cloned and introduced into maize as *cis*-genic markers [86]. The *c1* gene serves as a reliable marker with good expression in the aleurone; however an introduced *p1* gene tends to get silenced. Furthermore, expression of *p1* under constitutive promoter seems to be toxic to the plant. Similarly, *B-Peru*, a regulatory gene involved in anthocyanin pigmentation in maize kernel, embryo, and vegetative tissues can also be expressed under a constitutive promoter and used as an effective marker gene in maize (Krishnaswamy L, unpublished). However, it is best to screen young well-nurtured plants. This is because certain stress conditions enhance anthocyanin pigmentation in the vegetative tissue which can be hard to distinguish from the pigmentation regulated by the introduced *B-Peru* gene. Furthermore, as the chlorophyll content decreases in the aging vegetative tissue background level of anthocyanin could become prominent, making it hard to contrast against *B-Peru*-regulated pigmentation.
7. When transgenic plants are generated with transposon constructs, it is critical to be aware of potential variation in the copy number of the transgene. Multiple copies of a transposable element inserted as tandem copies or multiple copies in same or neighboring loci could often lead to a variety of genome rearrangements. Further, it is best to use a two-element system: first transform the nonautonomous element and screen for single locus, single-copy insertion lines; subsequently introduce the stable transposase gene either through genetic cross or second transformation. Introducing a nonautonomous transposon element into a line that already expresses the transposase could result in transposition events even before the copy number of insertion can be determined.
8. When *iaaH* gene is used as a negative selection marker the seedlings are selected on a medium containing naphthalene acetamide (NAM). The *iaaH* gene encodes indole acetamide hydroxylase which metabolizes NAM to naphthalene acetic acid (NAA). NAA is an auxin analogue and therefore at higher concentrations it affects root morphology to form a hairy or a

knotted structure. It is observed that when seeds are sown close to each other on the medium, it is difficult to identify NAM-resistant roots. Therefore seeds should be well spaced on the medium. For screening Arabidopsis (*Col*) seedlings, a NAM concentration of 5 μM is suitable. However, it is best to test the conditions for each set of experimental reagents and conditions. Further, the NAM-selected plants must be validated by molecular analysis for the absence of the *iaaH* gene [72].

9. Fluorescent protein genes *GFP* and *dsRED* have been widely used as reporters in plants. However, *EGFP* (a version of *GFP* engineered for enhanced fluorescence and folding efficiency at 37 °C) [91] may show some toxicity in high expression levels (Paremeswaran S, personal communication). Presence of autofluorescent compounds in plants could potentially interfere with the signal derived from fluorescent proteins. This can be overcome by using young tissue for screening expression of the fluorescent protein or by employing spectral imaging.

References

1. Jones RN (2005) McClintock's controlling elements: the full story. Cytogenet Genome Res 109:90–103
2. Kolkman JM, Conrad LJ, Farmer PR et al (2005) Distribution of activator (Ac) throughout the maize genome for use in regional mutagenesis. Genetics 169:981–995
3. Bancroft I, Bhatt AM, Sjodin C et al (1992) Development of an efficient two-element transposon tagging system in *Arabidopsis thaliana*. Mol Gen Genet 233:449–461
4. Dean C, Sjodin C, Bancroft I et al (1991) Development of an efficient transposon tagging system in *Arabidopsis thaliana*. Symp Soc Exp Biol 45:63–75
5. Fedoroff NV, Smith DL (1993) A versatile system for detecting transposition in Arabidopsis. Plant J 3:273–289
6. Sundaresan V, Springer P, Volpe T et al (1995) Patterns of gene action in plant development revealed by enhancer trap and gene trap transposable elements. Genes Dev 9:1797–1810
7. Grevelding C, Becker D, Kunze R et al (1992) High rates of Ac/Ds germinal transposition in Arabidopsis suitable for gene isolation by insertional mutagenesis. Proc Natl Acad Sci USA 89:6085–6089
8. Chuck G, Robbins T, Nijjar C et al (1993) Tagging and cloning of a Petunia flower color gene with the maize transposable element activator. Plant Cell 5:371–378
9. Meissner R, Chague V, Zhu Q et al (2000) Technical advance: a high throughput system for transposon tagging and promoter trapping in tomato. Plant J 22:265–274
10. Chin HG, Choe MS, Lee SH et al (1999) Molecular analysis of rice plants harboring an Ac/Ds transposable element-mediated gene trapping system. Plant J 19:615–623
11. Izawa T, Ohnishi T, Nakano T et al (1997) Transposon tagging in rice. Plant Mol Biol 35:219–229
12. McClintock B (1951) Chromosome organization and genic expression. Cold Spring Harb Symp Quant Biol 16:13–47
13. Greenblatt IM, Brink RA (1962) Twin mutations in medium variegated pericarp maize. Genetics 47:489–501
14. Greenblatt IM (1984) A chromosome replication pattern deduced from pericarp phenotypes resulting from movements of the transposable element, modulator, in maize. Genetics 108:471–485
15. Rhoades MM (1938) Effect of the Dt gene on the mutability of the a(1) allele in maize. Genetics 23:377–397
16. Nuffer MG (1955) Dosage effect of multiple *Dt* loci on mutation of a in the maize endosperm. Science 121:399–400
17. Brink RA, Nilan RA (1952) The relation between light variegated and medium variegated pericarp in maize. Genetics 37:519–544
18. Barclay PC, Brink RA (1954) The relation between modulator and activator in maize. Proc Natl Acad Sci USA 40:1118–1126
19. Peterson PA (1953) A mutable *pale green* locus in maize. Genetics 38:682–683
20. McClintock B (1954) Mutations in maize and chromosomal aberrations in neurospora. Carnegie Inst Wash Yr Bk 53:254–260

21. Peterson PA (1965) Relationship between the *Sp*, and *En* control systems in maize. Am Nat 99:391–398
22. Zhennan X, Dooner HK (2005) Mx-rMx, a family of interacting transposons in the growing hAT superfamily of maize. Plant Cell 17:375–388
23. Robertson DS (1978) Characterization of a mutator system in maize. Mutat Res 51:21–28
24. Tan BC, Chen Z, Shen Y et al (2011) Identification of an active new mutator transposable element in maize. G3 (Bethesda) 1:293–302
25. Iida S, Meyer J, Arber W (1983) Prokaryotic IS elements. In: Shapiro JA (ed) Mobile genetic elements. Academic, New York, pp 159–221
26. Heffron F (1983) *Tn3* and its relatives. In: Shapiro JA (ed) Mobile genetic elements. Academic, New York, pp 223–226
27. Roeder GS, Fink GG (1983) Transposobale elements in yeast. In: Shapiro JA (ed) Mobile genetic elements. Academic, New York, pp 300–332
28. Rubin GM (1983) Dispersed repetitive DNAs in drosophila. In: Shapiro JA (ed) Mobile genetic elements. Academic, New York, pp 329–336
29. Fedoroff N, Wessler S, Shure M (1983) Isolation of the transposable maize controlling elements Ac and Ds. Cell 35:235–242
30. Schwarz-Sommer Z, Gierl A, Klösgen RB et al (1984) The Spm (En) transposable element controls the excision of a 2-kb DNA insert at the wx allele of *Zea mays.* EMBO J 3:1021–1028
31. Pereira A, Schwarz-Sommer Z, Gierl A et al (1985) Genetic and molecular analysis of the enhancer (En) transposable element system of *Zea mays.* EMBO J 4:17–23
32. Pereira A, Cuypers H, Gierl A et al (1986) Molecular analysis of the En/Spm transposable element system of *Zea mays.* EMBO J 5:835–841
33. McClintock B (1956) Controlling elements and the gene. Cold Spring Harb Symp Quant Biol 21:197–216
34. McClintock B (1953) Induction of instability at selected loci in maize. Genetics 38:579–599
35. McClintock B (1950) The origin and behavior of mutable loci in maize. Proc Natl Acad Sci USA 36:344–355, 326 Zhang et al
36. McClintock B (1949) Mutable loci in maize. Carnegie Inst Wash Yr Bk 48:142–154, 29
37. McClintock B (1953) Mutation in maize. Carnegie Inst Wash Yr Bk 52:227–237
38. McClintock B (1948) Mutable loci in maize. Carnegie Inst Wash Yr Bk 47:155–169
39. McClintock B (1952) Mutable loci in maize. Carnegie Inst Wash Yr Bk 51:212–219
40. Weil CF, Wessler SR (1993) Molecular evidence that chromosome breakage by *Ds* elements is caused by aberrant transposition. Plant Cell 5:515–522
41. Huang JT, Dooner HK (2008) Macrotransposition and other complex chromosomal restructuring in maize by closely linked transposons in direct orientation. Plant Cell 20:2019–2032
42. Martínez-Férez IM, Dooner HK (1997) Sesqui-Ds, the chromosome-breaking insertion at bz-ml, links double Ds to the original Ds element. Mol Gen Genet 255:580–586
43. Dooner HK, Belachew A (1991) Chromosome breakage by pairs of closely linked transposable elements of the Ac-Ds family in maize. Genetics 129:855–862
44. Ralston E, English J, Dooner HK (1989) Chromosome-breaking structure in maize involving a fractured Ac element. Proc Natl Acad Sci USA 86:9451–9455
45. Zhang J, Peterson T (1999) Genome rearrangements by nonlinear transposons in maize. Genetics 153:1403–1410
46. Zhang J, Peterson T (2005) A segmental deletion series generated by sister-chromatid transposition of Ac transposable elements in maize. Genetics 171:333–344
47. Zhang J, Peterson T (2004) Transposition of reversed Ac element ends generates chromosome rearrangements in maize. Genetics 167:1929–1937
48. Zhang J, Yu C, Pulletikurti V et al (2009) Alternative Ac/Ds transposition induces major chromosomal rearrangements in maize. Genes Dev 23:755–765
49. Zhang J, Zhang F, Peterson T (2006) Transposition of reversed Ac element ends generates novel chimeric genes in maize. PLoS Genet 2:e164
50. Yu C, Zhang J, Pulletikurti V et al (2010) Spatial configuration of transposable element Ac termini affects their ability to induce chromosomal breakage in maize. Plant Cell 22:744–754
51. Pulletikurti V, Yu C, Zhang J et al (2009) Cytological evidence that alternative transposition by *Ac* elements causes reciprocal translocations and inversions in *Zea mays* L. Maydica 54:457–462
52. Neuffer MG (1995) Chromosome breaking sites for genetic analysis in maize. Maydica 40:99–116
53. Neuffer MG (2010) Chromosome breaking *Ds* sites in maize, revisited. Part I, background, methods, Description, Maize Genet Coop Newsl 84. http://www.agron.missouri.edu/mnl/84/PDF/54neuffer.pdf
54. Wienand U, Sommer H, Schwarz Z et al (1982) A general method to identify plant structural genes among genomic DNA clones using transposable element induced mutations. Mol Gen Genet 187:195–201

55. Bonas U, Sommer H, Saedler H (1984) The 17-kb Taml element of Antirrhinum majus induces a 3-bp duplication upon integration into the chalcone synthase gene. EMBO J 3:1015–1019
56. Sommer H, Carpenter R, Harrison BJ, Saedler H (1985) The transposable element Tam3 of *Antirrhinum majus* generates a novel type of sequence alterations upon excision. Mol Gen Genet 199:225–231
57. Upadhyaya KC, Hans S, Hans S, Enno K, Heinz S, Heinz S (1985) The paramutagenic line *niv*-44 has a 5 kb insert, Tam 2, in the chalcone synthase gene of *Antirrhinum majus.* Mol Gen Genet 199:201–207
58. Coen ES, Carpenter R, Martin C (1986) Transposable elements generate novel spatial patterns of gene expression in *Antirrhinum majus.* Cell 47:285–296
59. Chopra S, Brendel V, Zhang J, Axtell JD, Peterson T (1999) Molecular characterization of a mutable pigmentation phenotype and isolation of the first active transposable element from Sorghum bicolor. Proc Natl Acad Sci USA 96:15330–15335
60. Snowden KC, Napoli CA (1998) Psl: a novel Spm-like transposable element from *Petunia hybrida.* Plant J 14:43–54
61. Gerats AG, Huits H, Vrijlandt E, Maraña C, Souer E, Beld M (1990) Molecular characterization of a nonautonomous transposable element (dTphl) of petunia. Plant Cell 2:1121–1128
62. Inagaki Y, Hisatomi Y, Suzuki T, Kasahara K, Iida S (1994) Isolation of a suppressor-mutator/enhancer-like transposable element, Tpnl, from Japanese morning glory bearing variegated flowers. Plant Cell 6: 375–383
63. Tsay YF, Frank MJ, Page T, Dean C, Crawford NM (1993) Identification of a mobile endogenous transposon in *Arabidopsis thaliana.* Science 260:342–344
64. Pouteau S, Spielmann A, Meyer C, Grandbastien MA, Caboche M (1991) Effects of Tntl tobacco retrotransposon insertion on target gene transcription. Mol Gen Genet 228:233–239
65. Miller SM, Schmitt R, Kirk DL (1993) Jordan, an active Volvox transposable element similar to higher plant transposons. Plant Cell 5:1125–1138
66. Sundar IK, Sakthivel N (2008) Advances in selectable marker genes for plant transformation. J Plant Physiol 165:1698–1716
67. Barampuram S, Zhang ZJ (2011) Recent advances in plant transformation. Methods Mol Biol 701:1–35
68. Jones JD, Carland FM, Maliga P, Dooner HK (1989) Visual detection of transposition of the maize element activator (ac) in tobacco seedlings. Science 244:204–207
69. Jones JD, Shlumukov L, Carland F et al (1992) Effective vectors for transformation, expression of heterologous genes, and assaying transposon excision in transgenic plants. Transgenic Res 1:285–297
70. Aarts MG, Dirkse WG, Stiekema WJ, Pereira A (1993) Transposon tagging of a male sterility gene in Arabidopsis. Nature 363:715–717
71. Long D, Martin M, Sundberg E et al (1993) The maize transposable element system Ac/Ds as a mutagen in Arabidopsis: identification of an albino mutation induced by Ds insertion. Proc Natl Acad Sci USA 90:10370–10374
72. Krishnaswamy L, Zhang J, Peterson T (2008) Reversed end Ds element: a novel tool for chromosome engineering in Arabidopsis. Plant Mol Biol 68:399–411
73. Panjabi P, Burma PK, Pental D (2006) Use of the transposable elements Ac/Ds in conjunction with Spm/dSpm for gene tagging allows extensive genome coverage with a limited number of starter lines: functional analysis of a four-element system in *Arabidopsis thaliana.* Mol Genet Genomics 276:533–543
74. Kumar CS, Wing RA, Sundaresan V (2005) Efficient insertional mutagenesis in rice using the maize En/Spm elements. Plant J 44: 879–892
75. Sundaresan V (1996) Horizontal spread of transposon mutagenesis: new uses for old elements. Trends Plant Sci 1:184–190
76. Springer PS (2000) Gene traps: tools for plant development and genomics. Plant Cell 12:1007–1020
77. Acosta-García G, Autran D, Vielle-Calzada JP (2004) Enhancer detection and gene trapping as tools for functional genomics in plants. Methods Mol Biol 267:397–414
78. Athma P, Peterson T (1991) Ac induces homologous recombination at the maize P locus. Genetics 128:163–173
79. Xiao YL, Peterson T (2000) Intrachromosomal homologous recombination in Arabidopsis induced by a maize transposon. Mol Gen Genet 263:22–29
80. Courage-Tebbe U, Doring HP, Fedoroff N, Starlinger P (1983) The controlling element *Ds* at the *Shrunken* locus in *Zea mays*: structure of the unstable *sh-m5933* allele and several revertants. Cell 34:383–393
81. Burr B, Burr FA (1982) *Ds* controlling elements of maize at the *shrunken* locus are large and dissimilar insertions. Cell 29:977–986
82. Chaleff D, Mauvais J, McCormick S et al (1981) Controlling elements in maize. Carnegie Inst Wash Yr Bk 80:158–174
83. Doring HP, Nelsen-Salz B, Garber R, Tillman E (1981) Double *Ds* elements are involved in

specific chromosome breakage. Mol Gen Genet 219:299–305

84. Weck E, Courage U, Doring HP et al (1984) Analysis of *sh-m6233*, a mutation induced by the transposable element *Ds* in the *sucrose synthase* gene of *Zea mays*. EMBO J 3: 1713–1716
85. Yu C, Zhang J, Peterson T (2011) Genome rearrangements in maize induced by alternative transposition of reversed ac/ds termini. Genetics 188:59–67
86. Zhang J, Yu C, Krishnaswamy L, Peterson T (2011) Transposable elements as catalysts for chromosome rearrangements. Methods Mol Biol 701:315–326
87. English J, Harrison K, Jones JD (1993) A genetic analysis of DNA sequence requirements for dissociation state I activity in tobacco. Plant Cell 5:501–514
88. Yu C, Han F, Zhang J, Birchler J, Peterson T (2012) A transgenic system for generation of transposon Ac/Ds-induced chromosome rearrangements in rice. Theor Appl Genet 125(7): 1449–1462. doi: 10.1007/s00122-012-1925-4
89. Zhang J, Peterson T, Peterson P (2009) Transposons Ac/Ds, En/Spm and their relatives in maize. In: Bennetzen J, Hake S (eds) Maize handbook. Springer, New York, pp 251–276
90. Chen J, Greenblatt IM, Dellaporta SL (1987) Transposition of Ac from the P locus of maize into unreplicated chromosomal sites. Genetics 117:109–116
91. Zhang G, Gurtu V, Kain SR (1996) An enhanced green fluorescent protein allows sensitive detection of gene transfer in mammalian cells. Biochem Biophys Res Commun 227:707–711

Chapter 5

Molecular Biology of Maize *Ac/Ds* Elements: An Overview

Katina Lazarow, My-Linh Doll, and Reinhard Kunze

Abstract

Maize *Activator* (*Ac*) is one of the prototype transposable elements of the hAT transposon superfamily, members of which were identified in plants, fungi, and animals. The autonomous *Ac* and nonautonomous *Dissociation* (*Ds*) elements are mobilized by the single transposase protein encoded by *Ac*. To date *Ac/Ds* transposons were shown to be functional in approximately 20 plant species and have become the most widely used transposable elements for gene tagging and functional genomics approaches in plants. In this chapter we review the biology, regulation, and transposition mechanism of *Ac/Ds* elements in maize and heterologous plants. We discuss the parameters that are known to influence the functionality and transposition efficiency of *Ac/Ds* transposons and need to be considered when designing *Ac* transposase expression constructs and *Ds* elements for application in heterologous plant species.

Key words *Zea mays*, hAT transposon superfamily, Insertion mutagenesis, Gene tagging, DDE transposase

1 Introduction

Activator (*Ac*) and *Dissociation* (*Ds*) elements were the first transposons that were ever discovered. Barbara McClintock observed that in a certain maize strain, chromosome breaks frequently occurred at a specific locus on the short arm of chromosome 9 and that this locus, which she termed *Dissociation* (*Ds*), occasionally translocated to a new location in the genome. Subsequently she recognized that chromosome breakage at, and mobility of, *Ds* was dependent on the presence of another locus, termed *Activator* (*Ac*), that was autonomously able to transpose to new genomic positions [1]. Today we know that transposable elements are not only ubiquitous constituents of all pro- and eukaryotic genomes but also important players in genome evolution [2, 3].

Thomas Peterson (ed.), *Plant Transposable Elements: Methods and Protocols*, Methods in Molecular Biology, vol. 1057, DOI 10.1007/978-1-62703-568-2_5,

2 The hAT Transposon Superfamily

Ac is one of the three prototype transposons *hobo*, *Ac*, and *Tam3* of the hAT transposon superfamily. Members of this family were identified in plants, fungi, vertebrates, and invertebrates, although in some species, including humans, apparently no active elements have survived (reviewed in refs. 4–6). All hAT elements transpose by a cut-and-paste mechanism, have short terminal inverted repeats (11–14 bp in the majority of hAT elements), and generate 8-bp target site duplications upon integration. hAT elements encode only a single protein, the transposase, that catalyzes the transposition reaction. The transposases of hAT elements have three conserved regions, the hAT-domains [7, 8], that contain the catalytically active center and protein oligomerization functions. The hAT transposases share a catalytic DDE amino acid triad in the active center with a large number of recombinases, integrases, and transposases in prokaryotes and eukaryotes (*see* Subheading 5).

3 Structure of the Maize Transposable Element *Activator* (*Ac*)

The autonomous *Ac* is a small and simple structured transposable element. It is 4,565 bp long, has 11-bp imperfect terminal inverted repeats (TIRs) and approximately 240-bp essential subterminal sequences at both ends, and contains a single gene that encodes the 807 amino acid transposase (TPase) (reviewed in ref. 6). By making internal deletions of a *Ds* element, 238 bp from the *Ac* 5′-end and 209 bp from the 3′-end were mapped as the minimal terminal sequences required for uncompromised transposition frequencies [9].

The central region of *Ac* contains the transcription unit for a single 3.5 kb messenger RNA which encodes the transposase (TPase) protein [10, 11]. The *Ac* mRNA has multiple transcription initiation sites located between 304 and 364 nt from the 5′-end, with the major start site at position 334, and terminates 265 nt from the 3′-end of the element. The *Ac* promoter thus overlaps with the 5′ terminal sequence elements including the subterminal TPase-binding motifs that are essential for transposition (*see* Subheading 7.2). The *Ac* transcription unit spans ca. 87 % of the element, including four short introns with a length of 107, 71, 89, and 387 nt, respectively (*see* Fig. 1).

4 Classes of *Dissociation* (*Ds*) Elements in Maize

Ds elements are nonautonomous transposons that can be mobilized by the *Ac* TPase. In contrast to functional and active *Ac* elements, which have been identified in only a few maize lines and in low

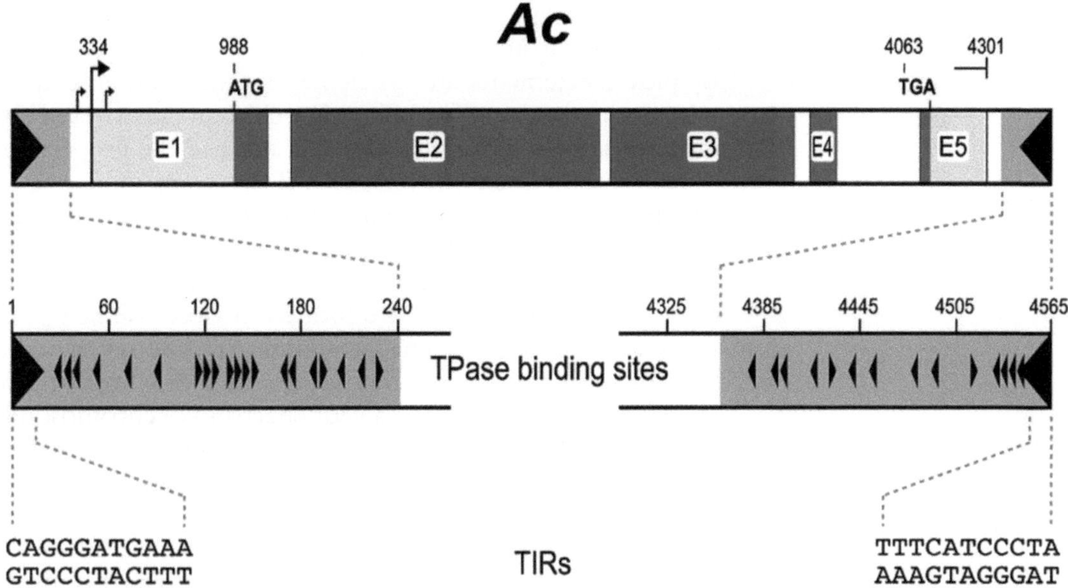

Fig. 1 The maize *Activator* (*Ac*) transposable element. *Top*: Diagram of the 4,565 bp *Ac* element. E1–E5: Exons 1–5. *Arrow* at position 334: Major transcription start site. ATG at position 988: TPase ORF start codon. TGA at position 4,063: TPase ORF stop codon. Position 4,301: Polyadenylation site. *Center*: Diagram of the distribution and orientation of the subterminal AAACGG TPase-binding sites. *Bottom*: Sequence of the *Ac* terminal inverted repeats

copy number, hundreds of *Ds* elements are spread throughout the maize genome. A recent survey identified 903 *Ds* (-related) elements in the genome of maize line B73, but different maize lines are highly polymorphic in genomic distribution and abundance of *Ds* elements. The majority of these elements apparently do not transpose in the presence of *Ac* and likely represent decaying transposons [12].

Ds elements are very divergent in architecture and can be classified in three categories (*see* below). With the exception of *Ds1*, all *Ds* elements have in common the TIRs with either a C or a T at the 5′-end and intact subterminal regions. Remarkably however, a similar mutation at the 3′-end of *Ac* that replaces the 3′-terminal A by a C causes a 3,800-fold reduction in excision frequency [13]. The *Ds* 5′- and 3′-ends are very similar in sequence composition but functionally distinct, as was shown by constructing artificial *Ds* elements. Elements flanked by two *Ac* 5′-ends or two *Ac* 3′-ends, and elements having a 5′-end and a 3′-end where the *Ac* TIRs were replaced by the *Tam3* TIRs, that are identical with the *Ac* TIRs in 7 of the 11 bp, cannot be mobilized by the *Ac* TPase [9, 14].

Simple *Ds* elements are derived from *Ac* by mutation of the TPase gene. These can either be internal deletions that remove parts of the TPase gene or more complex internal rearrangements where unrelated maize sequences were copied between the two *Ac* ends (reviewed in refs. 6, 15). Simple *Ds* elements can transpose, but do not cause chromosome breaks.

Complex *Ds* elements are *Ds* elements with multiple ends in alternative orientations. The first *Ds* element discovered by McClintock was such a complex element and had the property to induce chromosome breakage [16, 17]. If two complete *Ac* or *Ds* elements in direct or opposite orientation, or one complete and one "fractured" element consisting of a single residual transposon end in both alternative orientations, or a pair of individual 5′- and 3′-ends in reversed orientation are located in closely linked positions on a chromosome, aberrant transposition reactions can occur. Such reactions may involve *Ac/Ds* ends on the same chromatid or on sister chromatids and result in a wide spectrum of chromosomal rearrangements, including the generation of novel chimeric genes, deletions, and chromosome breakage-fusion-bridge cycles in maize [18–29]. These alternative constellations of *Ac/Ds* ends cause comparable aberrant transpositions also in transgenic plants [30–35]. Aberrant transposition reactions involving reversed transposon ends on two different molecules (e.g., sister chromatids after replication) are probably driven by differential binding of the TPase to methylated and hemimethylated *Ac/Ds* ends (*see* Subheading 8.2) [8, 36, 37].

Ds1 is unique in that it deviates from all other known *Ds* elements in structure and sequence organization [38]. The element is approximately 0.4 kb in size and has only 15 bp from the *Ac/Ds* 5′-end and 26 bp from the 3′-end in common with the other *Ac/Ds* elements. *Ds1* is thus lacking the subterminal regions that are characteristic for all other known *Ds* elements, but it contains several individual TPase-binding sites in its otherwise AT-rich center part which are required for mobilization by *Ac* [39]. Another unusual feature of *Ds1* is that it is transactivated not only by *Ac* but also the maize transposon *Uq*, which is unable to mobilize the other *Ds* elements [40]. However, the autonomous *Uq* has not been isolated and transposition of *Ds1* has not been molecularly investigated in depth and thus the mechanistic background of the unusual features of these elements remains somewhat enigmatic.

5 The *Ac* Transposase (TPase)

The *Ac* TPase protein has been partially characterized by mutant analysis in transgenic plants, transposition assays in plant protoplasts and yeast (*Saccharomyces cerevisiae*), and in vitro DNA binding analyses. However, a more detailed characterization is still missing because it has not yet been possible to establish an in vitro transposition assay and to obtain a soluble TPase preparation. The 807 amino acid *Ac* TPase contains three nuclear localization sequences (NLS) near its amino-terminal end (*see* Fig. 2).

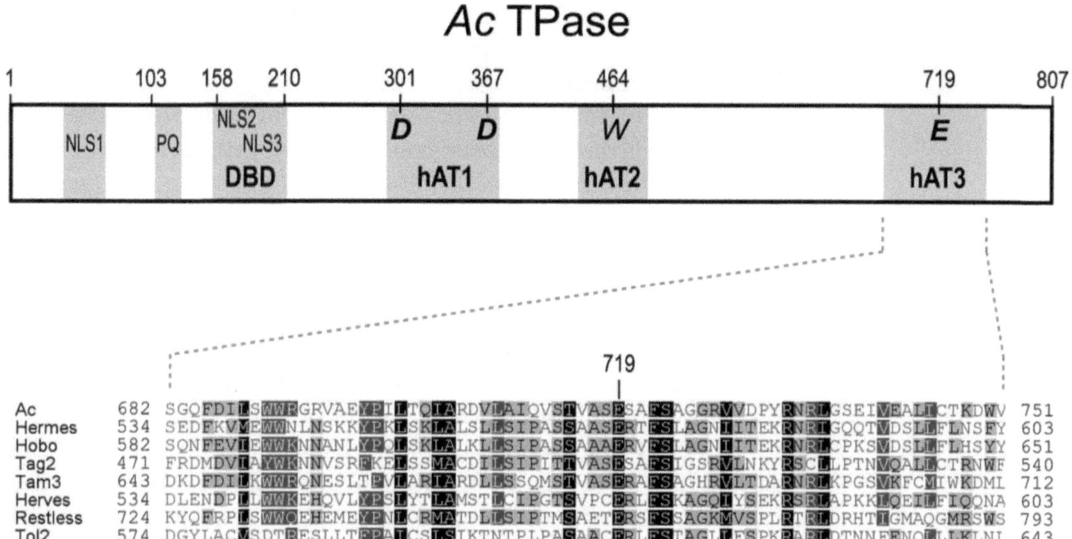

Fig. 2 The *Ac* transposase (TPase). *Top*: Diagram of the 807 amino acid *Ac* TPase protein. NLS 1–3: Nuclear localization signals 1–3. PQ: Tenfold PQ/PE-repeat. DBD: DNA-binding domain. hAT1-3: Conserved domains in the hAT-family transposases. D301, D367, E719: DDE catalytic motif. W464: Conserved tryptophan. *Bottom*: Alignment of the hAT3 domains from *Ac* (*Zea mays*), *Hermes* (*Musca domestica*), *hobo* (*Drosophila melanogaster*), *Tag2* (*Arabidopsis thaliana*), *Tam3* (*Antirrhinum majus*), *Herves* (*Anopheles gambiae*), *Restless* (*Tolypocladium inflatum*), and *Tol2* (*Oryzias latipes*)

For quantitative transport of the TPase into the nucleus all three NLS are required [41]. The second and third NLS motifs, $NLS_{159–178}$ and $NLS_{174–206}$, overlap with each other and with the bipartite DNA-binding domain that extends from residues 159 to 206. Binding to the subterminal sequence motifs is achieved by the C-terminal DNA-binding subdomain alone, whereas for recognition of the TIRs the N-terminal subdomain is additionally required [42, 43]. In contrast to several other hAT transposases that share the BED finger motif in their DNA-binding domains (e.g., *Tam3*, *hobo*, *Hermes*, *Slide*, *Tag1*) [44], the *Ac* TPase DNA-binding domain has no obvious sequence similarity to this domain. It has been noted that the DNA-binding domains of phylogenetically related transposases may display a marked structural diversity [45].

Between residues 109 and 129 the TPase contains a tenfold repeat of the dipeptides Pro-Gln and Pro-Glu (..PQPQPQPQPE PQPQPQPEPE..). This repeat is a unique feature of the *Ac* TPase; no other transposase contains a homologous amino acid sequence. The function of the PQ/PE dipeptide repeat is still unclear, but its deletion results in complete loss of TPase activity [10, 46]. Similar PQ repeats are found in various animal, plant, and viral proteins, but the function of the dipeptide repeats remains unknown. However, as in *Arabidopsis*, a number of putative transcription regulators contain PQ repeats, it is tempting to speculate that the PQ/PE dipeptide repeat in the *Ac* TPase is involved in DNA binding and autoregulation.

The alignment of the *Ac* TPase with the transposases of other hAT elements discloses three conserved regions, termed the "hAT regions". The hAT1 region encompasses approximately residues 293 to 376 of the *Ac* TPase, hAT2 extends from residues 442 to 490, and the most highly conserved C-terminal hAT3 region from 678 to 759 (*see* Fig. 2) [42, 47–50]. The hAT3 region constitutes a dimerization domain that is essential for TPase activity. In addition, between the DNA-binding domain and the hAT3 region one or more additional self-interaction interfaces are located that are responsible for TPase oligomerization in vivo [7].

The hAT transposons had long been thought to be a distinct superfamily because no prominent homologies between their transposase proteins with those of other transposon families had been recognized [51]. However, more recently is was found that also hAT transposases contain an acidic DDE motif in the catalytic center which has previously been shown to be involved in the catalytic activity of members of the retroviral integrase superfamily, the RAG recombinase of the V(D)J immunoglobulin gene recombination apparatus, and prokaryotic transposases [52, 53]. All DDE family proteins share an RNase H-like fold that brings the three catalytically active DDE residues into close proximity. Beyond this fold, the structures of their catalytic domains vary considerably [45]. Recently it was suggested that all eukaryotic cut-and-paste transposon superfamilies, including the hAT elements, have a common evolutionary origin [54].

The transposase of *Hermes* is the only transposase of the hAT transposon family whose structure has been determined [55]. Substitution of any amino acid of the DDE motif of *Hermes* severely affects DNA cleavage and strand transfer activity [53]. The alignment of the *Ac* TPase with the *Hermes* transposase identified the putative amino acids that constitute the DDE motif in the *Ac* TPase as D301, D367, and E719 (*see* Fig. 2). Consistent with their presumed activities, substitution of each of these residues completely abolishes *Ac* TPase activity in yeast [56].

Although the N-terminal 102 amino acids of the *Ac* TPase contain one of the three NLS (*see* Fig. 2) and the truncation of these residues consequently results in severely impaired nuclear transport, the N-terminus is not essential for the transposase activity [41, 57, 58]. The $TPase_{103-807}$ has been shown to efficiently mobilize *Ds* elements in plants, yeast, and vertebrates [58–65]. Astonishingly, the truncated TPase is more active in transactivation of *Ds* elements than the full-length protein (*see* Subheading 8.1). The *Hermes* transposase is somewhat similar to the *Ac* TPase in that its N-terminal 78 residues are not crucial for DNA binding or catalysis but involved in nuclear localization [55, 66].

6 *Ac* Transposition Mechanism

Except from sequence-specific and methylation-dependent binding of the *Ac* TPase to the TIRs and the unrelated, multiple subterminal binding sites of *Ac/Ds* elements, the distinct chemical steps of transposition have not been studied due to the lack of an in vitro transposition assay. However, the analysis of *Ac/Ds* excision and reinsertion patterns in maize, of genome rearrangements caused by aberrant *Ac/Ds* transposition, and of *Ds* transposition frequencies and footprints in plants and yeast yielded some insight into the transposition mechanism of *Ac/Ds* elements (reviewed in ref. 6).

Ac/Ds elements transpose like all hAT transposons by a cut-and-paste mechanism. When *Ac/Ds* elements excise, they leave behind a footprint that in the vast majority of analyzed excision sites in maize, transgenic plants, and yeast can readily be explained by a hairpin model. This mechanism was for the first time suggested for the *Tam3* element from snapdragon (*Antirrhinum majus*) [67]. Subsequently it was discovered that hairpin formation, opening, and repair during coding end-joining in V(D)J recombination [68, 69] employ the same biochemical mechanisms as in transposition of hAT elements, as was exemplified experimentally for the *Hermes* transposon [53].

6.1 *Ac/Ds* Excision and Footprint Formation

In the first step of transposase-mediated cleavage, a nick is introduced in the top strand one nucleotide distal to the 5′ end of the transposon (*see* Fig. 3). Following this first hydrolysis step, a nucleophilic attack of the free 3′-OH on the phosphodiester bond between the transposon 3′-end on the bottom strand is catalyzed by the transposase, resulting in an intramolecular transesterification and generation of a hairpin intermediate on the flanking DNA. As hairpin formation is sterically difficult, a recombinase or a transposase must be able to induce considerable DNA distortion to proceed with catalysis. For the V(D)J recombinase Rag1, it was demonstrated that a tryptophan residue, which is highly conserved in all hairpin-generating DDE recombinases, is involved in stabilizing the flipped-out base in the center of the hairpin [70]. Thus it is likely that in the *Ac* TPase the corresponding tryptophan 464 (*see* Fig. 2) performs this function.

The subsequent step is the opening of the hairpin. In V(D)J recombination, the Ku70–80/DNA-PK complex identifies the DNA damage by double-strand breakage and signals it to the non-homologous end joining (NHEJ) DNA-repair machinery of the cell. Next, Artemis is recruited by the Ku70–80/DNA-PK complex to the damage site and phosphorylated by the DNA-PKs, leading to the opening of the coding end hairpins [71]. In case of *Hermes*, it is not yet known which enzyme opens the hairpin, but it is assumed that a host function like Artemis is involved [53]. An Artemis-like protein is present in insect genomes, but it has not yet been molecularly studied.

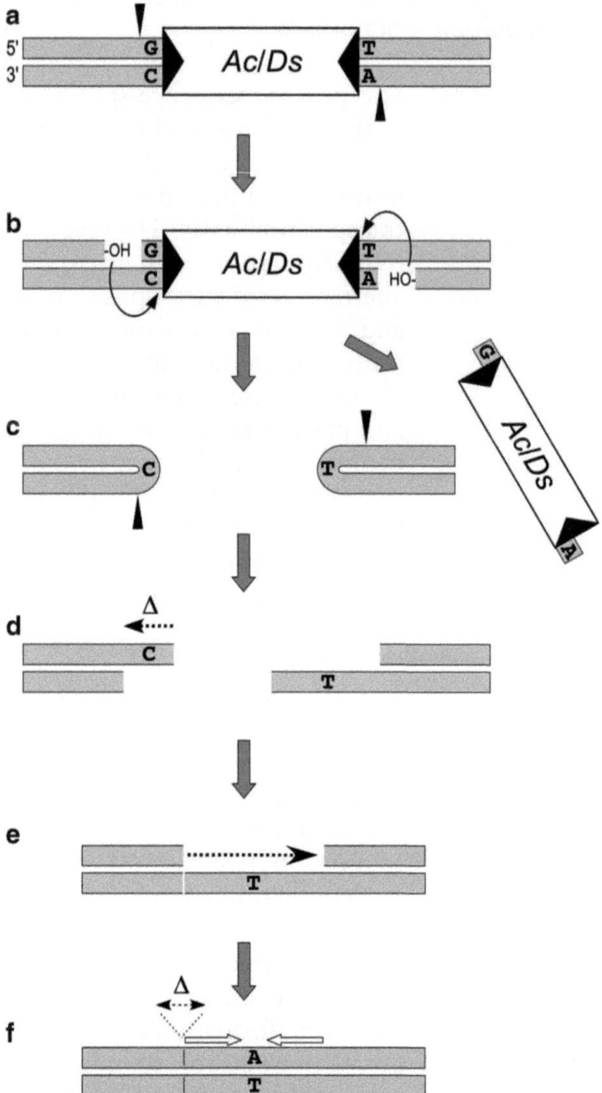

Fig. 3 Hairpin model for the hAT transposon excision and footprint formation reaction. (**a**) *Ac/Ds* excision is initiated by single-strand cleavage 1 bp distal to the 5′-ends of the transposon DNA. (**b**) Hairpins are generated at the flanking DNA by nucleophilic attack of the exposed 3′-hydroxyl groups to the phosphodiester bonds at the 3′-ends of the transposon DNA leading to the release of the transposon. (**c**) Hairpins are opened by single-strand cleavage at variable distance 3′ from the nucleotide at the center of the hairpin. (**d**) The resulting single-stranded 3′ overhangs may or may not be exonucleolytically degraded prior to rejoining (in this example the overhang at the left end is degraded). (**e**) Rejoining of the ends and fill-in of the remaining single-stranded gap. (**f**) The resulting *Ac/Ds* excision footprints carry at the former left and right borders either deletions or palindromic duplications centering around the complementary nucleotide of the one that previously flanked the 3′ transposon ends

The factors that accomplish opening of hAT transposon excision-generated hairpins, DNA repair, and religation in plants have not been identified yet. A study of *Ac/Ds* excision in yeast suggested an involvement of the Mre11/Rad50 complex, SAE2, NEJ1, and the Ku complex in repair of excision sites [63]. Recently a deep-sequencing survey of *Ac/Ds* excision footprints in *Arabidopsis* lines with defects in different DNA repair or DNA damage response genes (*Lig4-2*, *Lig6-1*, *Ku70*, *Atr2*, *Atm2*, and *PolL*) revealed that in all these mutants the preservation of the chromosomal DNA flanking the transposon is negatively affected, resulting in an enhanced frequency of deletions, insertions, and inversions at the excision site. However, it also turns out that excision site repair is not strictly dependent on the canonical NHEJ pathway factors Ku70 and DNA ligase IV (*Lig4*) or any of the investigated accessory NHEJ factors [72].

Ac/Ds excision sites in plants and also in yeast very frequently contain palindromic flanking sequence duplications that are centered around the complementary nucleotide of the base which previously was joined to the 3′ transposon ends. This footprint structure can be readily explained by a symmetric positioning of the endonucleolytic single-strand breaks, leading to hairpin opening on the DNA strands formerly flanking the 3′-ends of the transposon (*see* Fig. 3). For example, the most frequent *Ac/Ds* excision footprints at the *wx-m7* locus in maize and at four different insertion sites in transgenic *Arabidopsis* lines are presumably generated by hairpin opening at the phosphodiester bond on the top strand right behind the nucleotide in the center of the hairpin [73].

6.2 Ac/Ds Transposon Reinsertion

Upon reinsertion the exposed 3′-OH groups of the transposon ends join to the target DNA, a strategy highly conserved among mobile elements. The distinct chemical steps of transposition, nicking, hairpin formation, and target joining are supposedly performed by the same catalytic DDE active center [53], and for the prokaryotic *Tn*10 and *Mu* transposons, it was shown that all of these steps are catalyzed by a single transposase monomer at each transposon end [74–76]. It is unclear whether this is also the case for *Ac* and other hAT elements. The asymmetry in distribution of the multiple TPase-binding sites in both *Ac/Ds* termini could require an alternative mechanism with multiple transposase active sites at each end.

Like all hAT transposons, at the insertion site *Ac/Ds* elements generate an 8-bp target site duplication (TSD). The X-ray crystallographic analysis of the *Hermes* transposase revealed that this spacing of single-strand cuts during integration is imposed by the three-dimensional molecular architecture of the transpososome with the hexameric transposase at its core [55]. The *Ac* TPase is also active as an oligomeric protein [46]; however, whether *Ac* TPase or other hAT transposases assemble into hexamers is unknown.

The analysis of a large number of *Ac/Ds* insertion sites in maize and other plants has to date not revealed any obvious preferred sequence motif or strong consensus sequence. In a recent survey of *Ds* insertions throughout the maize genome, Vollbrecht and colleagues aligned 1,741 *Ds* insertion target sites plus 120 bp flanking nucleotides on each side and analyzed the multiple sequence alignment in depth. They discovered that the DNA in a 16–18 nucleotide window centered on the 8-bp target site differs in certain structural features relative to the flanking DNA. Most prominently, a pattern of alternating sites with more and less than average DNA deformability and a symmetric arrangement of paired hydrogen bond donor and acceptor motifs emerges [77]. This pattern would be consistent with DNA binding of two adjacent TPase dimers in a hexameric TPase oligomer as with *Hermes* transposase [55]. Interestingly, also in case of the *Drosophila melanogaster* P transposable element it was observed that the transposon has no obvious insertion sequence specificity, but rather that the insertion sites have a 14-nucleotide-long palindromic pattern centered on the 8-bp TSD as a common structural feature [78].

6.3 Ac/Ds Distribution in the Maize Genome

Ac/Ds elements have no strong target site consensus sequence, but their distribution in the genome is nonrandom. *Ac/Ds* insertions in the maize genome are highly overrepresented in hypomethylated gene-rich chromosomal regions [79–81]. A similar preference for insertion in gene-rich regions is also exhibited by *Mutator* and other class II-maize transposons, but not by the CACTA-family transposable elements [82]. Moreover, *Ac/Ds* elements show a strong bias toward integration into exons. Genome-wide analyses of *Ac* and *Ds* target sites in maize revealed that more than two-thirds of the transposons are residing in exons, approximately 20 % in introns, and only a few in the 5′- or 3′-UTRs [77, 83]. Preferential integration into exons is also seen with *Mutator*, but it is much less pronounced.

In early genetic and also later molecular studies, it was found that transposition of *Ac* in maize is biased toward genetically linked sites. More than half of the *Ac* transpositions from the maize *Bronze*, *P*, and *Waxy* genes result in intramolecular reinsertion within short distances ranging from only a few base pairs to 10 cM from the donor locus (reviewed in refs. 6, 84–88). Surprisingly, a recent survey of *Ds* transposition from the *r1* locus revealed a somewhat reduced preference (44 %) for transposition to linked sites compared to *Ac* (>60 %). Among the linked transpositions, similar to *Ac*, a strong preference for reinsertions within 10 cM of the donor locus is observed and, moreover, this study highlighted a directional bias of *Ds* transposition toward the centromere of the chromosome [77]. A comparable propensity for transposition to linked sites is not exhibited by other plant transposons and has been exploited to create allelic mutant series when a target locus is close to an *Ac* or a *Ds* donor site.

7 *Ac* Regulation

7.1 Transcriptional Regulation and Posttranscriptional Silencing

The *Ac* promoter and its regulation are not characterized in much detail. The *Ac* promoter lacks the canonical CAAT and TATA boxes, which may account for the multiple transcription initiation sites and its constitutively low activity. In maize, *Ac* is transcribed at similarly low levels in all organs and also in transgenic tobacco constitutively weak transcription is observed (reviewed in ref. 6). *Ac* contains an approximately 650-nucleotide-long 5′-untranslated leader sequence (5′-UTL) that, like the promoter, is rich in CpG-dinucleotides. In *Arabidopsis* the deletion of this 5′-UTL leads to increased activity of *Ac*, although the overall TPase mRNA level is not raised [89]. The 5′-UTL also has a strong negative effect on the expression of a downstream gene in tobacco suspension cells [90]. The functions of the 5′-UTL in maize are not exactly known, but it is likely involved in epigenetic downregulation of *Ac* transcription. Transposons have a high mutagenic potential and thus are a threat to genome integrity. Consequently, host genomes have evolved sophisticated mechanisms to counteract transposable element mobilization, with epigenetic silencing being the most important and common mechanism in plants and animals (reviewed in ref. 91). Although not molecularly proven, it is likely that Barbara McClintock once reactivated *Ac*, and also the CACTA-family *En/Spm* element, from an epigenetically silenced state by the application of DNA damage [1, 92]. Later it was demonstrated that functional reactivation of a highly methylated, transcriptionally silent *Ac* element is associated with demethylation of the promoter and 5′-UTL region and reappearance of transcription [36, 37, 93–96]. Consistently, a dedifferentiation-induced reduction in the level of DNA methylation coincided with an increase in *Ac* transposition in transgenic rice [97]. Moreover, transcriptional regulation by promoter methylation has been observed with *Ac-st2*, a derivative of *Ac* [98]. The *Ac* promoter is embedded in the 5′-*cis*-acting region of the element and thus also present in all *Ds* elements (except *Ds1*) and decaying remnants of *Ac/Ds* elements. The maize genome contains hundreds of *Ds* elements and *Ac/Ds* debris fragments, some of which are transcribed [10, 95]. It is conceivable that these transcripts contribute to the regulation of *Ac* activity in the plant genome by posttranscriptional and transcriptional gene silencing via formation of double-stranded RNAs as has been documented for the maize *Mutator* transposon [99, 100].

7.2 Posttranslational Regulation

In this overview, two posttranslational *Ac/Ds* regulation mechanisms are summarized that must be taken into account when performing large-scale mutagenesis experiments with *Ac/Ds*: autoregulation of TPase activity, and epigenetic regulation of DNA binding by TPase.

In her studies on *Ac*, McClintock observed that an increase in the number of *Ac* elements in the genome resulted in a decreased transposition frequency and a shift toward later transposition events in endosperm development. She termed this phenomenon the "negative dosage effect" [17]. Later genetic and molecular studies revealed that the activity of *Ac* is regulated by several mechanisms and that there is no linear correlation between transcription level, TPase concentration, and transposition frequency. Many genetic and molecular studies in maize, transgenic plants, and yeast support the model, according to which an increase of TPase protein expression above a certain level results in formation of insoluble and transpositionally inactive aggregates [46, 58, 62, 101–103]. This model can also explain the "negative dosage effect" (reviewed in ref. 6). Negative autoregulation by TPase overexpression has also been observed in several other transposable elements, among them the hAT transposon *Tol2* from medaka fish, and investigated in most detail in *Tc1/mariner* elements from invertebrates and vertebrates, where it has been termed "overproduction inhibition" (OPI) [104, 105].

Furthermore, posttranslational regulation of *Ac/Ds* can be complicated by the expression of mutant, inactive TPase proteins that may act as dominant inhibitors by formation of nonfunctional heterooligomers [46]. It is conceivable that decaying but still transcribed *Ac* elements in the maize genome can be a source of inhibitory TPase fragments. In different plant species, transposition frequency and timing in response to altering *Ac* TPase gene transcription have turned out to be highly variable and largely unpredictable. Presumably this variation is at least partially caused by mis-processing of the TPase gene transcripts and expression of truncated TPase fragments (*see* Subheading 8.1).

A second posttranslational regulation mechanism of *Ac* is based on the methylation-sensitivity of the TPase. The ~240 bp *cis*-acting regions at both *Ac/Ds* ends contain multiple subterminal AAACGG TPase-binding sites which include the canonical CpG methylation motif. C-methylation on the top strand of the (non-palindromic) binding motif dominantly blocks TPase binding, whereas methylation of the two cytosines on the complementary bottom strand promotes TPase binding. Consequently, the TPase binds to unmethylated and, with higher affinity, to bottom-strand hemimethylated binding sites, whereas holomethylated and top-strand hemimethylated binding sites are not recognized [8]. Genomic DNA methylation analysis of *Ac* in the maize *Waxy* locus (*wx-m9::Ac*) and a silenced derivative that behaves like a *Ds* element (*wx-m9::Ds-cy*) revealed that both transposons were fully C-methylated throughout the ~240 bp *cis*-acting 3′-end, including the TPase-binding sites. In contrast, at the 5′-end the *Ds-cy* was as fully methylated as at the 3′-end, whereas the active *Ac* element was not or only partially methylated [36, 37]. These findings led to a model that explains the "chromatid selectivity" of *Ac* transposition in maize: *Ac* transposes preferentially during or shortly after the S-phase of the cell cycle from only one of the two sister chromatids [86, 106]. Prior to replication *Ac/Ds* are holomethylated at both ends, preventing binding of TPase and thus transposition. After replication, the hemimethylation pattern of the *Ac/Ds* element on the two sister chromatids differs such that TPase can bind to both transposon ends on one, but not the other sister chromatid. Subsequently, the maintenance methylation system will reestablish holomethylation of the elements on both sister chromatids. Transposition assays in petunia protoplasts with artificially holo- and hemimethylated *Ds* elements corroborated this model [107]. Moreover, the methylation-dependent TPase interaction with the transposon ends can account for chromosome breakage and rearrangements at complete or fractured "double-Ds" elements or closely linked *Ds* elements in inverse orientation (reviewed in refs. 6, 26, 34, 37).

However, there is evidence that the methylation sensitivity of TPase binding is not the exclusive regulator of replication-dependent *Ac/Ds* transposition. In contrast to dicot cell cultures, where unmethylated *Ds* elements can excise from non-replicating vectors [107, 108], in transient assays in maize, wheat, barley, and rice cells *Ds* transposition is strictly dependent on vector replication although the transfected plasmids are unmethylated [109–112]. It is therefore conceivable that a cell cycle-dependent repressor of *Ac/Ds* transposition occurs exclusively in monocots or that the respective factor in dicots is unable to interact with *Ac/Ds*.

8 *Ac/Ds* as Gene Tagging and Functional Genomics Tools in Maize and Heterologous Plants

Only 3 years after its cloning, *Ac* was the first transposable element to be transformed into a heterologous host plant, tobacco, where it readily transposed [113]. Thereafter *Ac/Ds* were introduced in rapid succession into various other plant species (reviewed in ref. 6). *Ac/Ds* elements became the most frequently used transposons for application in heterologous plants because (a) the sequences required in *cis* for transposition of *Ds* are well defined; (b) the TPase is the only protein required for mobilization of *Ds* elements; and (c) it turned out that *Ac/Ds* elements were functional in all plant species tested. The latter seems to be a unique feature of *Ac/Ds*. The maize *En/Spm* and *Mu* transposons are very efficient mutagenesis tools in their native host; however, *Mu* is not functional in other species and application of *En/Spm* for insertion mutagenesis has only been feasible in rice and *Arabidopsis* (*see* Chapter 14).

8.1 Wild-Type and Mutant Ac TPase Expression in Heterologous Plants

In the early experiments the unmodified *Ac* element was transformed into a number of plant species. However, for gene tagging this approach is unfavorable. The mutations are unstable as *Ac* can transpose off the target gene, *Ac* is not suitable for inserting foreign "cargo genes", and it turned out that the transposition frequency of *Ac* can vary dramatically in the progenitors of the primary transformants and in different species. For example, in tobacco and tomato, high *Ac* transposition frequencies were observed [114, 115] whereas in *Arabidopsis*, broccoli, lettuce, and flax the element shows very low activity [116–120]. The first steps to improve *Ac/Ds* as insertion mutagenesis and functional genomics tools were the design of two-element systems, consisting of an immobile TPase source and a nonautonomous *Ds* transposon carrying a marker gene [121, 122], and the use of alternative promoters for *Ac* TPase expression [123–125]. For application of *Ac/Ds* as gene delivery tools, it is of advantage that *Ds* elements have no strict size limit and the identity and sequence composition of the internal "cargo" DNA is not critical. In maize, *Ds* elements more than 20 kb in size are not significantly impaired in transposition frequency and even bigger "macrotransposons" are mobilized by *Ac* TPase [26, 27, 126].

The very low *Ac/Ds* transposition frequencies in some plant species prompted the analysis of TPase expression in these plants and attempts to enhance transposition by using stronger promoters. In *Arabidopsis* the fusion of stronger promoters to the TPase open reading frame generally leads to increased *Ds* excision frequency and germinal transmission; however, the increase in transposition is much lower than the increase in transcript levels and in individual transgenic lines those frequencies are highly variable [123, 127, 128]. A similar response of *Ds* excision frequency upon increased TPase gene transcription is seen in petunia protoplasts [58, 125, 129]. In lettuce callus cells a fusion of the strong T-DNA 2′-promoter to the TPase cDNA enhanced *Ds* excision [118]. A significant increase in *Ds* transactivation frequencies is achieved in *Arabidopsis* by the deletion of the long 5′-untranslated leader sequence [89, 122]. In tobacco suspension cells it was found that this sequence has a strong negative effect on the expression of a downstream gene [90]. However, in flax removal of the UTL only marginally increased transposition [129].

In tobacco and apparently also in wheat the *Ac* promoter is, like in maize, weakly active and the *Ac* transcript is correctly processed [10, 130, 131]. In contrast, in *Arabidopsis* and sugar beet part of the *Ac* transcripts are misprocessed by alternative splicing at cryptic splice sites and by premature termination at cryptic polyadenylation signals [132–135]. Thus, it is conceivable that in these and other plant species with low *Ac/Ds* transposition frequencies even upon overexpression by the strong CaMV 35S promoter, only very little active TPase is synthesized and that additional truncated TPase fragments are generated, which may negatively interfere with TPase function.

A mutational analysis of the *Ac* TPase revealed that the N-terminal 102 amino acids contain a nuclear localization signal but are not essential for TPase activity [57]. Astonishingly, the truncated $\text{TPase}_{103\text{–}807}$ induces significantly higher *Ds* excision frequencies in transfected petunia protoplasts, more frequent and developmentally earlier *Ds* excision events in transgenic tobacco plants, and higher *Ds* transposition frequencies in transgenic barley lines than the full-length TPase [46, 58, 60, 61, 125, 136]. In a transient assay for excision of a *Ds1* element from a maize streak virus in agroinfected maize cells this higher activity of the truncated $\text{TPase}_{103\text{–}807}$ was not observed [137]. However, as *Ds1* differs in structure and in its response to the transactivating *Uq* element (*see* Subheading 4) it cannot be excluded that it also responds differently to the truncated *Ac* TPase. Moreover, as the expression levels are not known in these experiments it is possible that $\text{TPase}_{103\text{–}807}$ concentration was higher than that of the full-length protein and "OPI" became effective.

As in maize, OPI has been observed with the full-length *Ac* TPase in tobacco, petunia cells, wheat callus, and yeast. OPI is not limited to the full-length TPase but has also been detected with the N-terminally truncated $\text{TPase}_{103\text{–}807}$ in the three systems that were scrutinized on this aspect (petunia cells, tobacco, yeast) [46, 58, 62, 102, 131].

Recently a novel *Ac* $\text{TPase}_{103\text{–}807}$ derivative carrying four amino acid substitutions was reported that catalyzes 100-fold more frequent *Ds* excisions in yeast and a ≥6-fold higher *Ds* excision rate in *Arabidopsis* than the wild-type protein [56]. Although the reinsertion frequency and thus the forward mutation activity of this mutant TPase in plants are not known yet, it is a promising candidate to improve transposon mutagenesis strategies.

8.2 Genomic Distribution of Transposed Ac/Ds Elements in Heterologous Plants

By now *Ac/Ds* elements have been transiently or stably introduced into ca. 20 plant species. However, sufficiently large insertion mutant collections for use in functional genomics and reverse genetics approaches and a meaningful evaluation of the genomic distribution and insertion preferences of transposed elements exist, besides in maize, only in *Arabidopsis*, rice, and, at much smaller scale, barley. Table 1 shows a selection of recent publications on the application of *Ac/Ds* transposable elements in different plant species.

Particularly in large genomes with a low gene density and in plant species that are recalcitrant to large-scale transformation, the preference of *Ac/Ds* elements for insertion into hypomethylated, genic regions could be beneficial for gene tagging. Fortunately, this property of *Ac/Ds* elements appears to be retained in at least some heterologous plants as was reported for *Arabidopsis* [138–141], tobacco [142], tomato [143, 144], barley [145–147], and rice [97, 148–152].

Table 1
A selection of recent publications on *Ac/Ds* in maize and heterologous plants

Zea mays	
Kolkman et al. [81]	Collection of 158 W22 lines with mapped *Ac* insertions for regional mutagenesis
Ahern et al. [171]	Review on methods and resources for regional mutagenesis using >1,700 *Ds* starter lines
Vollbrecht et al. [77]	Distribution and target preferences of >1,500 *Ds* insertions compared
Wang et al. [172]	208 long-distance *Ac* transposants
Arabidopsis thaliana	
Kuromori et al. [140], Ito et al. [138]	RIKEN collection of 18,000 *Ds* insertion lines
Kuromori et al. [173]	Phenotyping of 4,000 RIKEN *Ds* insertion lines
Zhang et al. [174]	31 *Ds*-GUS launch pad lines and 44 *Ds*-loxP launch pad lines for deletional mutagenesis
Nishal et al. [165]	40 gene-trap *Ds* launch pad lines with heat shock-inducible *Ac* TPase
Muskett et al. [175]	89 gene-trap *Ds* launch pad lines
Panjabi et al. [176]	11 *Ds*(dSpm) launch pad lines, combination of *Ac/Ds* and *En/Spm* transposons
Sundaresan et al. [166], Parinov et al. [139]	567 enhancer-trap and gene-trap *Ds* launch pad lines
Oryza sativa	
Greco et al. [149], van Enckevort et al. [152]	1,380 enhancer-trap *Ds* lines
Kolesnik et al. [151]	4,400 gene-trap *Ds* lines
Kim et al. [150]	>1,000 gene-trap *Ds* lines
Park et al. [177]	>5,000 gene-trap *Ds* lines in japonica and indica subspecies backgrounds
Eamens et al. [178], Upadhyaya et al. [163]	>600 bidirectional gene-trap and plasmid-rescue *Ds* launch pad lines; transient *Ac* TPase expression
Qu et al. [179]	638 activation-tag *Ds* lines
Qu et al. [180]	Novel *Ac* TPase and *Ds* vectors
Luan et al. [181]	>2,800 gene-trap *Ds* lines
Hordeum vulgare	
Cooper et al. [145]	19 mapped *Ds* launch pad lines
Singh et al. [147], Zhao et al. [146]	>100 mapped *Ds* launch pad lines
Lazarow et al. [61]	34 gene-trap *Ds* lines
Ayliffe et al. [182]	Novel activation-tag *Ds* vectors
Glycine max	
Mathieu et al. [183]	900 *Ds* insertion lines; activation-tag, gene- and enhancer-trap *Ds* vectors

Like in maize, *Ac/Ds* elements transpose also in heterologous organisms frequently into genetically linked chromosomal target sites. Linked transposition has been reported in tobacco [153, 154], tomato [144, 155, 156], *Arabidopsis* [138, 140, 157–161], barley [162], and rice [163]. However, in *Brassica oleracea* a more dispersed reinsertion of *Ac/Ds* elements was observed [164]. Preferential transposition into closely linked target loci has been exploited in maize for local saturation mutagenesis [84, 87] and is also applicable in heterologous plants (*see* for example ref. 165), given that a starter line is at hand with a transposon donor site in close proximity to the target locus.

For genome-wide random gene tagging approaches this property of *Ac/Ds* elements has the disadvantage that the frequency of insertion into unlinked genes is severely depleted. To circumvent this problem, large-scale *Ac/Ds*-based insertion mutagenesis experiments ideally should be initiated with a collection of lines carrying "*Ds* launch pads" distributed on all chromosomes in less than approximately 30 cM distance. For some plant species such transposon launch pad lines have been developed (*see* Table 1). Alternatively, enrichment for unlinked *Ds* insertions may be achieved by counter-selection against a conditionally toxic gene flanking the *Ds* launch pad [166]. In maize, enrichment for novel, unlinked *Ds* transposants was achieved by visual selection against colored kernels that carry the empty launch pad with a revertant pigmentation gene [77]. As *Mu* and *En/Spm* transposons have no or only a weaker bias for transposition to linked targets, they have been used to generate mutant collections in their native host maize. *En/Spm* transposons were also successfully used to construct mutant collections in *Arabidopsis* and rice [167–170], but in other species apparently neither *Mu* nor *En/Spm* are sufficiently active for large-scale insertion mutagenesis.

Since the first transposon tagging experiments in maize in the early 1980s, a wealth of knowledge about the transposition mechanism, regulation, and insertion characteristics of *Ac/Ds* elements in its native host and in heterologous organisms has accumulated. This knowledge enabled the development of custom-tailored *Ac/Ds* transposon derivatives for specific applications (e.g., knockout insertion mutagenesis; promoter-, enhancer-, and gene-trapping; activation tagging). In plants where large-scale T-DNA insertion mutagenesis is not feasible or that have very large genomes, transposon mutagenesis may be a valuable alternative or complementary method.

8.3 Design of *Ac/Ds* Element Components for Use in Heterologous Plants

When designing *Ac* TPase expression construct and *Ds* elements for application in heterologous plants one should pay attention to the following parameters which are known to influence the functionality and efficiency of the system.

1. *Design of the* Ac *TPase expression cassette*:
   (a) The optimal expression level for the TPase coding region must be empirically determined. Due to unpredictable degrees of *Ac* transcript mis-processing and translation efficiency (e.g., due to unfavorable codon usage), it can vary dramatically in different plant species. The native *Ac* promoter is too weak in most plants. However, a strong promoter, in particular if it is active in very early developmental stages, may lead to TPase OPI and a reduced frequency of independent germinal transposition events.
   (b) In heterologous plants the *Ac* cDNA lacking the 5′-UTR, and not the genomic TPase gene, should be used for TPase expression.
   (c) In several plants the N-terminally truncated $TPase_{103-807}$ mobilizes *Ds* elements more efficiently than the wild-type protein. Though it has not yet been examined in large-scale experiments, the truncated TPase therefore may also be advantageous in transposon tagging approaches.
2. *Design of the* Ds *element*:
   (a) The minimal *Ac*-terminal sequences required for uncompromised transposition frequency are 238 bp from the 5′-end and 209 bp from the 3′-end.
   (b) There seems to be no strict limit for the cargo DNA between the *Ds* ends. In maize transposition reactions were observed with *Ds* ends separated by more than 100 kb. However, in transgenic plants there is no experience with *Ds* elements larger than approximately 10 kb.
   (c) The 5′-end of *Ds* harbors in addition to the TPase-binding sites also the *Ac* promoter. Although this promoter is very weak, one should be aware that downstream sequences might be transcribed.
   (d) As *Ac/Ds* transposons are methylation-prone, particularly in the absence of TPase, transgenic plants should be examined in subsequent generations whether methylation of transposon ends has spread into the cargo genes.

Acknowledgments

This work was supported by Deutsche Forschungsgemeinschaft (DFG) grant KU-715/9 and the Dahlem Centre of Plant Sciences (DCPS).

References

1. McClintock B (1951) Chromosome organization and genic expression. Cold Spring Harb Symp Quant Biol 16:13–47
2. Oliver KR, Greene WK (2009) Transposable elements: powerful facilitators of evolution. Bioessays 31:703–714
3. Zeh DW, Zeh JA, Ishida Y (2009) Transposable elements and an epigenetic basis for punctuated equilibria. Bioessays 31:715–726
4. Arensburger P et al (2011) Phylogenetic and functional characterization of the hAT transposon superfamily. Genetics 188:45–57
5. Kempken F, Windhofer F (2001) The hAT family: a versatile transposon group common to plants, fungi, animals, and man. Chromosoma 110:1–9
6. Kunze R, Weil CF (2002) The hAT and CACTA superfamilies of plant transposons. In: Craig NL, Craigie R, Gellert M et al (eds) Mobile DNA II. ASM Press, Washington, pp 565–610
7. Essers L, Adolphs RH, Kunze R (2000) A highly conserved domain of the maize *Activator* transposase is involved in dimerization. Plant Cell 12:211–224
8. Kunze R, Starlinger P (1989) The putative transposase of transposable element *Ac* from *Zea mays* L. interacts with subterminal sequences of *Ac*. EMBO J 8:3177–3185
9. Coupland G et al (1989) Sequences near the termini are required for transposition of the maize transposon *Ac* in transgenic tobacco plants. Proc Natl Acad Sci USA 86: 9385–9388
10. Kunze R et al (1987) Transcription of transposable element *Activator* (*Ac*) of *Zea mays* L. EMBO J 6:1555–1563
11. Coupland G et al (1988) Characterization of the maize transposable element *Ac* by internal deletions. EMBO J 7:3653–3659
12. Du C et al (2011) The complete Ac/Ds transposon family of maize. BMC Genomics 12:588
13. Xiao YL, Peterson T (2002) Ac transposition is impaired by a small terminal deletion. Mol Genet Genomics 266:720–731
14. Chatterjee S, Starlinger P (1995) The role of subterminal sites of transposable element Ds of *Zea mays* in excision. Mol Gen Genet 249:281–288
15. Döring H-P, Starlinger P (1986) Molecular genetics of transposable elements in plants. Annu Rev Genet 20:175–200
16. Martinez-Ferez IM, Dooner HK (1997) Sesqui-Ds, the chromosome-breaking insertion at bz-m1, links double Ds to the original Ds element. Mol Gen Genet 255:580–586
17. McClintock B (1948) Mutable loci in maize. Carnegie Inst Wash Yr Bk 47:155–169
18. Döring H-P et al (1989) Double *Ds* elements are involved in specific chromosome breakage. Mol Gen Genet 219:299–305
19. Dooner HK, Belachew A (1991) Chromosome breakage by pairs of closely linked transposable elements of the *Ac-Ds* family in maize. Genetics 129:855–862
20. Weil CF, Wessler SR (1993) Molecular evidence that chromosome breakage by Ds elements is caused by aberrant transposition. Plant Cell 5:515–522
21. Zhang J, Peterson T (1999) Genome rearrangements by nonlinear transposons in maize. Genetics 153:1403–1410
22. Zhang J, Peterson T (2004) Transposition of reversed Ac element ends generates chromosome rearrangements in maize. Genetics 167: 1929–1937
23. Zhang J, Peterson T (2005) A segmental deletion series generated by sister-chromatid transposition of Ac transposable elements in maize. Genetics 171:333–344
24. Zhang J, Zhang F, Peterson T (2006) Transposition of reversed Ac element ends generates novel chimeric genes in maize. PLoS Genet 2:e164
25. Zhang J et al (2009) Alternative Ac/Ds transposition induces major chromosomal rearrangements in maize. Genes Dev 23: 755–765
26. Yu C et al (2010) Spatial configuration of transposable element Ac termini affects their ability to induce chromosomal breakage in maize. Plant Cell 22:744–754
27. Huang JT, Dooner HK (2008) Macrotransposition and other complex chromosomal restructuring in maize by closely linked transposons in direct orientation. Plant Cell 20:2019–2032
28. Ralston EJ, English J, Dooner HK (1989) Chromosome-breaking structure in maize involved in a fractured Ac element. Proc Natl Acad Sci USA 86:9451–9455
29. Yu C, Zhang J, Peterson T (2011) Genome rearrangements in maize induced by alternative transposition of reversed *Ac/Ds* termini. Genetics 188:59–67
30. Krishnaswamy L, Zhang J, Peterson T (2008) Reversed end Ds element: a novel tool for chromosome engineering in Arabidopsis. Plant Mol Biol 68:399–411
31. Xuan YH et al (2011) Transposon Ac/Ds-induced chromosomal rearrangements at the rice OsRLG5 locus. Nucleic Acids Res 39:e149
32. Belzile F, Yoder JI (1994) Unstable transmission and frequent rearrangement of two closely linked transposed *Ac* elements in transgenic tomato. Genome 37:832–839

33. English J, Harrison K, Jones JDG (1993) A genetic analysis of DNA sequence requirements for dissociation state I activity in tobacco. Plant Cell 5:501–514
34. English JJ, Harrison K, Jones JDG (1995) Aberrant transpositions of maize double Ds-like elements usually involve Ds ends on sister chromatids. Plant Cell 7:1235–1247
35. Yu C et al (2012) A transgenic system for generation of transposon Ac/Ds-induced chromosome rearrangements in rice. Theor Appl Genet 125:1449–1462
36. Wang L, Heinlein M, Kunze R (1996) Methylation pattern of activator (Ac) transposase binding sites in maize endosperm. Plant Cell 8:747–758
37. Wang L, Kunze R (1998) Transposase binding site methylation in the epigenetically inactivated Ac derivative Ds-cy. Plant J 13:577–582
38. Sutton WD et al (1984) Molecular analysis of *Ds* controlling element mutations at the *Adh1* locus of maize. Science 223:1265–1268
39. Bravo-Angel AM et al (1995) The binding motifs for *Ac* transposase are absolutely required for excision of *Ds1*. Mol Gen Genet 248:527–534
40. Caldwell EEO, Peterson PA (1992) The *Ac* and *Uq* transposable element systems in maize: interactions among components. Genetics 131:723–731
41. Boehm U et al (1995) One of three nuclear localization signals of maize *Activator* (*Ac*) transposase overlaps the DNA-binding domain. Plant J 7:441–451
42. Feldmar S, Kunze R (1991) The ORFa protein, the putative transposase of maize transposable element *Ac*, has a basic DNA binding domain. EMBO J 10:4003–4010
43. Becker H-A, Kunze R (1997) Maize *Activator* transposase has a bipartite DNA binding domain that recognizes subterminal motifs and the terminal inverted repeats. Mol Gen Genet 254:219–230
44. Aravind L (2000) The BED finger, a novel DNA-binding domain in chromatin-boundary-element-binding proteins and transposases. Trends Biochem Sci 25:421–423
45. Nesmelova IV, Hackett PB (2010) DDE transposases: structural similarity and diversity. Adv Drug Deliv Rev 62:1187–1195
46. Kunze R et al (1993) Dominant transposition-deficient mutants of maize *Activator* (*Ac*) transposase. Proc Natl Acad Sci USA 90: 7094–7098
47. Calvi BR et al (1991) Evidence for a common evolutionary origin of inverted repeat transposons in *Drosophila* and plants: *hobo*, *Activator* and *Tam3*. Cell 66:465–471
48. Hehl R et al (1991) Structural analysis of *Tam3*, a transposable element from *Antirrhinum majus*, reveals homologies to the *Ac* element from maize. Plant Mol Biol 16:369–371
49. Essers L, Kunze R (1995) Transposable elements *Bg* (*Zea mays*) and *Tag1* (*Arabidopsis thaliana*) encode protein sequences with homology to *Ac*-like transposases. Maize Genet Coop Newsl 69:39–41
50. Kunze R, Saedler H, Lönnig W-E (1997) Plant transposable elements. In: Callow JA (ed) Adv Bot Res, vol 27. Academic, London, pp 331–470
51. Capy P et al (1997) Do the integrases of LTR-retrotransposons and class II element transposases have a common ancestor? Genetica 100:63–72
52. Haren L, Ton-Hoang B, Chandler M (1999) Integrating DNA: transposases and retroviral integrases. Annu Rev Microbiol 53:245–281
53. Zhou L et al (2004) Transposition of hAT elements links transposable elements and V(D)J recombination. Nature 432:995–1001
54. Yuan YW, Wessler SR (2011) The catalytic domain of all eukaryotic cut-and-paste transposase superfamilies. Proc Natl Acad Sci USA 108:7884–7889
55. Hickman AB et al (2005) Molecular architecture of a eukaryotic DNA transposase. Nat Struct Mol Biol 12:715–721
56. Lazarow K et al (2012) A hyperactive transposase of the maize transposable element activator (ac). Genetics 191:747–756
57. Li M-g, Starlinger P (1990) Mutational analysis of the N terminus of the protein of maize transposable element *Ac*. Proc Natl Acad Sci USA 87:6044–6048
58. Heinlein M, Brattig T, Kunze R (1994) In vivo aggregation of maize activator (Ac) transposase in nuclei of maize endosperm and petunia protoplasts. Plant J 5:705–714
59. Emelyanov A et al (2006) Trans-kingdom transposition of the maize dissociation element. Genetics 174:1095–1104
60. Kunze R et al (1995) Somatic and germinal activities of maize activator (Ac) transposase mutants in transgenic tobacco. Plant J 8:45–54
61. Lazarow K, Lütticke S (2009) An Ac/Ds-mediated gene trap system for functional genomics in barley. BMC Genomics 10:55
62. Weil CF, Kunze R (2000) Transposition of maize Ac/Ds transposable elements in the yeast *Saccharomyces cerevisiae*. Nat Genet 26:187–190
63. Yu J et al (2004) Microhomology-dependent end joining and repair of transposon-induced DNA hairpins by host factors in *Saccharomyces cerevisiae*. Mol Cell Biol 24:1351–1364
64. Boon Ng GH, Gong Z (2011) Maize Ac/Ds transposon system leads to highly efficient germline transmission of transgenes in medaka (*Oryzias latipes*). Biochimie 93:1858–1864

65. Froschauer A et al (2012) Effective generation of transgenic reporter and gene trap lines of the medaka (*Oryzias latipes*) using the Ac/Ds transposon system. Transgenic Res 21: 149–162
66. Michel K, Atkinson PW (2003) Nuclear localization of the Hermes transposase depends on basic amino acid residues at the N-terminus of the protein. J Cell Biochem 89:778–790
67. Coen ES, Carpenter R, Martin C (1986) Transposable elements generate novel spatial patterns of gene expression in *Antirrhinum majus*. Cell 47:285–296
68. Roth DB et al (1992) V(D)J recombination: broken DNA molecules with covalently sealed (hairpin) coding ends in *scid* mouse thymocytes. Cell 70:983–991
69. McBlane JF et al (1995) Cleavage at a V(D)J recombination signal requires only RAG1 and RAG2 proteins and occurs in two steps. Cell 83:387–395
70. Lu CP et al (2006) Amino acid residues in Rag1 crucial for DNA hairpin formation. Nat Struct Mol Biol 13:1010–1015
71. Ma Y et al (2002) Hairpin opening and overhang processing by an Artemis/DNA-dependent protein kinase complex in nonhomologous end joining and V(D)J recombination. Cell 108:781–794
72. Huefner ND et al (2011) Breadth by depth: expanding our understanding of the repair of transposon-induced DNA double strand breaks via deep-sequencing. DNA Repair (Amst) 10:1023–1033
73. Rinehart TA, Dean C, Weil CF (1997) Comparative analysis of non-random DNA repair following Ac transposon excision in maize and Arabidopsis. Plant J 12:1419–1427
74. Namgoong SY, Harshey RM (1998) The same two monomers within a MuA tetramer provide the DDE domains for the strand cleavage and strand transfer steps of transposition. EMBO J 17:3775–3785
75. Kennedy AK, Haniford DB, Mizuuchi K (2000) Single active site catalysis of the successive phosphoryl transfer steps by DNA transposases: insights from phosphorothioate stereoselectivity. Cell 101:295–305
76. Williams TL et al (1999) Organization and dynamics of the Mu transpososome: recombination by communication between two active sites. Genes Dev 13:2725–2737
77. Vollbrecht E et al (2010) Genome-wide distribution of transposed dissociation elements in maize. Plant Cell 22:1667–1685
78. Liao GC, Rehm EJ, Rubin GM (2000) Insertion site preferences of the P transposable element in *Drosophila melanogaster*. Proc Natl Acad Sci USA 97:3347–3351
79. Bennetzen JL et al (1994) Active maize genes are unmodified and flanked by diverse classes of modified, highly repetitive DNA. Genome 37:565–576
80. Chen J, Greenblatt IM, Dellaporta SL (1987) Transposition of *Ac* from the *P* locus of maize into unreplicated chromosomal sites. Genetics 117:109–116
81. Kolkman JM et al (2005) Distribution of activator (Ac) throughout the maize genome for use in regional mutagenesis. Genetics 169:981–995
82. Schnable PS et al (2009) The B73 maize genome: complexity, diversity, and dynamics. Science 326:1112–1115
83. Cowperthwaite M et al (2002) Use of the transposon Ac as a gene-searching engine in the maize genome. Plant Cell 14:713–726
84. Athma P, Grotewold E, Peterson T (1992) Insertional mutagenesis of the maize *P* gene by intragenic transposition of *Ac*. Genetics 131:199–209
85. Dooner HK, Belachew A (1989) Transposition pattern of the maize element *Ac* from the *bz-m2(Ac)* allele. Genetics 122: 447–457
86. Greenblatt IM (1984) A chromosomal replication pattern deduced from pericarp phenotypes resulting from movements of the transposable element, *Modulator*, in maize. Genetics 108:471–485
87. Moreno MA et al (1992) Reconstitutional mutagenesis of the maize P gene by short-range Ac transpositions. Genetics 131: 939–956
88. Van Schaik NW, Brink RA (1959) Transpositions of modulator, a component of the variegated pericarp allele in maize. Genetics 44:725–738
89. Lawson EJR et al (1994) Modification of the 5′ untranslated leader region of the maize *Activator* element leads to increased activity in Arabidopsis. Mol Gen Genet 245:608–615
90. Scortecci KC et al (1999) Negative effect of the 5′-untranslated leader sequence on Ac transposon promoter expression. Plant Mol Biol 40:935–944
91. Lisch D (2009) Epigenetic regulation of transposable elements in plants. Annu Rev Plant Biol 60:43–66
92. McClintock B (1984) The significance of responses of the genome to challenge. Science 226:792–801
93. Brettell RIS, Dennis ES (1991) Reactivation of a silent Ac following tissue culture is associated with heritable alterations in its methylation pattern. Mol Gen Genet 229:365–372
94. Brutnell TP, Dellaporta SL (1994) Somatic inactivation and reactivation of Ac associated with changes in cytosine methylation and transposase expression. Genetics 138:213–225
95. Kunze R, Starlinger P, Schwartz D (1988) DNA methylation of the maize transposable

element *Ac* interferes with its transcription. Mol Gen Genet 214:325–327
96. Fusswinkel H et al (1991) Detection and abundance of mRNA and protein encoded by transposable element *Activator* (*Ac*) in maize. Mol Gen Genet 225:186–192
97. Kohli A et al (2004) Dedifferentiation-mediated changes in transposition behavior make the activator transposon an ideal tool for functional genomics in rice. Mol Breeding 13:177–191
98. Brutnell TP, May BP, Dellaporta SL (1997) The Ac-st2 element of maize exhibits a positive dosage effect and epigenetic regulation. Genetics 147:823–834
99. Slotkin RK, Freeling M, Lisch D (2003) Mu killer causes the heritable inactivation of the mutator family of transposable elements in *Zea mays*. Genetics 165:781–797
100. Slotkin RK, Freeling M, Lisch D (2005) Heritable transposon silencing initiated by a naturally occurring transposon inverted duplication. Nat Genet 37:641–644
101. Heinlein M (1996) Excision patterns of activator (Ac) and dissociation (Ds) elements in *Zea mays* L: implications for the regulation of transposition. Genetics 144:1851–1869
102. Scofield SR, English JJ, Jones JDG (1993) High level expression of the activator transposase gene inhibits the excision of dissociation in tobacco cotyledons. Cell 75:507–517
103. Takumi S et al (1999) Variations in the maize Ac transposase transcript level and the Ds excision frequency in transgenic wheat callus lines. Genome 42:1234–1241
104. Grabundzija I et al (2010) Comparative analysis of transposable element vector systems in human cells. Mol Ther 18:1200–1209
105. Lohe AR, Hartl DL (1996) Autoregulation of *mariner* transposase activity by overproduction and dominant-negative complementation. Mol Biol Evol 13:549–555
106. Chen J, Greenblatt IM, Dellaporta SL (1992) Molecular analysis of *Ac* transposition and DNA replication. Genetics 130:665–676
107. Ros F, Kunze R (2001) Regulation of activator/dissociation transposition by replication and DNA methylation. Genetics 157: 1723–1733
108. Houba-Hérin N, Domin M, Pedron J (1994) Transposition of a *Ds* element from a plasmid into the plant genome in *Nicotiana plumbaginifolia* protoplast-derived cells. Plant J 6:55–66
109. Laufs J et al (1990) Wheat dwarf virus Ac/Ds vectors: expression and excision of transposable elements introduced into various cereals by a viral replicon. Proc Natl Acad Sci USA 87:7752–7756
110. McElroy D et al (1997) Development of a simple transient assay for Ac/Ds activity in cells of intact barley tissue. Plant J 11:157–165
111. Sugimoto K et al (1994) Transposition of the maize Ds element from a viral vector to the rice genome. Plant J 5:863–871
112. Wirtz U, Osborne B, Baker B (1997) Ds excision from extrachromosomal geminivirus vector DNA is coupled to vector DNA replication in maize. Plant J 11:125–135
113. Baker B et al (1986) Transposition of the maize controlling element "*Activator*" in tobacco. Proc Natl Acad Sci USA 83: 4844–4848
114. Baker B et al (1987) Phenotypic assay for excision of the maize controlling element Ac in tobacco. EMBO J 6:1547–1554
115. Yoder JI (1990) Rapid proliferation of the maize transposable element *Activator* in transgenic tomato. Plant Cell 2:723–730
116. Roberts MR et al (1990) Excision of the maize transposable element *Ac* in flax callus. Plant Cell Rep 9:406–409
117. Schmidt R, Willmitzer L (1989) The maize autonomous element activator (Ac) shows a minimal germinal excision frequency of 0.2–0.5 % in transgenic *Arabidopsis thaliana* plants. Mol Gen Genet 220:17–24
118. Yang CH, Ellis JG, Michelmore RW (1993) Infrequent transposition of *Ac* in lettuce, *Lactuca sativa*. Plant Mol Biol 22:793–805
119. McKenzie N, Wen LY, Dale J (2002) Tissue-culture enhanced transposition of the maize transposable element dissociation in *Brassica oleracea* var. 'Italica'. Theor Appl Genet 105:23–33
120. Dean C et al (1992) Behavior of the maize transposable element Ac in *Arabidopsis thaliana*. Plant J 2:69–81
121. Hehl R, Baker B (1989) Induced transposition of *Ds* by a stable *Ac* in crosses of transgenic tobacco plants. Mol Gen Genet 217: 53–59
122. Bancroft I et al (1992) Development of an efficient two-element transposon tagging system in *Arabidopsis thaliana*. Mol Gen Genet 233:449–461
123. Swinburne J et al (1992) Elevated levels of *Activator* transposase mRNA are associated with high frequencies of *Dissociation* excision in *Arabidopsis*. Plant Cell 4:583–595
124. Scofield SR et al (1992) Promoter fusions to the *Activator* transposase gene cause distinct patterns of *Dissociation* excision in tobacco cotyledons. Plant Cell 4:573–582
125. Becker D et al (1992) Control of excision frequency of maize transposable element *Ds* in *Petunia* protoplasts. Proc Natl Acad Sci USA 89:5552–5556

126. Dowe MF Jr, Roman GW, Klein AS (1990) Excision and transposition of two *Ds* transposons from the *bronze mutable* 4 derivative 6856 allele of *Zea mays* L. Mol Gen Genet 221:475–485
127. Balcells L, Coupland G (1994) The presence of enhancers adjacent to the *Ac* promoter increases the abundance of transposase mRNA and alters the timing of *Ds* excision in Arabidopsis. Plant Mol Biol 24:789–798
128. Long D et al (1993) Analysis of the frequency of inheritance of transposed Ds elements in Arabidopsis after activation by a CaMV 35S promoter fusion to the Ac transposase gene. Mol Gen Genet 241:627–636
129. Finnegan EJ et al (1993) Behaviour of modified Ac elements in flax callus and regenerated plants. Plant Mol Biol 22:625–633
130. Finnegan EJ et al (1988) Transcription of the maize transposable element *Ac* in maize seedlings and in transgenic tobacco. Mol Gen Genet 212:505–509
131. Takumi S et al (1999) Trans-activation of a maize Ds transposable element in transgenic wheat plants expressing the Ac transposase gene. Theor Appl Genet 98:947–953
132. Grevelding C et al (1992) High rates of *Ac/Ds* germinal transposition in *Arabidopsis* suitable for gene isolation by insertional mutagenesis. Proc Natl Acad Sci USA 89: 6085–6089
133. Jarvis P, Belzile F, Dean C (1997) Inefficient and incorrect processing of the *Ac* transposase transcript in *iae1* and wild-type *Arabidopsis thaliana*. Plant J 11:921–931
134. Lisson R et al (2010) Alternative splicing of the maize Ac transposase transcript in transgenic sugar beet (*Beta vulgaris* L.). Plant Mol Biol 74:19–32
135. Martin DJ et al (1997) Alternative processing of the maize Ac transcript in Arabidopsis. Plant J 11:933–943
136. Houba-Hérin N et al (1990) Excision of a *Ds*-like maize transposable element (*Ac*Δ) in a transient assay in *Petunia* is enhanced by a truncated coding region of the transposable element *Ac*. Mol Gen Genet 224:17–23
137. Shen WH, Ramos C, Hohn B (1998) Excision of Ds1 from the genome of maize streak virus in response to different transposase-encoding genes. Plant Mol Biol 36:387–392
138. Ito T et al (2005) A resource of 5,814 dissociation transposon-tagged and sequence-indexed lines of Arabidopsis transposed from start loci on chromosome 5. Plant Cell Physiol 46:1149–1153
139. Parinov S et al (1999) Analysis of flanking sequences from dissociation insertion lines: a database for reverse genetics in Arabidopsis. Plant Cell 11:2263–2270
140. Kuromori T et al (2004) A collection of 11 800 single-copy Ds transposon insertion lines in Arabidopsis. Plant J 37:897–905
141. Pan X, Li Y, Stein L (2005) Site preferences of insertional mutagenesis agents in Arabidopsis. Plant Physiol 137:168–175
142. Zhang BD et al (1999) Cloning of the DNA fragments flanking Ds insertion sites in tobacco genome. Acta Phytophysiol Sinica 25:7–14
143. Meissner R et al (2000) Technical advance: a high throughput system for transposon tagging and promoter trapping in tomato. Plant J 22:265–274
144. Carroll BJ et al (1995) Germinal transpositions of the maize element *Dissociation* from T-DNA loci in tomato. Genetics 139: 407–420
145. Cooper LD et al (2004) Mapping Ds insertions in barley using a sequence-based approach. Mol Genet Genomics 272: 181–193
146. Zhao T et al (2006) Mapped Ds/T-DNA launch pads for functional genomics in barley. Plant J 47:811–826
147. Singh J et al (2006) High-frequency Ds remobilization over multiple generations in barley facilitates gene tagging in large genome cereals. Plant Mol Biol 62:937–950
148. Enoki H et al (1999) Ac as a tool for the functional genomics of rice. Plant J 19:605–613
149. Greco R et al (2003) Transpositional behaviour of an Ac/Ds system for reverse genetics in rice. Theor Appl Genet 108:10–24
150. Kim CM et al (2004) Rapid, large-scale generation of Ds transposant lines and analysis of the Ds insertion sites in rice. Plant J 39: 252–263
151. Kolesnik T et al (2004) Establishing an efficient Ac/Ds tagging system in rice: large-scale analysis of Ds flanking sequences. Plant J 37:301–314
152. van Enckevort LJ et al (2005) EU-OSTID: a collection of transposon insertional mutants for functional genomics in rice. Plant Mol Biol 59:99–110
153. Jones JDG et al (1990) Preferential transposition of the maize element *Activator* to linked chromosomal locations in tobacco. Plant Cell 2:701–707
154. Dooner HK et al (1991) Variable patterns of transposition of the maize element *Activator* in tobacco. Plant Cell 3:473–482
155. Stuurman J et al (1996) Single-site manipulation of tomato chromosomes in vitro and in vivo using Cre-lox site-specific recombination. Plant Mol Biol 32:901–913
156. Stuurman J, Nijkamp HJJ, van Haaren MJJ (1998) Molecular insertion-site selectivity of Ds in tomato. Plant J 14:215–223

157. Keller J, Lim E, Dooner HK (1993) Preferential transposition of *Ac* to linked sites in *Arabidopsis*. Theor Appl Genet 86: 585–588
158. Bancroft I, Dean C (1993) Transposition pattern of the maize element Ds in *Arabidopsis thaliana*. Genetics 134:1221–1229
159. Machida C et al (1997) Characterization of the transposition pattern of the Ac element in *Arabidopsis thaliana* using endonuclease I-SceI. Proc Natl Acad Sci USA 94: 8675–8680
160. Ito T et al (2002) A new resource of locally transposed dissociation elements for screening gene-knockout lines in silico on the Arabidopsis genome. Plant Physiol 129: 1695–1699
161. Smith D et al (1996) Characterization and mapping of Ds-GUS-T-DNA lines for targeted insertional mutagenesis. Plant J 10:721–732
162. Koprek T et al (2000) An efficient method for dispersing Ds elements in the barley genome as a tool for determining gene function. Plant J 24:253–263
163. Upadhyaya NM et al (2006) Dissociation (Ds) constructs, mapped Ds launch pads and a transiently-expressed transposase system suitable for localized insertional mutagenesis in rice. Theor Appl Genet 112: 1326–1341
164. Mckenzie N, Dale PJ (2004) Mapping of transposable element dissociation inserts in *Brassica oleracea* following plant regeneration from streptomycin selection of callus. Theor Appl Genet 109:333–341
165. Nishal B, Tantikanjana T, Sundaresan V (2005) An inducible targeted tagging system for localized saturation mutagenesis in Arabidopsis. Plant Physiol 137:3–12
166. Sundaresan V et al (1995) Patterns of gene action in plant development revealed by enhancer trap and gene trap transposable elements. Genes Dev 9:1797–1810
167. Marsch-Martinez N et al (2002) Activation tagging using the En-I maize transposon system in Arabidopsis. Plant Physiol 129: 1544–1556
168. Schneider A et al (2005) A transposon-based activation-tagging population in *Arabidopsis thaliana* (TAMARA) and its application in the identification of dominant developmental and metabolic mutations. FEBS Lett 579: 4622–4628
169. Greco R et al (2004) Transcription and somatic transposition of the maize En/Spm transposon system in rice. Mol Genet Genomics 270:514–523
170. Kumar CS, Wing RA, Sundaresan V (2005) Efficient insertional mutagenesis in rice using the maize En/Spm elements. Plant J 44: 879–892
171. Ahern KR et al (2009) Regional mutagenesis using dissociation in maize. Methods 49: 248–254
172. Wang F et al (2010) An Ac transposon system based on maize chromosome 4S for isolating long-distance-transposed Ac tags in the maize genome. Genetica 138:1261–1270
173. Kuromori T et al (2006) A trial of phenome analysis using 4000 Ds-insertional mutants in gene-coding regions of Arabidopsis. Plant J 47:640–651
174. Zhang S et al (2003) Resources for targeted insertional and deletional mutagenesis in Arabidopsis. Plant Mol Biol 53:133–150
175. Muskett PR et al (2003) A resource of mapped dissociation launch pads for targeted insertional mutagenesis in the Arabidopsis genome. Plant Physiol 132:506–516
176. Panjabi P, Burma PK, Pental D (2006) Use of the transposable element Ac/Ds in conjunction with Spm/dSpm for gene tagging allows extensive genome coverage with a limited number of starter lines: functional analysis of a four-element system in *Arabidopsis thaliana*. Mol Genet Genomics 276:533–543
177. Park SH et al (2007) Analysis of gene-trap Ds rice populations in Korea. Plant Mol Biol 65:373–384
178. Eamens AL et al (2004) A bidirectional gene trap construct suitable for T-DNA and Ds-mediated insertional mutagenesis in rice (*Oryza sativa* L.). Plant Biotechnol J 2: 367–380
179. Qu S et al (2008) A versatile transposon-based activation tag vector system for functional genomics in cereals and other monocot plants. Plant Physiol 146:189–199
180. Qu S et al (2009) Construction and application of efficient Ac-Ds transposon tagging vectors in rice. J Integr Plant Biol 51:982–992
181. Luan WJ et al (2008) An efficient field screening procedure for identifying transposants for constructing an Ac/Ds-based insertional-mutant library of rice. Genome 51:41–49
182. Ayliffe MA et al (2007) A barley activation tagging system. Plant Mol Biol 64:329–347
183. Mathieu M et al (2009) Establishment of a soybean (Glycine max Merr. L) transposon-based mutagenesis repository. Planta 229:279–289

Chapter 6

Gene Tagging with Engineered *Ds* Elements in Maize

Yubin Li, Gregorio Segal, Qinghua Wang, and Hugo K. Dooner

Abstract

We describe here protocols for isolating genes in maize using *Dissociation (Ds)* transposons marked with a green fluorescent protein (*GFP*) transgene. The introduced marker enables the phenotypic scoring of the nonautonomous element and the anchoring of unique primers on the element to facilitate the isolation of the adjacent DNA by PCR. Transposons such as *Ds* transpose preferentially to sites closely linked to the *Ds*-launching platform. Based on this transposition behavior, a genetic resource is being created to mobilize a modified *Ds* element from different starting sites in the genome. Enough transgenic lines are being generated to cover most of the maize genome, allowing the targeted tagging of most genes from a *Ds-launching* platform located nearby.

Keywords Transposon tagging, Maize, *Ac–Ds*, GFP-marked *Ds* element, Mapped transposon-launching platforms

1 Introduction

1.1 Tagging Concept

Transposon tagging refers to the use of transposons as forward genetic tools to isolate genes on the basis of their mutant phenotype or as reverse genetics tools to isolate insertion mutations in known genes [1, 2]. In maize, the transposons of choice have been native elements of the *Ac/Ds* [3] and *Mutator* [4] families. In general, two tagging strategies have been used with the *Ac/Ds* transposon system. Non-targeted tagging relies on the ability to select progeny carrying unlinked transposed elements. In contrast, targeted tagging is based on the well-known property of *Ac* and *Ds* to transpose preferentially to closely linked sites [5–9]. Here, we discuss tagging with an engineered *Ds** element carrying an introduced marker that enables the phenotypic tracing of the nonautonomous element and the anchoring of unique primers on the element to facilitate the isolation of the tagged gene by PCR. Since enough transgenic lines are being generated to cover most of the maize genome, we emphasize the tagging of target genes from a *Ds**-launching platform located nearby.

Thomas Peterson (ed.), *Plant Transposable Elements: Methods and Protocols*, Methods in Molecular Biology, vol. 1057, DOI 10.1007/978-1-62703-568-2_6,

1.2 Advantages of an Engineered (Marked) Ds Element Over Native Elements

A *Ds** element marked with a transgene, such as the jellyfish gene encoding GFP [10], has the same genetic advantages as *Ac* [8], plus it can be stabilized by segregating away the *Ac* element used originally to mobilize it. These advantages are the following: (a) *Ds** transpositions can be selected phenotypically, making it possible to use convenient seed phenotypes to readily assemble a large collection of independent transposed *Ds** (*trDs**) elements. (b) The marked *Ds** element has a dominant phenotype, so *trDs**s can be identified in heterozygous condition and then homozygosed to screen for mutant phenotypes. (c) The dominant phenotype permits a quick co-segregation analysis of the *trDs** and a new mutation. (d) The visible phenotype allows the mapping of a *trDs** insertion in the genome independently of the molecular mapping of its adjacent sequence. (e) Like *Ac*, *Ds* tends to transpose to sites that are closely linked to the donor site, making it possible to "saturate" defined regions of the genome with *trDs* insertions. We refer to the collection of *trDs** elements from any one launching pad as a *Ds** insertion library. Placing *Ds** in launching pads at multiple regularly spaced locations will allow selective targeting of the entire genome. (f) The unique transgene sequence (e.g., *GFP*) enables the design of specific primers for the PCR isolation of the DNA adjacent to the *trDs** (*tds* or *transposed Ds* site).

A comprehensive set of lines that will serve as starting points for the production of future *Ds** insertion libraries is presently being generated (Li et al., unpublished). Because elements of the *Ac/Ds* family transpose preferentially to linked sites, about one-third of all transpositions are within 7 cM on either side of the donor site [6–8]. Therefore, based on the 1,727-cM genetic map of [11], the minimum set would consist of 124 lines carrying a uniquely marked element at equally spaced locations in the genome (*see* **Note 1**). In this set, most genes will be within 7 cM of a launching platform and will be, therefore, realistic targets in localized transposon mutagenesis attempts. The active *Ds**-launching platforms are being mapped by matching the sequence of their adjacent DNA, isolated by inverse polymerase chain reaction (iPCR), to the maize genome. The lines have been produced by *Agrobacterium*-mediated transformation of the Hi II hybrid [12] and have been confirmed to carry a single copy of the T-DNA platform. The *Ds** transposon delivery system in the T-DNA is based on the *c1-m2* allele [13], which carries a 2.5-kb *Ds* element in the third exon of the *c1* gene [14] and produces a spotted seed (c-m) phenotype in the presence of *Ac*. The selection scheme for *trDs** elements (*see* Subheading 1.3) takes advantage of the fact that both parents of Hi II are *c1* and produce colorless seed, so it is easy to identify *Ds* transpositions as fully colored seed. A side benefit of this transposant selection system is that most maize lines are *c1* [15, 16], so researchers can cross the *c1-m2(Ds*) Ac* stocks directly to their lines in order to generate *Ds**-tagged mutations of their gene of interest (*goi*).

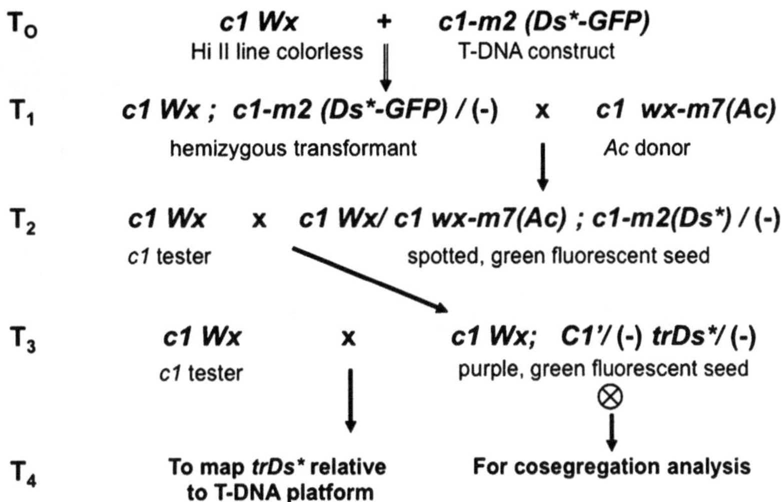

Fig. 1 Genetic scheme to generate: (1) Active *c1-m2(Ds*)* platforms (T_0–T_2) and (2) C′ purple revertants carrying *trDs*-GFP* elements (T_3–T_4)

1.3 Genetic Scheme for Identifying trDs*

Figure 1 outlines the genetic scheme used to generate (a) the T-DNA platforms carrying an active *Ds** element marked with a *GFP* gene expressed behind a seed-specific *α-zein* promoter (generations T_0–T_1) and (b) a *Ds** insertion library from a specific T-DNA platform (generations T_2–T_4). In short, the researcher obtains a specific line from the Maize Stock Center and uses it to pollinate *c1* tester plants (T_2). Transpositions of *Ds** from the modified *c1-m2* gene are selected as purple seeds that retain green fluorescence (T_3). These selections can be self-pollinated to produce *Ds** insertion homozygotes and backcrossed to the *c1* tester to map the location of the *trDs** (T_4).

2 Materials

2.1 Constructs

TAG21 is a pTF102-derived binary vector [12] in which the *gus* gene has been replaced by a modified *c1-m2* allele (Fig. 2a: Li et al., unpublished). In this engineered mutable allele, the *Ds** element carries P(*α-zein*)/*GFP*, a construct in which the *GFP* gene is driven by the *α-zein* promoter [17], conditioning green fluorescence in the endosperm. Germinal transpositions of *Ds** (*α-zein/GFP*) are easily identified in single-copy transgenotes as fully purple kernels that retain green fluorescence (Fig. 2b). In most transformants, the c-m spotted phenotype resembles that of the native *c1-m2* allele and segregates as a single Mendelian trait, indicating that the transgene has integrated at a single locus (Fig. 2c).

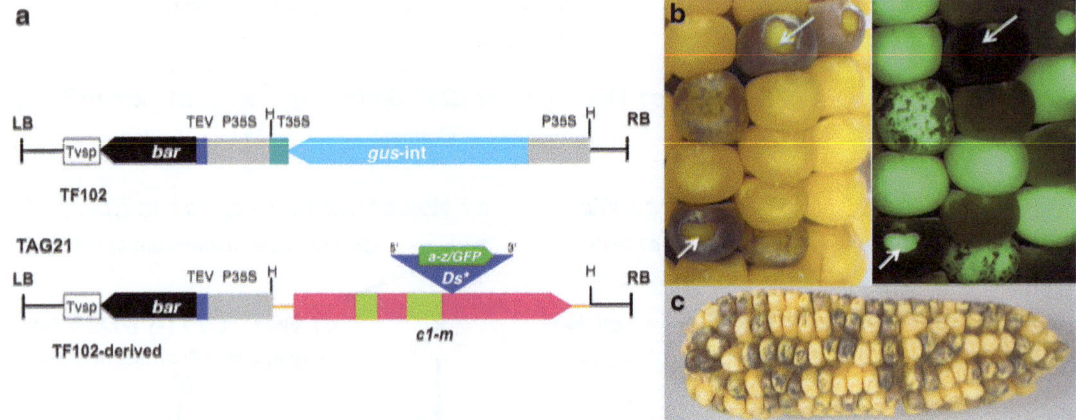

Fig. 2 **(a)** T-DNA construct used in *Agrobacterium* transformation of Hi II (*c1*) embryos. The *gus* HindIII (H) fragment of pTF102 was replaced with the *c1-m Ds** excision reporter shown in TAG 21 (*LB* left border, *RB* right border). (**b**) Purple kernel selections that retain (*white arrow*) or lose (*blue arrow*) *Ds**(α-*zein*)/*GFP*. (**c**) A testcross ear of an *Agrobacterium*-generated transformant, segregating 1 spotted (c-m):1 colorless (c)

2.2 Stocks

Most transgenic lines produce purple revertants (C′) in sufficiently high numbers (3–4 %) to serve as potential sources of transposed *Ds** elements. About half of C′ revertants carry new *trDs** based on retention of green fluorescence and GFP-hybridizing bands in Southern blots, a result consistent with the one-half fraction seen in other screens based on *Ac/Ds* excision [7]. The germinally active single-copy T-DNA platforms are mapped by isolating adjacent DNA via iPCR, sequencing it, and comparing it with the maize genome sequence. The genomic location of 82 platforms has been mapped so far (Fig. 3) and that of 150 other lines is presently being mapped. Stocks for all these *Ds**-launching platforms will be available from the Maize Genetics Stock Center.

2.3 Molecular Biology Solutions

All of the buffers, reagents, and other solutions used in the enzyme reactions (*see* Subheading 3) are found in kits provided by the manufacturers of the respective enzymes. Other reagents are prepared as follows.

1. DNA extraction buffer:
   (a) 0.1 M NaCl.
   (b) 50 mM EDTA (pH 8.0).
   (c) 50 mM Tris–HCl (pH 8.0).
   (d) 0.1 M DIECA (diethyldithiocarbamic acid sodium salt).
   (e) 1 % N-Laurylsarcosine Na-salt.
   (f) 8.5 mM β-mercaptoethanol.
2. CTAB solution (2 %):
   (a) 2 % CTAB (N-hexadecyl-N,N,N trimethylammonium bromide).
   (b) 50 mM Tris–HCl (pH 7.5), 10 mM EDTA (pH 8.0).

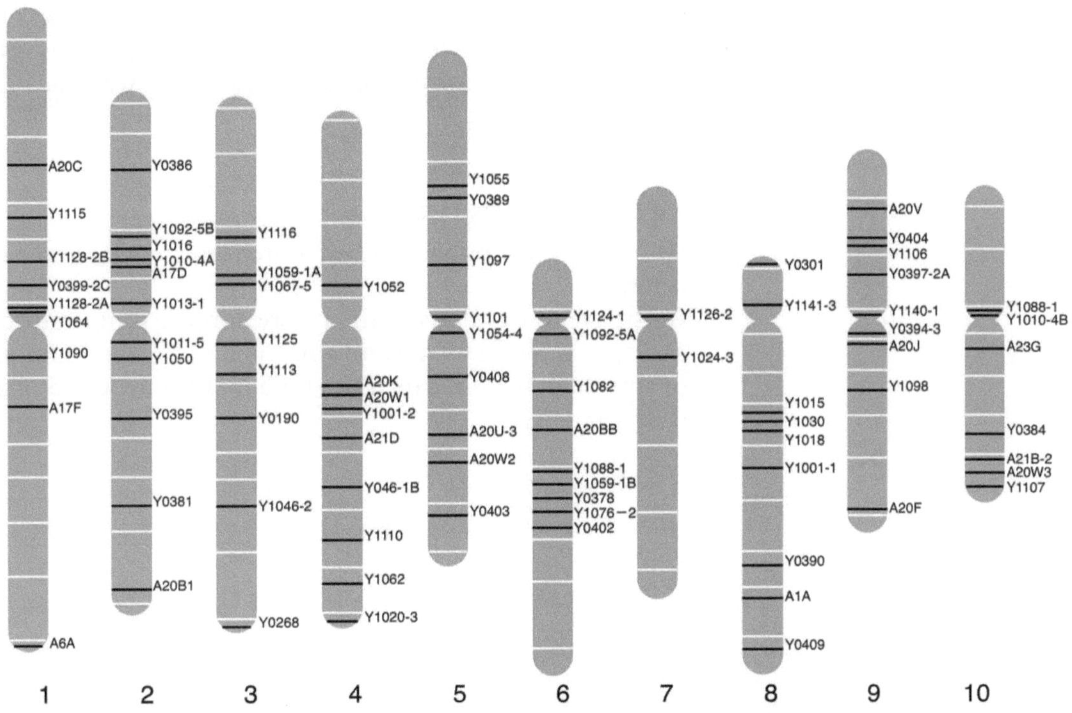

Fig. 3 Location of 82 T-DNA platforms currently mapped in the genome. The black marks identify the location of the mapped T-DNA platforms; the white marks, the location of the bin core markers. The space between the two core markers corresponds to roughly 20 cM in the genetic map

3. SOC medium:
   (a) 950 ml double-distilled water.
   (b) 20 g Bacto-tryptone.
   (c) 5 g Bacto-yeast extract.
   (d) 0.5 g NaCl.
   (e) 2.5 ml of 1 M KCl.

   Adjust the pH to 7.0 with 10 N NaOH and then make up to a final volume of 1 L using double-distilled water. Autoclave to sterilize, and then store the medium at room temperature (21–25 °C). Add 20 ml sterile 1 M glucose immediately before use.
4. LB + ampicillin plates:
   (a) 10 g Bacto-tryptone.
   (b) 5 g Bacto-yeast extract.
   (c) 10 g NaCl.
   (d) 15 g Bacto-agar to double-distilled water (such that the final volume is 1 L).

Autoclave to sterilize, cool to 55 °C, add 1 ml 100 mg/ml ampicillin (amp), and then pour into sterile 10 cm petri dishes. Store at 4 °C and dispose of unused plates after 3 weeks.

5. LB + ampicillin medium:
   (a) 10 g Bacto-tryptone.
   (b) 5 g Bacto-yeast extract.
   (c) 10 g NaCl to double-distilled water (such that the final volume is 1 L).
   Autoclave to sterilize and then store at room temperature. To 1 L add 1 ml 100 mg/ml ampicillin. After the addition of ampicillin store the medium for a maximum period of 2 weeks at 4 °C.

2.4 Molecular Biology Reagents

1. 1 kb DNA ladder.
2. Agarose (electrophoresis grade).
3. Ampicillin.
4. Bacto-agar.
5. β-mercaptoethanol.
6. Bacto-tryptone.
7. Bacto-yeast extract.
8. BigDye® Terminator v3.1 Cycle Sequencing Kit.
9. Buffer P1.
10. Buffer P2.
11. Buffer P3.
12. CTAB (N-hexadecyl-N,N,N trimethylammonium bromide).
13. DIECA (diethyldithiocarbamic acid sodium salt).
14. EDTA.
15. Efficiency™ DH5α™ Competent Cells.
16. Epicentre End-It™ DNA End-Repair kit.
17. Ethanol (99.7 % (vol/vol), AnalaR)).
18. Ethidium bromide solution (10 mg/ml).
19. Expand High Fidelity PCR System, dNTPack.
20. Glucose.
21. HPLC-purified oligonucleotides (*see* Table 2).
22. NEB buffer 2.
23. N-Laurylsarcosine Na-salt.
24. pGEM®-T Easy Vector.
25. Potassium chloride, KCl.
26. Qiagen Multiplex PCR kit.
27. QIAquick Gel Extraction Kit.
28. QIAquick PCR Purification Kit.

29. RNase A.
30. Sodium chloride, NaCl.
31. Sodium hydroxide, NaOH.
32. T4 DNA ligase (400 U/μl; supplied with 10× buffer).

2.5 Supplies and Equipment

The following supplies and equipment are used in Subheading 3.

1. Nebulizer Kit.
2. 1.5-ml microtubes.
3. PCR tubes and lids (strips of 8).
4. Thin-walled 96-well PCR plates.
5. Deep-well 96-well plates.
6. Millipore filter plates.
7. Bio-Rad S1000™ thermal cycler.
8. 37 °C incubator.
9. Heating block.
10. Gel-running equipment including gel tank and power supply.
11. Micro-centrifuge.
12. Sorvall RC6+ centrifuge.
13. MicroAmp® Optical 96-Well Reaction Plate.
14. 3730 DNA Analyzer.

3 Methods

3.1 Isolation of C' Revertants with trDs-GFP

The first step in setting up a targeted gene tagging experiment is to identify one or two transposon-launching platforms located as near as possible to the *goi* and to request the corresponding stocks from the Maize Genetics Stock Center in Champagne-Urbana, IL (http://maizecoop.cropsci.uiuc.edu/). As in any gene tagging experiment, a large number of independent transpositions is desirable. Using the parent with the *Ds**-launching platform as male is much more efficient for the following reasons: (a) The average germinal transposition frequency is about two times higher in the male than in the female. (b) Postmeiotic or gametophytic transpositions, which lead to non-corresponding kernels carrying a C′ endosperm and a c-m embryo (false selections), occur much more frequently in the female gametophyte than in the male gametophyte. and (c) Premeiotic transpositions, which lead to clusters of individuals with the same transposition event (jackpots), are rare in the male, allowing the same tassel to be used in multiple pollinations over a number of days without undue concern about recovering the same event multiple times. The average T_2 testcross ear (Fig. 1) will have 1–2 *trDs** elements. Thus, a collection of 1,000 independent transposition events can be obtained from about 700 ears. That number of ears would be produced in 700 pollinations of *c1* tester plants by *c1-m2(Ds-GFP)* plants (T2 in Fig. 1). Since each male plant can be used multiple times, no more than 50 male plants will be needed if the *c1* females are hand pollinated (*see* **Note 2**).

1. Plant 500–1,000 seeds of *c1* tester, either from the investigator's own stocks or obtained from the Maize Stock Center, to serve as female parents, and 50–100 seeds of the heterozygous *c1-m2(Ds*)/(-); wx-m7/Wx* stock, obtained from the Maize Stock Center, to serve as male parents (*see* **Note 3**): If the latter seed stock is limited, selfing the hemizygotes first to increase the seed would also result in an increase in the number of *c1-m2(Ds*)* alleles from 1 to 1.3 per individual among the spotted seed. In this case, the fine spotted seed from the selfed ear (*Ac* homozygotes) should be avoided because the frequency of transposition is reduced at higher doses of *Ac* [18, 19].
2. As the plants mature, protect all ears of the *c1* tester line with ear shoot bags (e.g., Lawson 217) prior to silk emergence.
3. Pollinate all *c1* plants with pollen of the *Ds** stock (collected in tassel bags, such as Lawson 404) and cover pollinated ears with the same bags. The number of the *Ds** parent plant can be recorded on the bag if the investigator wants to keep track of the parental origin of all C′ selections, although premeiotic events are rare (*see* **Note 4**).
4. After harvesting and drying the ears to ≤13 % moisture, screen all ears for fully purple kernels (i.e., C′ revertants). Then, inspect the C′ selections for green fluorescence under a blue light source (400–500 nm wavelength). Gently sand the top surface of individual C′ kernels to expose the underlying endosperm, as is done when staining for wx starch (Fig. 2b), and score the presence of reinserted *trDs** elements on the basis of green fluorescence of the exposed endosperm. Use Dark Reader Hand Lamp (HL32T, blue filter to remove green/red light) and Dark Reader viewing glasses, equipped with an amber filter to remove blue light, both of which are purchased from Clare Chemical Research, Inc (*see* **Note 5**).
5. Grow out the purple, green fluorescent seed selections (C′ GFP), preferably in the greenhouse or a similar protected environment, number them, and collect one leaf from each 4-week-old plant for future DNA analysis. Freeze the leaf in liquid nitrogen (or dry ice) and store the frozen leaf in a regular (noncycling) freezer until ready to extract DNA.
6. Self-pollinate and/or cross the C′ GFP selections to a *c1* tester. The option here will depend on several factors, such as what the researcher wants to accomplish at this stage, the time frame of the experiment, and the amount of greenhouse growing space.

3.2 Mutant Screen

1. To isolate a *Ds**-tagged allele of a gene defined genetically (i.e., with a known mutant phenotype), self-pollinate the T_3 C′ GFP selections in order to homozygose the *trDs** element and cross them as males to plants carrying the known reference mutant allele of the gene in order to identify new, putative *Ds**-tagged, alleles in the T_4 generation (Fig. 1). Alternatively, if a line with the reference mutant allele is available in a *c1* mutant version (very likely, as most colorless seed stocks are *c1* mutants), use that line as female in crosses to the *c1-m2(Ds*)/*(-) parent (Fig. 1, T_2). In that case, the screen for new mutations putatively tagged by *Ds** could be conducted in the T_3, rather than the T_4 (*see* **Note 6**).
2. To isolate a *Ds**-tagged allele of a gene defined only molecularly (*goi*) (i.e., with an unknown mutant phenotype), self-pollinate all selections and proceed to analyze them molecularly (*see* Subheading 3.6) to identify a *Ds** insertion in *goi*.

3.3 Mutant Analysis

1. To obtain preliminary information that a mutant phenotype is, indeed, caused by the insertion of a *trDs** element, one can perform a co-segregation analysis of the self-progeny of the pertinent *Ds** transposant, preferably in the absence of *Ac* (*see* **Note** 7). If the *trDs** has caused a new dominant mutation, all the mutants should be green fluorescent and vice versa. If the *trDs** has caused a new recessive mutation, all the mutants should be green fluorescent and all the green fluorescent non-mutant should segregate the mutation upon selfing. Secondary transpositions of *Ds** would result in exceptions to these outcomes. This is a correlation type of analysis and one can only conclude that the mutation is likely to be transposon tagged, not that it is tagged, regardless of the size of the experiment.
2. To confirm that a mutant phenotype is caused by the insertion of a *trDs** element, one conducts a reversion analysis. Mutant individuals carrying *Ac*, preferably in heterozygous condition, are self-pollinated and their self-progenies are screened for wild-type revertants. If the mutation was *Ds** tagged, the revertants should lose *Ds** from the mutant gene. This is best ascertained by a molecular analysis of the revertants, which usually involves PCR and sequencing, although Southern blots were traditionally used for this purpose. Since germinal transpositions of *Ac* and *Ds* are in the 1–5 % range, a population of several hundred individuals should be screened for reversion.

3.4 Isolation of the tds Site from a Mutant Allele by iPCR

The insertion site of a *trDs** element (*tds* site) in a *Ds**-mutagenized gene is readily isolated by iPCR amplification from total genomic DNA using unique primers from the *GFP* transgene and nonspecific primers from the *Ds* 5′ or 3′ ends. *Ds* primers are inevitably nonspecific because the different classes of *Ds* elements in the maize genome are highly redundant.

1. Digest 5 μg of maize genomic DNA (*see* Subheading 3.7) with a restriction enzyme that does not cut the known sequence between the ends of the iPCR primers to be used. Use several enzymes to be sure of obtaining a fragment of appropriate length. The reaction should contain:
   (a) Maize genomic DNA 5 μg.
   (b) 10× Restriction buffer 5 μl.
   (c) 10× BSA (if needed) 5 μl.
   (d) Restriction enzyme 20 U.
   (e) Distilled water to a final volume of 50 μl.

   Incubate at enzyme's optimum temperature for 5 h to overnight.
2. Stop the reaction by heating at 65 °C (or recommended temperature) for 15 min. Purify restriction products through a Qiagen PCR purification column or by standard ethanol precipitation.
3. Resuspend DNA in 100 μl water for intramolecular ligation by T4-DNA ligase:
   (a) Purified DNA digestion products 100 μl.
   (b) 10× T4 DNA ligase reaction buffer 20 μl.
   (c) T4 DNA ligase (from NEB: 400 U/μl) 1 μl.
   (d) Distilled water 79 μl.

   Incubate 200-μl ligation reaction at 16 °C overnight.
4. Stop the reaction by heating at 65 °C for 15 min. Purify ligation products through a Qiagen PCR purification column or by standard ethanol precipitation.
5. Resuspend DNA in 200 μl water, and use 1 μl as template for the 25-μl iPCR reaction.
6. The sequences and combinations of PCR primers are shown in Table 1.
7. Load 3 μl of the reaction on a 1× TAE 0.8 % agarose gel to visualize the amplified iPCR products (*see* **Note 8**). Purify iPCR products of a desired size and analyze on a 0.8 % agarose gel to confirm the recovery and to measure the approximate concentration for direct sequencing with an ABI BigDye Terminator V3.1 Cycle Sequencing Kit.
8. To determine the physical and genetic locations of *tds* sites, compare the flanking sequences with the B73 RefGen_v2 database provided in the MaizeGDB Genome Browser (http://gbrowse.maizegdb.org/cgi-bin/gbrowse/maize_v2/). BlastN results include their physical locations, such as chromosome number, contig number, and start and end positions in the pseudo

Table 1
Combinations of PCR primers to isolate *trDs flanking sequences**

*trDs** End	1st-round PCR	2nd-round PCR
5′-end	GFP-1/Ds-10	GFP-3/Ds-9
3′-end	GFP-2/Ds-13	GFP-4/Ds-14

Primer sequence:
GFP-1: 5′-GTCGCCACCATGGTGAGCAA-3′
GFP-2: 5′-GCGGCCGCTTTACTTGTACA-3′
GFP-3: 5′-CGTAAACGGCCACAAGTTCA-3′
GFP-4: 5′-TCGTCCATGCCGAGAGTGAT-3′
Ds-9: 5′-CGGTTATACGATAACGGTCG-3′
Ds-10: 5′-ACCTCGGGTTCGAAATCGAT-3′
Ds-13: 5′-CCGGTATATCCCGTTTTCGT-3′
Ds-14: 5′-TTTCGTTTCCGTCCCGCAA-3′

molecule, as well as their genetic locations, such as bin number and recombination coordinate number. The MaizeGDB Genome Browser provides additional helpful information, such as coding capacity and EST support for the query sequence.

3.5 Isolation of the tds Site from a Mutant Allele by Adapter-Ligated PCR

Splinkerette PCR is an alternative to iPCR for the isolation of the host sequence adjacent to a *trDs** [20]. This method involves the ligation of adapters with a hairpin at one end that are designed to eliminate the amplification of sequences adjacent to the many native *Ds* elements.

1. Preparation of the splinkerette adaptor with HPLC-purified "long-strand adaptor" and "short-strand adaptor" oligonucleotides (*see* Table 2):
   (a) Dissolve the oligonucleotides in 5× NEB buffer 2 to a concentration of 50 μM.
   (b) Add 50 μl of each adaptor stock to a PCR tube and vortex to mix.
   (c) The adaptor mix is denatured and annealed by heating it to 95 °C for 5 min and then cooling to room temperature at the rate of 1 °C every 15 s in a thermal cycler. Store the adaptor mix at −20 °C.
2. Shear 2 μg genomic DNA to 2–4 kb DNA fragments with a nebulizer.
3. Blunt-end the DNA using Epicentre End-It™ DNA End-Repair kit.
4. Combine and mix the following components in a microfuge tube:
   (a) 1–34 μl genomic DNA to end-repair (300 ng).
   (b) 5 μl 10× End-repair buffer.
   (c) 5 μl dNTP mix.

Table 2
Oligonucleotides required for splinkerette PCR protocol

Name	Sequence (5′–3′)
Adaptors	
Long-strand adaptor	CGAAGAGTAACCGTTGCTAGGAGAGACC GTGGCTGAATGAGACTGGTGTCGACA CTAGTGG
Short-strand adaptor	CCACTAGTGTCGACACCAGTCTCTAATTT TTTTTTTCAAAAAAA
1st-round primers	
Splink1	CGAAGAGTAACCGTTGCTAGGAGAGACC
GFP-12 (5′-end)	CAGCTCCTCGCCCTTGCTCACCA
GFP-10 (3′-end)	TCCGCCCTGAGCAAAGACC
2nd-round primers	
Splink2	GTGGCTGAATGAGACTGGTGTCGAC
Ac5′-178 (5′-end)	GTGAAACGGTCGGGAAACTAGCTCTAC
Ds-13 (3′-end)	CCGGTATATCCCGTTTTCGT

(d) 5 μl ATP.

(e) *x*μl sterile water to a reaction volume of 49 μl.

(f) 1 μl End-repair enzyme mix.

Total reaction volume: 50 μl.

Incubate at room temperature for 45 min. Stop the reaction by heating at 70 °C for 10 min. Place the tube on ice.

5. Set up the following splinkerette adaptor ligation reaction:

(a) 50 μl blunt-end DNA.

(b) 1 μl Adaptor mix (25 μM).

(c) 10 μl 10× T4 DNA ligase buffer.

(d) 1 μl T4 DNA ligase.

(e) 38 μl sterile water.

Total reaction volume: 100 μl.

Incubate the ligation reaction at 16 °C overnight. Heat-inactivate the T4 DNA ligase at 65 °C for 15 min. Use the QIAquick Gel Extraction Kit to clean up the ligation reaction. Elute the DNA in 40 μl of 10 mM Tris–HCL (pH 8.5) and store it at −20 °C.

6. Use 5 μl of purified splinkerette adaptor-ligated genomic DNA as template for first-round PCR amplification.

The PCR program is as follows:

(a) 2 min at 94 °C.

(b) 35 cycles of 20″ at 94 °C.

(c) 30″ at 56 °C.

(d) 3′ at 72 °C.

(e) 7′ at 72 °C.

Use 1 μl of 1/1,000 diluted primary PCR product as template for second-round PCR amplification following the above PCR program. (*See* Table 2 for primers used in first- and second-round PCR).

7. Visualize the secondary PCR products on a 1× TAE 1 % agarose gel, size-select the 0.5–1 kb products, and recover them using QIAquick Gel Extraction Kit.
8. Use Taq DNA polymerase to perform A-overhang reaction of recovered PCR products at 72 °C for 30 min.
9. Ligate the PCR products with pGEM®-T Easy Vector and transform Subcloning Efficiency™ DH5α™ Competent Cells.
10. Culture the transformed colonies in deep-well 96-well plates. For each individual, pick 12 colonies. Purify plasmids from the colonies and sequence them with BigDye Terminator on an ABI 3730 DNA Analyzer.
11. Determine the physical and genetic locations of the *tds* site as described in **step 8** of the previous section.

3.6 Isolation of the tds Site from a Gene Defined Only Molecularly

The *tds* site of a gene defined only molecularly can be efficiently identified in DNA pools of C′ GFP revertants (*see* **Note 9**). The following is a simplified account of the analysis of DNA from 960 such individuals.

1. Germinate individual C′ GFP revertants and array in ten plates of 12 × 8 format.
2. For each plate, collect 2-week-old seedling leaves from all 96 individuals and pool roughly equal amounts of leaf tissue for DNA extraction.
3. Test the DNA pools by a multiplex PCR with two *GFP* primers from the 5′ and 3′ ends of *Ds** (*see* Table 1) and one primer from *goi*. Several *goi* primers may be needed to cover the full length of the desired gene. A PCR band corresponding to a new junction between *Ds** and *goi* will only be amplified in pools containing DNA from an individual with a new *trDs** insertion in *goi*.
4. Extract individual DNAs from all of the C′ revertants in the positive pool and array them in a 96-well plate.

5. Pool DNAs into 8 row pools and 12 column pools.
6. Test the 20 DNA pools by PCR. The intersection of the positive row and column pools identifies the well containing DNA from the desired individual.

3.7 Isolation of Genomic DNA from Maize Leaf Tissues

1. Grind the leaf tissue (~5 g) into fine powder (in liquid N_2), transfer into a 50 ml centrifuge tube, add 20 ml 1× DNA extraction buffer, mix well by shaking, and put on ice for 30 min.
2. Bring up the volume to 45 ml by adding chloroform, and mix well by inverting the tubes four to six times.
3. Spin at 3,300×*g* for 15 min.
4. Transfer the top phase into a fresh 50 ml centrifuge tube. Add the same volume of phenol, chloroform, isoamyl alcohol (25:24:1) and mix well by flipping the tubes four to six times.
5. Spin at 3,300 × *g* for 15 min for separation of phases.
6. Transfer the aqueous phase (above the white interface layer) into a fresh 50 ml centrifuge tube. Add the same volume of chloroform and mix well by flipping the tubes four to six times.
7. Spin at 3,300 × *g* for 15 min for separation of phases.
8. Transfer the aqueous phase (above the white interface layer) into a fresh 50 ml centrifuge tube. Add the same volume of 100 % ethanol and mix well by flipping the tubes four to six times.
9. Spin at 3,300 × *g* for 15 min.
10. Pour away the supernatant and air-dry the pellet.
11. Resuspend the pellet in 5 ml 1× TE (pH 8.0) with RNaseA (final concentration of 25 μg/ml), and incubate at 37 °C for 30 min or until the pellet is totally dissolved.
12. Add 5.7 ml 2 % CTAB mixture (*see* Subheading 2.3) and 0.7 ml 5 M NaCl in the RNaseA-treated DNA samples.
13. Vortex gently to allow the DNA to precipitate out of solution (*see* **Note 10**).
14. Drip off the supernatant, keeping the precipitated DNA in the centrifuge tube.
15. Add 10 ml 70 % ethanol (0.15 M NaCl) and wash the precipitated DNA for 15 min in order to remove salts.
16. Pour off the wash solution and wash again with 10 ml 70 % ethanol for 15 min. Use a 1-ml pipette tip to hold the DNA against the tube wall and pour off the ethanol wash.
17. Pipette the DNA which has adhered to the tip into a 2 ml Eppendorf tube, allow it to air-dry, and dissolve in 1× TE (pH 8.0).

4 Notes

1. More recently, the estimated genetic size of the maize genome has grown to about 2,000 cM [21], raising the number of equidistant *Ds*-launching platforms required to cover it to 140.
2. Of 1,000 *Ds** transpositions, one-third or about 333 are expected to fall within 7 cM on either side of the *c1-m2(Ds*)* donor locus. Assuming a genetic map of 1,727 cM [11] and a gene number of 32,000 [21], there are roughly 259 genes in the average 14-cM interval. So, if transposition was randomly distributed in the 14 cM flanking the donor locus, each gene would be hit 1.3 times among 333 transpositions. That assumption is clearly not valid as the probability of hitting a specific gene is higher close to the donor locus [6–8]. The distributions of linked and unlinked transpositions have been examined for a limited set of transpositions from only a handful of loci, the largest number studied to date being the 1,228 *Ac* transpositions from *wx-m7(Ac)* mapped by Cowperthwaite et al. [8]. In that study, 387 transpositions fell within 7 cM of the *wx* donor locus, of which 249 fell within 4.6 cM of *wx*. Therefore, doubling or tripling the number of desirable transpositions would be advisable for genes known to be >5 cM away from the *c1-m2(Ds*)* platform being used. Even 3,000 *Ds** transpositions could be generated from no more than 2,000 pollinations, a still reasonably small transposon tagging experiment.
3. Several plantings of the *c1* tester line should be made by the investigator to ensure that enough *c1* female plants "nick" (i.e., flower concurrently) with the *c1-m2(Ds*) Ac* male plants. The latter are in a Hi II x W22 hybrid background, so they flower earlier than most inbreds. In the average NJ summer, they shed pollen between 62 and 68 days after planting. Most of our experience is based on *c1-m2(Ds*) Ac* that have flowered in late July after a mid to late May planting.
4. It is advisable to spread the pollen from individual *c1-m2(Ds*) Ac* plants over several days to dilute out potential premeiotic transpositions, which lead to clusters of C′ revertants carrying the same *trDs** element.
5. An alternative, but more expensive, piece of equipment that serves the same purpose as the Dark Reader Hand Lamp is a fluorescent dissecting microscope, such as the Leica MZ16 that was used for Fig. 2b. Kernel green fluorescence can also be visualized under the blue light provided by a conventional LCD projector.
6. Screening only the T_3 C′ GFP selections from the cross to the reference mutation is an extremely efficient way of identifying mutations with an adult plant phenotype and can be done in

the greenhouse. A mutation with a seed phenotype can be directly screened in the testcross ears, making sure that any mutant seeds are green fluorescent, so as to eliminate unwanted self-contaminants. A mutation with a seedling phenotype can be screened on sand benches in the greenhouse by germinating only the T_3 C′ GFP selections, if space is limited, or the entire T_3 GFP population, if space is plentiful. The latter is not an efficient method, but has the advantage that all *Ds** transpositions would be included in the screen: those co-segregating with the *C1'* excision product (purple kernels), those co-segregating with the *c1-m2(Ds*)* parental allele (spotted), and those co-segregating with *c1-m2(Ds*)*, but without *Ac* (colorless). The last two can arise from replicative transpositions of *Ds** [22].

7. It is preferable to conduct the co-segregation analysis in the absence of *Ac* because secondary transpositions of *Ds** will affect the outcome. The *wx-m7(Ac)* allele, which is the source of *Ac*, will most likely be segregating in the T_3 selfed ear and most of the *Wx/Wx* progeny in that ear will have lost *Ac* by segregation.
8. It is always very informative to run an agarose gel to check the first-round PCR products, and the pattern of the amplicons on the gel can serve as guidance for the performance of the second-round PCR.
9. If the sequence, but not the mutant phenotype, is known, the PCR identification of a *Ds** insertion into the desired sequence is best carried out in pools of individuals carrying a *trDs**. An example of a pooling strategy employed to identify a transposon-tagged sequence can be found in Das and Martienssen [23].
10. Do not spin down DNA by centrifugation because the pelleted DNA is hard to dissolve.

Acknowledgment

We thank Jun Huang and Limei He for helpful suggestions. This work is supported by National Science Foundation Plant Genome Program project DBI-0929350.

References

1. Walbot V (1992) Strategies for mutagenesis and gene cloning using transposon tagging and T-DNA insertional mutagnesis. Ann Rev Plant Physiol Plant Mol Biol 43:49–82
2. Brutnell TP (2002) Transposon tagging in maize. Funct Integr Genomics 2:4–12
3. Dellaporta SL, Moreno MA (1994) Gene tagging with *Ac/Ds* elements in maize. In: Freeling M, Walbot V (eds) The maize handbook. Springer, New York, pp 219–233
4. Chomet PS (1994) Transposon tagging with Mutator. In: Freeling M, Walbot V (eds)

The maize handbook. Springer, New York, pp 243–248
5. Van Schaik N, Brink RA (1959) Transposition of *Modulator*, a component of the variegated pericarp in maize. Genetics 44:725–738
6. Greenblatt IM (1984) A chromosome replication pattern deduced from pericarp phenotypes resulting from movement of the transposable element *Modulator* in maize. Genetics 108:471–485
7. Dooner HK, Belachew A (1989) Transposition pattern of the maize element *Ac* from the *bz-m2(Ac)* allele. Genetics 122:447–457
8. Cowperthwaite M et al (2002) Use of the transposon *Ac* as a gene-searching engine in the maize genome. Plant Cell 14:713–726
9. Vollbrecht E et al (2010) Genome-wide distribution of transposed *Dissociation* elements in maize. Plant Cell 22:1667–1685
10. Chalfie M et al (1994) Green fluorescent protein as a marker for gene expression. Science 263:802–805
11. Davis GL et al (1999) A maize map standard with sequenced core markers, grass genome reference points and 932 expressed sequence tagged sites (ESTs) in a 1736-locus map. Genetics 152:1137–1172
12. Frame BR et al (2002) Agrobacterium tumefaciens-mediated transformation of maize embryos using a standard binary vector system. Plant Physiol 129:13–22
13. McClintock B (1948) Mutable loci in maize. Carnegie Inst Wash Yearbook 47:155–169
14. Neuffer MG, Coe EH, Wessler S (1997) The mutants of maize. Cold Spring Harbor Laboratory Press, Cold Spring Harbor, NY
15. Chen SM, Coe EH Jr (1977) Control of anthocyanin synthesis by the C locus in maize. Biochem Genet 15:333–346
16. Hanson MA et al (1996) Evolution of anthocyanin biosynthesis in maize kernels: the role of regulatory and enzymatic loci. Genetics 143:1395–1407
17. Segal G, Song R, Messing J (2003) A new opaque variant of maize by a single dominant RNA-interference-inducing transgene. Genetics 165:387–397
18. Brink RA, Nilan RA (1952) The relation between light variegated and medium variegated pericarp in maize. Genetics 37:519–544
19. McClintock B (1952) Chromosome organization and gene expression. Cold Spring Harbor Symp Quant Biol 16:13–47
20. Devon RS, Porteous DJ, Brookes AJ (1995) Splinkerettes—improved vectorettes for greater efficiency in PCR walking. Nucleic Acid Res 23:1644–1645
21. Schnable PS et al (2009) The B73 maize genome: complexity, diversity, and dynamics. Science 326:1112–1115
22. Greenblatt IM, Brink RA (1962) Twin mutations in medium variegated pericarp maize. Genetics 47:489–501
23. Das L, Martienssen R (1995) Site-selected transposon mutagenesis at the *hcf106* locus in maize. Plant Cell 7:287–294

Chapter 7

Plant Regeneration Methods for Rapid Generation of a Large Scale *Ds* Transposant Population in Rice

Yuan Hu Xuan, Jin Huang, Gihwan Yi, Dong-Soo Park, Soo Kwon Park, Moo Young Eun, Doh Won Yun, Gang-Seob Lee, Tae Ho Kim, and Chang-deok Han

Abstract

To mutagenize rice genomes, a two-element system is utilized. This system comprises an immobile *Ac* element driven by the CaMV 35S promoter, and a gene trap *Ds* carrying a partial intron with alternative splice acceptors fused to the GUS coding region. Rapid, large-scale generation of a *Ds* transposant population was achieved using a plant regeneration procedure involving the tissue culture of seed-derived calli carrying *Ac* and *Ds* elements. During tissue cultures, *Ds* mobility accompanies changes in methylation patterns of a terminal region of *Ds*, where over 70 % of plants contained independent *Ds* insertions. In the transposon population, around 12 % of plants expressed GUS at the early seedling stage. A flanking-sequence-tag (FST) database has been established by cloning over 19,968 *Ds* insertion sites and the *Ds* map shows relatively uniform distribution across the rice chromosomes.

Key words Rice, *Ac/Ds* transposable elements, Gene trap, Plant regeneration, Tissue culture, Mutagenesis

1 Introduction

Enhancer/gene traps are advanced versions of the conventional transposon-mediated insertional mutagenesis techniques [1, 2]. A gene trap contains multiple splicing acceptor sites fused to the coding region of a reporter gene. In our two-element system, *Ac* cDNA was used as a transposase source under the control of the CaMV 35S promoter (Fig. 1a) [3]. The gene trap *Ds* carried a partial intron with alternative splice acceptors in front of the reporter gene GUS (Fig. 1b). The intron was a modified form of the fourth intron of the *Arabidopsis* GPA1 (the alpha subunit of G-protein alpha subunit) gene [4].

In maize and *Arabidopsis*, *Ac/Ds* as an insertional vehicle is mainly propagated by genetic manipulations such as crossing and

Thomas Peterson (ed.), *Plant Transposable Elements: Methods and Protocols*, Methods in Molecular Biology, vol. 1057, DOI 10.1007/978-1-62703-568-2_7, © Springer Science+Business Media New York 2013

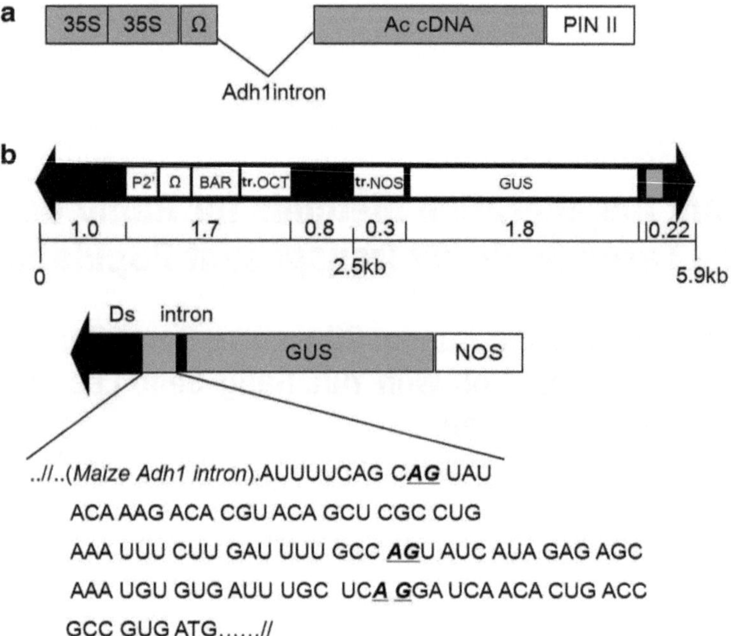

Fig. 1 Ac construct and gene trap Ds. (**a**) A doubly enhanced CaMV 35S promoter, untranslated sequence, and maize *Adh1* intron I are fused to the cDNA of an Ac protein coding sequence. The transcription terminator is from a proteinase inhibitor II (PINII) gene of potato. (**b**) Gene trap *Ds* carrying the reporter gene GUS and the selection marker bar. A *bar* coding region is expressed under the control of a promoter and a terminator from the 2′ gene and an octopine synthetase from the Ti plasmid, respectively. An intron fused to GUS carries three alternative splice acceptor sites (***AG*** sites) and is situated after the 222nd base from the 3′-end of *Ds*

selfing, and is utilized as a low-copy mutagen [5]. Many *Ds* elements become immobile even though *Ac* transposase was expressed. In maize, the germinal transposition of *Ac* is usually 5–10 % in a given population [5]. Therefore, the efficiency of an *Ac*/*Ds* tagging system largely depends on selection schemes for transposants. In our systems, rapid and large-scale generation of a *Ds* transposant population was achieved using a regeneration procedure involving the tissue culture of seed-derived calli carrying *Ac* and *Ds* elements. In rice, methods for growing plantlets from seed-derived calli, derived from dry seeds, are sufficiently robust to produce a large number of plants within a short period of time [6, 7]. During tissue culture, *Ds* mobility is dramatically increased and is accompanied by changes in methylation patterns of a terminal region of *Ds* and alteration in the steady-state level of *Ac* mRNA expressed by a constitutive CaMV 35S promoter [6].

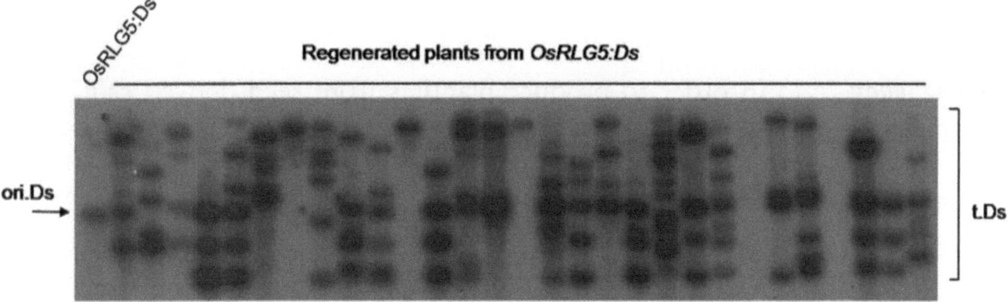

Fig. 2 Polymorphic display of *Ds* elements in regenerated plants. Southern blot hybridization was performed to identify *Ds* transpositions in plants regenerated from *OsRLG5::Ds* seeds. *Eco*RI-digested genomic DNA samples were hybridized with a 1.8 kb DNA fragment from the GUS coding region. The original *Ds* (ori.Ds) and transposed *Ds* (t.Ds) were identified from *OsRLG5:Ds* and its regenerated population, respectively

In the regenerated population, over 70 % of plants carried independent *Ds* insertions (Fig. 2). This population does not require a selection scheme for transposants or the control or maintenance of *Ac/Ds* activity since transposed *Ds* elements in regenerated plants were stably transmitted in subsequent generations [7]. Also, enhanced *Ds* mobility during tissue cultures leads to the high frequencies of new allelic mutations and revertants (Fig. 2) [8]. More importantly, the regeneration strategy for *Ds* mutagenesis can be applied to various genetic backgrounds of both japonica and indica varieties.

The *Ds* population was analyzed for major agronomic traits in the glasshouse and field. Ten to twenty samples of each line were scored for 26 different traits ranging from the seedling to maturity stages [9]. The efficiency of gene identification via GUS staining was also estimated. Overall, 12 % of the lines exhibited GUS expression detectable in various organs and tissues of 5-day-old seedlings. Since *Ds* could be inserted in either transcriptional orientation at equal frequency, these results suggest that approximately 25 % of the population contained *Ds* insertions into loci expressed during the early-seedling stage. Transposed *Ds* elements and genomic flanking sequences were cloned using iPCR and TAIL-PCR, and the *Ds* insertion sites were mapped on rice chromosomes, with iPCR and TAIL-PCR used to clone DNA sequences flanking the *Ds* insertion sites [10, 11]. TIGR Rice Genome pseudo molecules (ver.5) (http://www.tigr.org/tdb/e2k1/osa1/pseudomolecules) were used to map 19,968 *Ds*-flanking DNA sequences. Except for the vicinities of the original *Ds* sites, distribution of the transposed *Ds* elements is relatively uniform among the 12 chromosomes [12].

2 Materials

2.1 Tissue Culture for Regeneration of Plants

In order to produce plantlets from seed-derived calli, sequential incubations were performed using four types of tissue culture media: (a) NB medium for callus induction, (b) N6-7-CH medium for pre-regeneration, (c) N6S3-CH-I medium for regeneration I, (d) N6S3-CH-II medium for regeneration II [6, 12].

1. NB medium:

   5 % (v/v) N6 macro-element (20×) stock (for 1 L (20×) stock):

   (a) 56.6 g KNO_3.

   (b) 9.26 g $(NH_4)_2SO_4$.

   (c) 8 g KH_2PO_4.

   (d) 1.8 g $MgSO_4$.

   (e) 3.3 g $CaCl_2 \cdot 2H_2O$.

   1 % (v/v) B5 micro-element 100× stock (for 1 L (100×) stock):

   (a) 1 g $MnSO_4 \cdot 4H_2O$.

   (b) 200 mg $ZnSO_4 \cdot 7H_2O$.

   (c) 75 mg KI.

   (d) 25 mg $NaMoO_4 \cdot 2H_2O$.

   (e) 300 mg H_3BO_3.

   (f) 2.5 mg $CuSO_4 \cdot H_2O$.

   (g) 2.5 mg $CoCl_2 \cdot 6H_2O$.

   0.1 % (v/v) B5 vitamin (1,000×) stock (for 1 mL (1,000×) stock):

   (a) 100 mg myo-inositol.

   (b) 1 mg nicotine acid.

   (c) 1 mg pyridoxine–HCl.

   (d) 10 mg thiamine–HCl.

   0.11 mM MS medium Fe-EDTA.

   9 μM 2,4-D.

   87.6 mM sucrose.

   4.3 mM L-proline.

   3.4 mM L-glutamine.

   0.03 % (v/v) casein enzymatic hydrolysate.

   Adjust pH to 5.8 and add 2.5 g/L phytagel.

2. N6-7-CH medium:

   5 % (v/v) N6 macro-element (20×) stock (for 1 L (20×) stock):

(a) 56.6 g KNO_3.

(b) 9.26 g $(NH_4)_2SO_4$.

(c) 8 g KH_2PO_4.

(d) 1.8 g $MgSO_4$.

(e) 3.3 g $CaCl_2 \cdot 2H_2O$.

1 % (v/v) B5 micro-element (100×) stock (for 1 L (100×) stock):

(a) 1 g $MnSO_4 \cdot 4H_2O$.

(b) 200 mg $ZnSO_4 \cdot 7H_2O$.

(c) 75 mg KI.

(d) 25 mg $NaMoO_4 \cdot 2H_2O$.

(e) 300 mg H_3BO_3.

(f) 2.5 mg $CuSO_4 \cdot H_2O$.

(g) 2.5 mg $CoCl_2 \cdot 6H_2O$.

0.1 % (v/v) B5 vitamin (1,000×) stock (for 1 mL (1,000×) stock):

(a) 100 mg myo-inositol.

(b) 1 mg nicotine acid.

(c) 1 mg pyridoxine–HCl.

(d) 10 mg thiamine–HCl.

0.11 mM MS medium Fe-EDTA.

9 μM 2,4-D.

61.6 mM sucrose.

16.5 mM sorbitol.

0.2 % (v/v) casein enzymatic hydrolysate.

Adjust pH to 5.8 and add 2.5 g/L phytagel.

After autoclaving, add 0.5 mg/L 6-benzyladenine.

3. N6S3-CH-I medium:

5 % (v/v) N6 macro-element (20×) stock (for 1 L (20×) stock):

(a) 56.6 g KNO_3.

(b) 9.26 g $(NH_4)_2SO_4$.

(c) 8 g KH_2PO_4.

(d) 1.8 g $MgSO_4$.

(e) 3.3 g $CaCl_2 \cdot 2H_2O$.

1 % (v/v) B5 micro-element (100×) stock (for 1 L (100×) stock):

(a) 1 g $MnSO_4 \cdot 4H_2O$.

(b) 200 mg $ZnSO_4 \cdot 7H_2O$.

(c) 75 mg KI.

(d) 25 mg $NaMoO_4 \cdot 2H_2O$.

(e) 300 mg H_3BO_3.

(f) 2.5 mg $CuSO_4 \cdot H_2O$.

(g) 2.5 mg $CoCl_2 \cdot 6H_2O$.

0.1 % (v/v) B5 vitamin (1,000×) stock (for 1 mL (1,000×) stock):

(a) 100 mg myo-inositol.

(b) 1 mg nicotine acid.

(c) 1 mg pyridoxine–HCl.

(d) 10 mg thiamine–HCl.

0.11 mM MS medium Fe-EDTA.

9 μM 2,4-D.

Amino acids (amino acids for 1 L medium: 877 mg glutamine, 228 mg arginine, 75 mg glycine, 266 mg aspartic acid).

61.6 mM sucrose.

16.5 mM sorbitol.

0.2 % (v/v) casein enzymatic hydrolysate.

Adjust pH to 5.8 and add 5 g/L phytagel (*see* **Note 1**).

After autoclaving, add 1 mg/L naphthalene acetic acids and 5 mg/L kinetin.

4. N6S3-CH-II medium:

5 % (v/v) N6 macro-element (20×) stock (for 1 L (20×) stock):

(a) 56.6 g KNO_3.

(b) 9.26 g $(NH_4)_2SO_4$.

(c) 8 g KH_2PO_4.

(d) 1.8 g $MgSO_4$.

(e) 3.3 g $CaCl_2 \cdot 2H_2O$.

1 % (v/v) B5 micro-element (100×) stock (for 1 L (100×) stock):

(a) 1 g $MnSO_4 \cdot 4H_2O$.

(b) 200 mg $ZnSO_4 \cdot 7H_2O$.

(c) 75 mg KI.

(d) 25 mg $NaMoO_4 \cdot 2H_2O$.

(e) 300 mg H_3BO_3.

(f) 2.5 mg $CuSO_4 \cdot H_2O$.

(g) 2.5 mg $CoCl_2 \cdot 6H_2O$.

0.1 % (v/v) B5 vitamin (1,000×) stock (for 1 mL (1,000×) stock):

(a) 100 mg myo-inositol.

(b) 1 mg nicotine acid.

(c) 1 mg pyridoxine–HCl.

(d) 10 mg thiamine–HCl.

0.11 mM MS medium Fe-EDTA.

9 μM 2,4-D.

Amino acids (amino acids for 1 L medium: 877 mg glutamine, 228 mg arginine, 75 mg glycine, 266 mg aspartic acid).

87.6 mM sucrose.

0.2 % (v/v) casein enzymatic hydrolysate.

Adjust pH to 5.8 and add 5 g/L phytagel.

After autoclaving, add 0.5 mg/L naphthalene acetic acids and 2 mg/L kinetin (*see* **Note 2**).

2.2 Genomic DNA Extraction (See Note 3)

1. Lysis buffer: 350 mM NaCl, 100 mM Tris–HCl (pH 7.6), 7 M urea, 50 mM EDTA, 2 % SDS.
2. Phenol:chloroform:isoamyl alcohol (PCI) to 25:24:1 (v/v/v). Phenol: Analytical quality phenol (Merck) is melted at 60 °C and mixed with 1 g 8-hydroxyquinoline and 0.1 M Tris–HCl, pH 8.0. The solution is mixed by shaking and maintained at 4 °C until the two phases are completely separated. The supernatant is discarded. Neutralization is repeated several times with 1 volume of 0.1 M Tris–HCl (pH 8.0) until the pH of the phenol is close to 7.0.
3. Isopropanol.
4. 70 % ethanol.
5. TE: 10 mM Tris–HCl (pH 8.0), 1 mM EDTA (pH 7.5).
6. RNase A (10 g/L).

2.3 Inverse PCR (iPCR)

1. 1 μg genomic DNA.
2. 4-base cutters: *Nla*III, *Hae*III, or *Bfa*I.
3. T4 DNA ligase (NEB).
4. *rTaq* polymerase (TaKaRa): 1 U per reaction.
5. Specific PCR primers. The iPCR primers used to amplify sequences flanking the 5′ ends of *Ds* are as follows:

   (a) Ds5-1, CCGTTTACCGTTTTGTATATCCCG.

   (b) Ds5-2, CGTTCCGTTTTCGTTTTTTACC.

   (c) Ds5-3, GTACGGAATTCTC CCATCCT.

   (d) Ds502, ATACGATAACGGTCGG.

(e) Ds5I-1, TAATCGGGATGATCCCGTTCGTT.
(f) Ds5I-2, ATGACTG CAATATGGCCAGC.
(g) Ds5I-3, TTCTAATTCGGGATGACTGC.

The primers used to amplify the 3′ ends of *Ds* are the same as those for the TAIL-PCR described below. In addition, the following primers are used for iPCR amplification of the 3′ ends of *Ds*:

(a) Ds3I-105, AAACGAACGGGATAAATACGG.
(b) Ds3I-150, GGTTAAAGTCGAAATCGGACG.

2.4 TAIL-PCR

1. *rTaq* polymerase (TaKaRa): 1 U per reaction (*see* **Note 4**).
2. Specific PCR primers. The Ds 5′-primers are the same as those used for iPCR. The Ds 3′-primers are as follows:
   (a) Ds3-1, CCGACCGGATCGTATCGGT.
   (b) Ds3-2, TTAACCCGACCGGATCGTATCGGTTTCG.
   (c) Ds3-3, GTTTCGTTACCGGTATATCCCGTTTCG.
   (d) Ds3-4, GTTAAATATGAAAATGAAAACGGTAGAGG.
   (e) Ds3-5, ATGAAA ACGGTAGAGG.
   (f) Ds3-6, ACCGTTTTCATCC.

   Arbitrary degenerate primers are as follows:
   (a) cggc1, GVCTYCGWSSGC.
   (b) SAD11, NTCAGSTWTSGWGWT.
   (c) SW41, AGWGHAGSAHCADAAS.
   (d) BAD5, WTCCASNTGSNACG.
   (e) DRM-CG2, GCNGNWCGWCGWG.
   (f) CST1, GTANTCGWAWNCST.
   (g) CTG1, GWWGGTSCWASWCTG.
   (h) AMS2(=GAG3), GWSIDRAMSCTGCTC.
   (i) geeky1, GKYKGCKGCNGC.
   (j) DRM-AG1, GNGWSASTNGAGC.
   (k) BAD8, GTGASNTGSWATGG.
   (l) DRM-NC1, GSCNCSGWNCC.
   (m) AD10, TTGIAGNACIANAGG.
   (n) AD20, TCTTICGNACITNGGA.
   (o) W4, AGWGNAGWANCANAGA.
   (p) AD-1, NTCGASTWTSGWGTT.
   (q) AD-2A, NGTCGASWGANAWGAA.

   (r) AD-2B, NGTCGASWGANAWGTT.
   (s) AD-2C, NGTCGASWGANAWAGA.
   (t) AD-2D, NGTCGASWGANAWTGT.
   (u) AD-3, WGTGNAGWANCANAGA.
   (v) AD-4, STTNTASTNCTNTGC.
   (w) AD-5, WCAGNTGWTNGTNCTG.

3. GUS Solution:
   (a) 50 mM sodium phosphate (pH 8.0).
   (b) 10 mM EDTA (pH 7.5).
   (c) 0.1 % Triton X-100.
   (d) 2 mM potassium ferrocyanide.
   (e) 2 mM potassium ferricyanide.
   (f) 0.6 % chloramphenicol (stock: 34 mg/mL).
   (g) 1 mg/mL gluc powder dissolved in DMFO. Since the GUS solution is light sensitive, it should be wrapped with lighttight foil. The solution can be used twice. Between uses, it should be kept at 4 °C.

3 Methods

3.1 Tissue Culture for Transgenic Plants

1. Dry mature seeds are shaken in the sterilization solution for 30–40 min at RT.
2. Seeds are washed with autoclaved dH_2O 15–20 times of 2-min intervals.
3. Seeds are dried on paper towels to remove water, and then dried for 30 min inside a clean bench hood (*see* **Note 5**).
4. Sterilized seeds are sown on NB media for callus induction, which takes 4 weeks in the dark.
5. Calli are transferred onto N6-7-CH medium for 10 days in the dark (*see* **Note 6**).
6. After 10 days pre-regeneration, calli are transferred to N6S3-CH-I medium and maintained for 1 month under a 16/8-h light/dark cycle (*see* **Note 7**).
7. After green spots are well developed, calli are transferred onto N6S3-CH-II medium and maintained for an additional 1 month under a 16/8-h light/dark cycle.
8. Before being transferred to the greenhouse, regenerated plants are transplanted into bottles that contain half-strength MS media [7] (*see* **Note 8**).

3.2 Preparation to Grow Rice Plants in a Paddy

Daily high and low temperatures in a greenhouse at Gyeongsang National University were typically 30 °C during the day and 20 °C at night. For vegetative growth, the light/dark cycle in the greenhouse was 14/10 h. To induce flowering, the light period was reduced to 10 h. For ratoon cultures, the culms were cut out at 10–20 cm above the ground. Cut plants were grown on fresh soil for the next round of culture (*see* **Note 9**). Agronomic traits such as plant architecture, color, flowering date, and height were examined in the GM fields at the National Institute of Crop Science (NICS) (Milyang, Korea) and at the National Academy of Agriculture Science (NAAS) (Gunwi, Korea) (*see* **Note 10**). The following procedures were used for seed germination and seedling culture before transplantation into the outdoor paddy fields:

1. Seedling trays are filled with soil and watered until completely wet.
2. Seeds are sown in rows and covered with 5 mm of dry soil.
3. Seedling trays are covered with dark plastic films and maintained at about 30 °C until coleoptiles emerge 10–15 mm above the surface.
4. Trays are transferred to wet beds and covered with white light fabrics until coleoptiles turn completely green.
5. Young seedlings are nurtured to 10–15 cm with occasional watering (never being flooded).
6. Trays are moved to seedling beds in the paddy field and submerged up to less than one-third of the plant height in the water.
7. Seedlings are grown until the three- or four-leaf stage before being transplanted (*see* **Note 11**).

3.3 Genomic DNA Extraction

1. Leaves are frozen in liquid nitrogen and ground with a mortar and pestle (glass pestle or machine) into powder.
2. 750 μl of lysis buffer are added and vortexed well, and the solution is maintained at 37 °C for 20 min with shaking.
3. An equal volume of PCI is added to the sample solution and vortexed vigorously.
4. The upper (aqueous) phase is saved after a 10-min centrifugation.
5. PCI extraction is repeated one or two times (*see* **Note 12**).
6. Either 0.6 or the same volume of isopropanol is added to the supernatant and kept at RT or −20 °C for 10 min.
7. DNA is pelleted by 15-min centrifugation.
8. DNA pellets are washed with 75 % EtOH, centrifuged for 5 min, and the EtOH is discarded.
9. Air-dried DNA pellets are dissolved with 30–50 μl TE-RNase and kept at RT overnight or at 37 °C for 1 h.

3.4 iPCR

1. 1 μg of genomic DNA is digested with 10 U of *Nla*III, *Hae*III, or *Bfa*I for 2 h (*see* **Note 13**).
2. After phenol extraction, DNA is ethanol-precipitated and dissolved in 400 μl of ligation buffer.
3. Digested DNA is self-ligated at 16 °C for 12 h by adding 1 U T4 DNA ligase.
4. After heat inactivation, DNA is ethanol-precipitated and dissolved in 20 μl TE.
5. The first round of PCR is performed in a volume of 50 μl using 5 μl dissolved DNA. PCR reaction consists of the following:
   (a) 10 mM Tris–HCl (pH 8.3).
   (b) 50 mM KCl.
   (c) 1 mM $MgCl_2$.
   (d) dNTPs (0.2 mM each).
   (e) 0.2 pM Ds primer.
   (f) 1 U Taq polymerase (TaKaRa).
6. Nested PCR is performed in a volume of 50 μl using 50 times dilution of the first PCR products as template.
7. A 3 % agarose gel is run to compare DNA sizes of the first and second PCR products.

The cycling conditions for iPCRs are as shown below:

First-round reaction:

1. 94 °C for 3 min.
2. (35 cycles):
   (a) 94 °C for 15 s.
   (b) 55 °C for 1 min.
   (c) 72 °C for 2 min.
3. 72 °C for 10 min.

Second-round reaction:

1. 94 °C for 3 min.
2. (30 cycles):
   (a) 94 °C for 15 s.
   (b) 55 °C for 1 min.
   (c) 72 °C for 2 min 30 s.
3. 72 °C for 10 min.

3.5 TAIL-PCR

For TAIL-PCR, three *Ds* end-specific primers with one of the arbitrary degenerate primers (*see* **Note 14**) were used to amplify *Ds* flanking sequences. The tertiary products should be 80 bp shorter for 5′ flanking fragments (using Ds5-1, Ds5-2, and Ds5-3) and 33 bp shorter for 3′ flanking fragments (using Ds3-155, Ds3-37, and Ds3-4 primer set) than the secondary products.

1. The primary PCR reaction is carried out in a volume of 20 μl and consists of the following:
   (a) 10 mM Tris–HCl (pH 8.3).
   (b) 50 mM KCl.
   (c) 1 mM $MgCl_2$.
   (d) dNTPs (0.2 mM each).
   (e) 0.2 pM Ds primer.
   (f) 3 pM AD primer.
   (g) 1 U Taq polymerase (TaKaRa).
2. The secondary PCR is carried out in a volume of 20 μl using 50 times dilution of the primary PCR products as template in a reaction consisting of the following:
   (a) 10 mM Tris–HCl (pH 8.3).
   (b) 50 mM KCl.
   (c) 1 mM $MgCl_2$.
   (d) dNTPs (0.2 mM each).
   (e) 0.2 pM Ds primer.
   (f) 2 pM AD primer.
   (g) 1 U Taq polymerase.
3. The tertiary PCR is performed in a volume of 30 μl using 50 times dilution of the secondary PCR products as template in a reaction consisting of the following:
   (a) 10 mM Tris–HCl (pH 8.3).
   (b) 50 mM KCl.
   (c) 1 mM $MgCl_2$.
   (d) dNTPs (0.2 mM each).
   (e) 0.13 pM Ds primer.
   (f) 1.3 pM AD primer.
   (g) 1 U Taq polymerase.
4. The primary, secondary, and tertiary PCR products are run in parallel lanes on a 3 % agarose gel to compare the sizes of DNA products.

The cycling conditions for the three sequential TAIL-PCR reactions are as below:

Primary reaction:

1. 94 °C for 3 min.
2. (5 cycles):
   (a) 94 °C for 1 min.
   (b) 65 °C for 1 min.
   (c) 72 °C for 2 min 30 s.
3. 94 °C for 1 min; 35 °C for 1 min; ramp to 72 °C in 2 min 30 s.
4. 72 °C for 2 min 30 s.
5. 94 °C for 30 s; 68 °C for 1 min; 72 °C for 2 min 30 s.
6. 94 °C for 30 s; 68 °C for 1 min; 72 °C for 2 min 30 s.
7. (15 cycles):
   (a) 94 °C for 30 s.
   (b) 44 °C for 1 min.
   (c) 72 °C for 2 min 30 s.
8. 72 °C for 5 min.

Secondary reaction:

1. 94 °C for 3 min.
2. (15 cycles):
   (a) 94 °C for 30 s; 65 °C for 1 min; 72 °C for 2 min 30 s.
   (b) 94 °C for 30 s; 65 °C for 1 min; 72 °C for 2 min 30 s.
   (c) 94 °C for 30 s; 44 °C for 1 min; 72 °C for 2 min 30 s.
3. 72 °C for 5 min.

Tertiary reaction:

1. 94 °C for 3 min.
2. (35 cycles):
   (a) 94 °C for 1 min.
   (b) 37 °C for 1 min.
   (c) 72 °C for 1 min.
3. 72 °C for 5 min.

3.6 GUS Assay

1. GUS solutions are aliquoted into Eppendorf or Falcon tubes and kept on ice.
2. Sample tissues are immediately submerged into GUS solution.
3. Lighttight sample tubes wrapped with foil are incubated at 37 °C for up to 2 days.

4. Samples are washed with 70 % ethanol. GUS solutions are saved at 4 °C for later use.
5. Samples can be stored in 70 % ethanol at 4 °C (*see* **Note 15**).

4 Notes

1. For plant regeneration, N6S3-CH-I medium should contain 5 g/L of phytagel. If it contains less phytagel, plates will be too moist, preventing green calli from being further developed.
2. Currently, we use MSR16 media instead of N6S3-CH-I medium and N6S3-CH-II medium for plantlet regeneration. The components of MSR16 media are as follows: 4.43 g/L MS media (+vitamin) (Duchefa), 100 mg/L myo-inositol, 50 g/L sucrose, 20 g/L sorbitol, 0.1 mg/L NAA, 2.0 mg/L kinetin. Adjust pH to 5.6–5.8 and add phytagel 5 g/L.
3. For faster DNA extraction, the following rapid grinding methods may be employed:
   (a) Mix 0.5-cm leaf pieces with two or three tungsten carbide beads (3-mm diameter, Qiagen, 69997) in a 2-mL Eppendorf tube (Axygen, MCT-200).
   (b) Punch tube caps with a pin and place tubes into a 1.5-mL polystyrene tube rack (Sigma, R4511).
   (c) Fill the bottom of racks with liquid nitrogen.
   (d) Vortex frozen tubes at full speed using a regular vortex machine. For larger scale preparation, 50-mL tubes (Falcon) and 5-mm porcelain beads (Kurabo, B-500) can be used in place of Eppendorf tubes and 3-mm tungsten beads.
4. We found that the quality of *Taq* polymerase influences the success rate of the first and second rounds of TAIL-PCR. For example, *rTaq*, a high-quality DNA polymerase from TaKaRa, performed better than other *Taq* polymerases from the same company. However, for the tertiary reaction, all of the DNA polymerases work equally well.
5. Air-drying of seeds after surface sterilization may help prevent cross-contamination from seeds that are not completely sterilized. When treating with sodium hypochlorite solutions, sterilization efficiency can be enhanced if 100 seeds are aliquoted in separate vessels.
6. Usually, calli were grown on N6-7-CH for 10 days before transfer to N6S3-CH-I. However, sometimes calli failed to generate green spots on N6S3-CH-I medium, even though the calli appeared to be healthy. In such cases, the incubation time on N6-7-CH medium can be extended up to 14–15 days,

which helps to increase the regeneration frequency on N6S3-CH-I.

7. It is recommended that calli of less than 1 mm in diameter be transferred to regeneration medium. This could reduce the chance of producing plants carrying the same *Ds* insertion from the same calli.
8. Two to three days before transfer to soil, an appropriate amount of water is added to the bottles, which should greatly help the plantlets survive on soil.
9. If the plants do not produce enough seeds, they can be maintained by ratoon culture. This process involves cutting culms and transferring stumped plants onto fresh soils for the next round of culture. Seed sets of ratooned plants are usually less than those of the original plants. To obtain sufficient seeds from ratoon culture, plants can be split into several pieces.
10. Chlorophyll-deficient mutants need close attention in the early growth stage. Dwarf and poor-growth mutants are best nurtured in a field isolated from normal plants.
11. When young leaves are emerging, removing excess water promotes vigorous rooting, which eventually leads to fast and healthy growth of young plantlets.
12. For iPCR or TAIL-PCR requiring high-quality genomic DNA, we performed PCI extraction with genomic DNA at least twice.
13. For iPCR, we usually used *Nla*III. The efficiency of iPCR was lower if genomic DNA was digested with *Nla*III for more than 3 h. Genomic DNA digestion should be completed within 2–2.5 h. For a template of the first PCR reaction, we sometimes used DNA recut with *Cla*I or *Bam*HI. However, if the *Cla*I or *Bam*HI enzyme sites are suspected to be located between two *Ds* end primers, self-ligated DNA is directly used for the first PCR reaction. In this case, we used ten times diluted (rather than 50 times diluted) first-reaction PCR product.
14. We used several arbitrary degenerate (AD) primers for TAIL-PCR reactions. Each AD primer might have different success rates in obtaining PCR products, which could depend on sequence compositions adjacent to *Ds* insertions. Using different combinations of AD and *Ds* end primers, you can greatly increase the success rate in cloning *Ds* insertion sites.
15. Samples are maintained at 4 °C after being photographed. If necessary, before being photographed, samples can be placed in a couple of drops of clearing solution containing 20 % lactic acid and 20 % glycerol in 1× PBS. This process could improve the clarity of the photos.

Acknowledgement

This research was supported by grants from the Next-Generation BioGreen 21 Program (PJ008215 and PJ008168), Rural Development Administration, Republic of Korea. Y.H.X. was supported by a scholarship from the BK21 program.

References

1. Skarnes WC (1990) Entrapment vectors: a new tool for mammalian genetics. Biotechnology 8:827–831
2. Sundaresan V (1996) Horizontal spread of treansposon mutagenesis: new uses for old elements. Trends Plant Sci 1:184–190
3. Chin HG et al (1999) Molecular analysis of rice plants harboring an *Ac/Ds* transposable element-mediated gene trapping system. Plant J 19:615–624
4. Sundaresan V et al (1995) Patterns of gene action in plant development revealed by enhancer trap and gene trap transposable elements. Genes Dev 9:1797–1810
5. Walbot V (2000) Saturation mutagenesis using maize transposons. Curr Opin Plant Biol 3:103–107
6. Kim CM et al (2002) Reprogramming of the activity of *Ac/Ds* transposon family during plant regeneration in rice. Mol Cell 14:231–237
7. Kim CM et al (2004) Rapid, large-scale generation of *Ds* transposant lines and analysis of the *Ds* insertion sites in rice. Plant J 39:252–263
8. Park SJ et al (2006) Analysis of intragenic *Ds* transpositions and excision events generating novel allelic variation in rice. Mol Cell 2:284–293
9. Park DS et al (2009) Genetic variation through *Dissociation* (*Ds*) insertional mutagenesis system for rice in Korea: progress and current status. Mol Breed 24:1–15
10. Liu YG et al (1995) Efficient isolation and mapping of *Arabidopsis thaliana* T-DNA insert junctions by thermal asymmetric interlaced PCR. Plant J 8:457–463
11. Pereira A, Aarts MGM (1998) Transposon tagging with the *En-I* system. In: Martinez-Zapater J, Salinas J (eds) Methods in molecular biology, vol 82, Arabidopsis protocols. Humana, Totowa, NJ, pp 329–338
12. Park SH et al (2007) Analysis of gene-trap *Ds* rice populations in Korea. Plant Mol Biol 65:373–384
13. Hiei Y et al (1994) Efficient transformation of rice (*Oryza sativa* L.) mediated by *Agrobacterium* and sequence analysis of the boundaries of the T-DNA. Plant J 6:271–282

Chapter 8

Isolation of Sequences Flanking *Ac* Insertion Sites by *Ac* Casting

Dafang Wang and Thomas Peterson

Abstract

Localizing *Ac* insertions is a fundamental task in studying *Ac*-induced mutation and chromosomal rearrangements involving *Ac* elements. Researchers may sometimes be faced with the situation in which the sequence flanking one side of an *Ac/Ds* element is known, but the other flank is unknown. Or, a researcher may have a small sequence surrounding the *Ac/Ds* insertion site and needs to obtain additional flanking genomic sequences. One way to rapidly clone unknown *Ac/Ds* flanking sequences is via a PCR-based method termed *Ac* casting. This approach utilizes the somatic transposition activity of *Ac* during plant development, and provides an efficient means for short-range genome walking. Here we describe the principle of *Ac* casting, and show how it can be applied to isolate *Ac* macrotransposon insertion sites.

Key words Mutagenesis, *Ac* donor, *Ac* insertion, *Ac* casting, Somatic transposition, Nested PCR

1 Introduction

Ac casting, first described by Singh et al. [1], was successfully applied to isolate genomic DNA flanking *Ac* insertions in the maize *pink scutellum1/viviparous7* (*ps1/vp7*) locus. Compared with the conventional four-step inverse PCR approach (digestion, ligation, DNA precipitation, and nested-PCR), *Ac* casting only requires nested-PCR and is more convenient and efficient. Using *Ac* casting, Zhang et al. [2] identified a series of deletions and translocations induced by *Ac* alternative transposition at *p1* locus, and Yu et al. [3] isolated the breakpoint of small inversions generated by *Ac* reversed-end transposition.

Ac casting takes advantage of spontaneous somatic *Ac* transposition which generates a pool of templates flanking the *Ac* donor site. *Ac* transposition occurs during or shortly after DNA replication [4, 5] with high frequency of transposition from the donor site to a genetically linked site. For example, previous reports have identified local *Ac* transpositions ranging from 6-basepairs to

Thomas Peterson (ed.), *Plant Transposable Elements: Methods and Protocols*, Methods in Molecular Biology, vol. 1057,
DOI 10.1007/978-1-62703-568-2_8, © Springer Science+Business Media New York 2013

15-kilobases between the donor and target sites [6–8]. These distances are suitable for amplification by PCR, thus providing the basis for the *Ac* casting approach described below.

The efficiency of *Ac* casting is closely related to the frequency of *Ac* transposition in somatic cells, and this somatic transposition frequency varies by genetic background and *Ac* dosage [9]. For example, the frequency of somatic transposition of a single *Ac* insertion at the maize *p1* locus is approximately 5 %, i.e., five independent *Ac* excision events per 100 maize leaf cells. However, transposition is less in stocks homozygous for *Ac* due to a negative dosage effect (i.e., *Ac* activity is reduced as *Ac* copy number is increased) [10, 11]. Finally, the timing of *Ac* transposition varies with development of the host plant [12]. Although these factors may in principle affect the efficiency of *Ac* casting, the technique is so simple and rapid that researchers risk little by trying the method with whatever material is most convenient.

2 Method (An Example to Isolate Sequences Flanking *Ac* Macrotransposon Insertions)

2.1 Tissue Collection, DNA Prep, and PCR Analysis

Leaf tissue (*see* **Note 1**) was collected and ground in liquid nitrogen (*see* **Note 2**). Total DNA was prepared by using a modified cetyltrimethylammonium bromide (CTAB) extraction protocol [13]. HotMaster Taq polymerase from Eppendorf (Hamburg, Germany) was used in the PCR reaction. The PCR master mix (1× HotMaster™ Taq Buffer with 2.5 mM Mg^{2+}, 0.25 mM dNTP, 0.2 mM primer, and HotMaster™ Taq DNA polymerase) was heated at 94 °C for 2 min, followed by 35 cycles of 94 °C for 20 s, 60 °C for 30 s, and 65 °C for 2 min, followed by a final cycle of extension at 65 °C for 8 min.

2.2 Primer Selection

Due to the low copy number of the template, nested-PCR was used to obtain sufficient PCR products for sequencing or other downstream applications (*see* **Note 3**). Two primer pairs ("p1-f2" with "Ac 3f-1," and "p1-f2" with "Ac 3f-2") are complementary to *Ac* and *Ac* donor flanking sequences (*see* Fig. 1a). Primer "p1-f1" and "Ac3f-1" were used for the first round of PCR. The resulting PCR product was used as template and amplified by primer "p1-f2" and "Ac 3f-2." Sequences of the representative primers are listed in Table 1.

2.3 Result Analysis and Confirmation

The PCR products were subject to gel electrophoresis analysis, and multiple bands were observed (*see* Fig. 2a). This type of result is common, because various sizes of PCR products result from

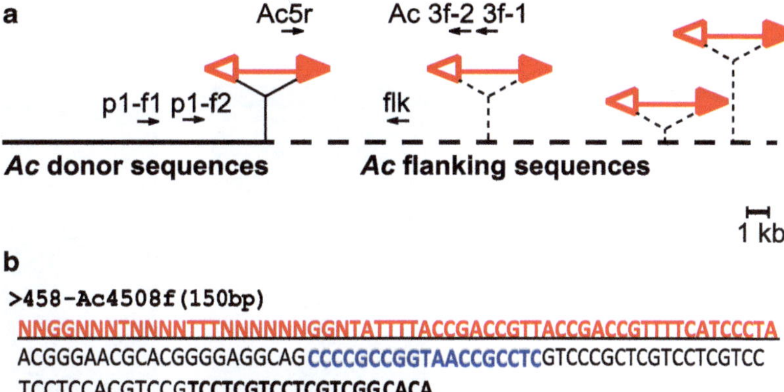

Fig. 1 *Ac* casting strategy and PCR analysis. (**a**) *Ac* casting primer locations. *Triangles* indicate the 4,565-bp *Ac* element, and the *solid red* and *open arrowheads* show the *Ac* 5′ and 3′ ends, respectively. The donor *Ac* element on left has a solid stem inserted into the known *p1* sequences; new *Ac* insertions resulting from somatic transpositions are shown with dashed stems inserted into the flanking sequence. The locations of primers used in *Ac* casting are shown by *short horizontal arrows*. (**b**) Raw data from sequencing *Ac* casting product. The sequence was obtained using *Ac* primer Ac3f-2 which reads from the 3′ end of the newly transposed *Ac* (*red underlined*), through the newly-obtained *Ac* flanking sequences and into the *Ac* donor sequences. (The direction of read is from right to left in (**a**).) The black bold underlined letters indicate the sequence homologous to primer p1-f2. *Letters in blue* indicate the position of primer "flk" which can be paired with *Ac* primer "Ac5r" to test the *Ac* casting result

Table 1
Sequences of the representative primers used in *Ac* casting

Primer	Primer sequences
Ac5r	CCCGTTTCCGTTCCGTTTTCGT
Ac3f-1	GATTACCGTATTTATCCCGTTCGTTTTC
Ac3f-2	ATGAAAATGAAAACGGTAGAGG
P1-f1(458)	AGAAGGGCCAGGCTCGTTCG
P1-f2(458)	GTGCCCGACGAGGACGAGGA
P1-f1(460)	CGCCCTCGAATCGAAAGCAT
P1-f2(460)	GAGCATCGGATTCGGGGACG
Flk-458	CCCCGCCGGTAACCGCCTC
Flk-460	CGCCTGCTTCGTCCTGCTCAA

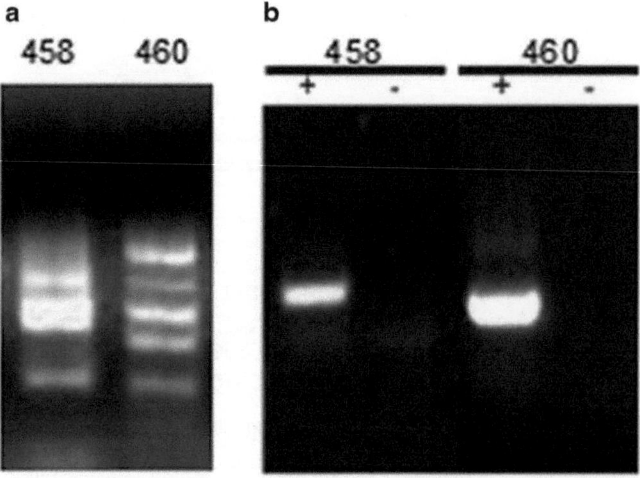

Fig. 2 *Ac* casting agarose gel analysis and confirmation. (**a**) *Ac* casting products were run on 0.8 % agarose gel at voltage of 100 V for 40 min in 1×TAE buffer (40 mM Tris–acetate, 1 mM ethylenediaminetetraacetic acid) (*see* **Note 7**). Lanes labeled "458" and "460" indicate two independent *Ac* macrotransposon alleles. Bands in lane 458 are fewer in number but more intense, suggesting that the plant tissue contained a smaller number of larger insertion sectors. However, bands in the sample 460 are greater in number but less intense, suggesting that the plant tissue contained a larger number of smaller insertion sectors. (**b**) Gel photo showing confirmation of the *Ac* casting products. For each sample, a specific primer complementary to the predicted flanking sequence was used in combination with primer Ac5r; specific bands of the expected size for each allele confirm the results. *Lanes marked "+"* contain templates from the allele of interest, while *lanes marked "−"* are negative controls which contain DNA from a sibling allele lacking *Ac*

multiple independent *Ac* somatic insertions into the flanking sequences. If PCR fails to produce bands, or bands are smeared (*see* **Note 4**). The bands were purified from the gel for sequencing (*see* **Note 5**). Based on the sequences obtained, a new primer "flk" reverse-complementary to the putative flanking sequence was designed for use in a final PCR to confirm the tentative insertion site (*see* **Note 6**). The confirmation step utilizes the new primer "flk" together with primer "Ac5r" (homologous to *Ac*) in PCR, which should yield a product whose size can be predicted based on the sequence obtained from *Ac* casting. The correct confirmation product does not depend on somatic transposition, and hence should be produced using any template containing the original test allele. DNA from a sibling plant lacking the *Ac* insertion serves as a negative control and should yield no band in the confirmation step.

3 Notes

1. The frequency of *Ac* transposition may vary among individual plants, and within different tissues of each plant. In order to increase the probability that the leaf tissue samples include enough somatic transposition events, one can collect samples from a number of young leaves and pool them together for DNA extraction. It has been observed that different tissue collections representing the same allele can generate different patterns of *Ac* casting products (data not shown). Finally, one can prepare DNA from larger amounts (0.3–1 g) of leaf samples and aliquot the DNA for multiple *Ac* casting attempts. In general, sampling more leaves gives greater chance for success.
2. Handle frozen samples quickly. Make sure that the frozen tissue/powder does not thaw before beta-mercaptoethanol is added. If the DNA becomes degraded it will not give a good PCR result.
3. Nested-PCR has the advantage of providing increased specificity by using two pairs of primers. But the disadvantage is that contamination from sequential PCR reactions can accumulate and spoil the results. Be extremely careful when you transfer tubes, and always include negative controls in both the first round and second round of PCR. If the negative controls show the same pattern of bands as the test samples, one can conclude that some component of the PCR has been contaminated.
4. Troubleshooting:

   (a) PCR fails to produce any bands:

   - Check your PCR reaction: Is the annealing temperature too high? Is the elongation time too short?
   - Try a new design for your experiment. The example given here applies for transpositions to one side of the *Ac* donor; however, *Ac* may have an equal probability of transposition to the opposite side. In that scenario, try using two pairs of primers in the nested PCR, including the **Ac** 3′ forward primers and the reverse primers from the donor sequences.
   - Repeat PCR using samples from different leaves or seedlings, as described in (*see* **Note 1**).

   (b) PCR bands are smeared:

   - Check gel resolution (*see* **Note 7**).
   - Check whether PCR condition is optimal: Is annealing temperature too low?
   - Check for contamination. Always include a negative control as stated in **Note 3**.

- Check whether template is too concentrated. Try a 1/10 or 1/100 dilution of your first PCR product as template for the second PCR.

5. Independent transposition events may produce multiple bands in a single *Ac* casting experiment. We choose one to three strong bands and purify the DNA from the gel for sequencing. The rationale is that strong bands are indicative of early transpositions and these are more likely to be real.
6. Evaluate your sequence before designing new primer(s) for confirmation PCR (*see* Fig. 1b). The sequence from your *Ac* casting should contain three parts: your previously known *Ac* flanking sequences; the newly obtained putative flanking sequences; and a portion of sequence from the transposed *Ac* element (determined by the location of your primer). Sequences which contain all three parts are less likely to be false positives, and may be advanced to the confirmation step.
7. Because *Ac* casting commonly gives multiple bands, a good separation of bands on gel electrophoresis is important. For greater gel resolution, use a higher agarose percentage (>0.8 %) and lower voltage (5–7 V/cm). High voltages can heat and melt the gel, but voltages too low can result in diffusion of the bands.

References

1. Singh M et al (2003) Activator mutagenesis of the Pink scutellum1/viviparous7 locus of maize. Plant Cell 15:874–884
2. Zhang J et al (2009) Alternative Ac/Ds transposition induces major chromosomal rearrangements in maize. Genes Dev 23:755–765
3. Yu C et al (2009) Spatial configuration of Ac termini affects their ability to induce chromosomal breakage. Plant Cell 22:744–754
4. Chen J, Greenblatt IM, Dellaporta SL (1987) Transposition of Ac from the *P* locus of maize into unreplicated chromosomal sites. Genetics 117:109–116
5. Chen J, Greenblatt IM, Dellaporta SL (1992) Molecular analysis of Ac transposition and DNA replication. Genetics 130:665–676
6. Athma P, Grotewold E, Peterson T (1992) Insertional mutagenesis of the maize *P* gene by intragenic transposition of Ac. Genetics 131:199–209
7. Moreno MA et al (1992) Reconstitutional mutagenesis of the maize P gene by short-range Ac transpositions. Genetics 131: 939–956
8. Weil CF et al (1992) Changes in state of the *wx-m5* allele of maize are due to intragenic transposition of Ds. Genetics 130:175–185
9. Greenblatt IM, Brink RA (1962) Twin mutations in medium variegated pericarp maize. Genetics 47:489–501
10. McClintock B (1948) Mutable loci in maize. Garnegie Inst Wash Year Book 47:155–169
11. McClintock B (1951) Mutable loci in maize. Garnegie Inst Wash Year Book 50: 174–181
12. Levy AA, Walbot V (1990) Regulation of the timing of transposable element excision during maize development. Science 248: 1534–1537
13. Porebski S, Bailey LG, Baum BR (1997) Modification of a CTAB DNA extraction protocol for plants containing high replication polysaccharide and polyphenol components. Plant Mol Biol Rep 15:8–15

Chapter 9

Regulation of the *Mutator* System of Transposons in Maize

Damon Lisch

Abstract

The *Mutator* system has proved to be an invaluable tool for elucidating gene function via insertional mutagenesis. Its high copy number, high transposition frequency, relative lack of insertion specificity, and ease of use has made it the preferred method for gene tagging in maize. Recent advances in high throughput sequencing of insertion sites, combined with the availability of large numbers of pre-mutagenized and sequence-indexed stocks, ensure that this resource will only be more useful in the years ahead. *Muk* is a locus that can silence *Mu*-active lines, making it possible to ameliorate the phenotypic effects of high numbers of active *Mu* transposons and reduce the copy number of these elements during introgressions.

Key words *Mutator*, *MuDR*, Gene tagging, Epigenetic, Maize, Transposon

1 Introduction

All transposable elements (TEs) are potentially mutagenic, but some of them are manifestly virulent. *Mutator* elements are certainly that. In lines with active *Mu* transposons mutation frequencies can exceed 50 times background, and forward mutation frequencies can be as high as 10^{-3} per locus per generation [1–4]. Because of this property, these transposons have been used to generate new mutations in maize with great effect [5]. Initially, lines with active *Mu* transposons were used in both directed and undirected screens to mutagenize and tag genes, and, over the past three decades, large numbers of genes have been mutagenized using this strategy [6, 7]. Advances in next generation *Mu* insertion flanking site sequencing suggest that there are few technical challenges remaining in identification of co-segregating insertions from forward screens [6]. In addition, a major effort has been made to mobilize a limited number of *Mu* transposons in a large number of plants and then sequence the insertion sites and propagate the resulting lineages for distribution to researchers [8].

Thomas Peterson (ed.), *Plant Transposable Elements: Methods and Protocols*, Methods in Molecular Biology, vol. 1057,
DOI 10.1007/978-1-62703-568-2_9,

To date, several thousand sequence-indexed lineages have been made available to maize researchers and many more should be available soon.

The *Mutator* system has been remarkably useful, and it will continue to be so. However, it is also fascinating in its own right. Investigation of the regulation of this particularly virulent TE system has revealed important clues concerning the means by which TE activity is regulated in plants [9, 10]. This investigation has led to the discovery of *Mu killer* (*Muk*), a locus that is competent to heritably epigenetically silence *MuDR*, the regulatory transposon for the *Mutator* system [11, 12]. Isolation of *Muk* has greatly facilitated our understanding of the means by which active TEs can be recognized and silenced, a topic that will be discussed at some length in this chapter. The availability of *Muk* has also made it possible to tame an otherwise unruly transposon system, making it possible to maximize the benefits of the availability of large numbers of insertional mutations.

2 Basic Biology and Behavior of *Mu* Elements

Although the biology of the *Mutator system* has been well covered elsewhere [9, 13], it is useful to outline what is known. The *Mutator* system is composed of an autonomous element, *MuDR*, and several nonautonomous elements (*Mu1*–*Mu13*) [13–15]. All *Mu* elements share similar, roughly 200 bp terminal inverted repeats (TIRs), but each class carries unique internal sequences. The system is controlled by autonomous *MuDR* elements, which carry two genes, *mudrA* and *mudrB* [16–20] (*see* Fig. 1). The *mudrA* gene encodes MURA, the putative transposase which has been shown to bind to *Mu* element TIRs [21]. Genes similar to *mudrA* are found in a wide variety of organisms, from bacteria, to yeast and mammals, suggesting that *Mu*-like elements (MULEs) are an ancient and widespread family of TEs [13, 22–25]. The second gene encoded by *MuDR*, *mudrB*, is much more mysterious. There are no homologs outside of the genus *Zea*, but in maize the protein encoded by this gene, MURB, appears to be required for insertion but not excision of *Mu* elements [26–28]. Given the propensity for *Mu* elements to capture host sequences [29, 30], it is likely that *mudrB* was a gene or gene fragment that was incorporated by *MuDR* at some point and mutated beyond recognition due to positive selection for its new function. Also present in all maize lines are sequences very similar to *MuDR* called *hMuDRs* that appear to be relatively inert and do not appear to contribute positively or negatively to *Mutator* activity [31].

The nonautonomous elements carry fragments of non-TE genes and are referred to generically as "Pack-MULEs" [30]. Like all nonautonomous elements, these elements only transpose in the

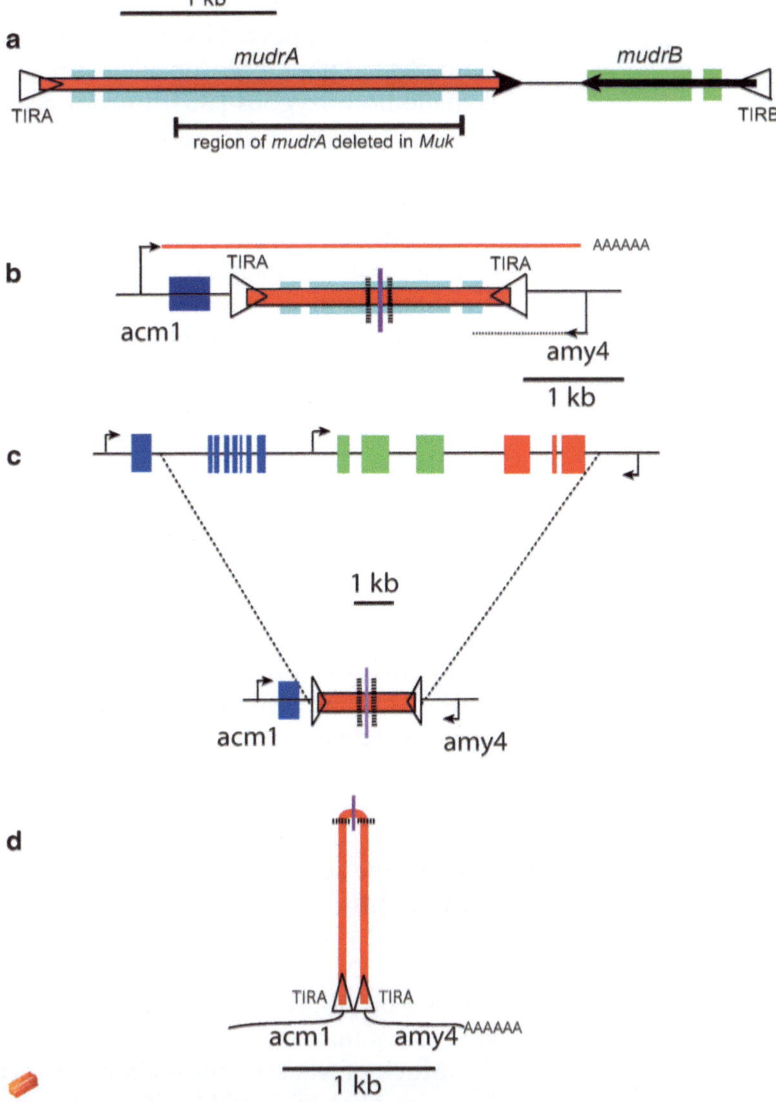

Fig. 1 (**a**) The structure of *MuDR*, including the terminal inverted repeats (TIRs) as well as the *mudrA* and *mudrB* genes. (**b**) The structure of *Muk*. (**c**) The deletion associated with the rearrangement that gave rise to *Muk*. (**d**) A model of the structure of the hairpin transcript produced by *Muk*

presence of the autonomous *MuDR* element. In an active *Mutator* gene tagging line, these nonautonomous elements can reach copy numbers in the hundreds, and the bulk of new mutations are probably caused by insertions of these elements [32].

Based on analysis of empty excision sites, somatic *Mu* element excision appears to largely result from NHEJ of double-stranded gaps rather than repair from the homologous chromosome [33, 34], a conclusion supported by the observation that reversion frequency is unaffected if an insertion allele is placed in trans to a deletion or is made homozygous [35]. In root cells, coordinate excision and transposition have been directly observed, demonstrating that somatic reversion events are associated with integration [36]. To some extent, the excision frequency of nonautonomous elements is dependent on the dose of *MuDR*, so two copies of *MuDR* cause roughly twice as many excisions of nonautonomous elements, and a reduced level of activity of a single (or many) *MuDR* element results in a concomitant reduction in excision frequency of those elements [11, 37, 38]. This feature makes excision frequency a useful measure of overall activity (but not necessarily of mutagenic potential, which is also dependent on the overall copy number of *Mu* elements).

In contrast to somatic events, which often involve reversion of mutant phenotypes due to the loss of the *Mu* insertion, germinal reversions are extremely rare [31]. The vast majority of activity in the germ line involves transposition without loss of the original element [32, 37]. Although it has been suggested that this is due to a switch from non-replicative to replicative transposition [31], current evidence supports a switch from NHEJ in late somatic tissue to repair using the sister chromatid in the germ line [39]. For instance, heritable deletions within *MuDR* elements are invariably flanked by short direct repeats, consistent with interrupted gap repair, and successful heritable duplication of *Mu* elements requires the activity of RAD51, a key component of the gap repair pathway [40, 41]. Germinal transposition rates for *Mu* elements can exceed 100 %, or an average of one new insertion event per *Mu* element [32]. Transposition is generally into unlinked sites and shows a pronounced bias towards low copy (genic) sequences that are recombinationally active [7, 35, 42, 43].

Much of what we know about the details concerning the regulation and behavior of *Mu* elements comes from analysis of what is referred to as minimal *Mutator* lines. These are lines that carry a very low number of *Mu* elements, as few as a single *MuDR* element and a single nonautonomous *Mu* element inserted into a color gene [19, 44, 45]. Our laboratory uses *a1-mum2*, an allele that contains a *Mu1* element inserted into the promoter of the *A1* gene [46]. In the presence of *MuDR*, *Mu1* excises from this *A1* allele, resulting in characteristic late somatic excision events (*see* Fig. 2). Minimal lines have been isolated at least twice from high copy lines, and in each case, the single remaining *MuDR* element was at the same position on chromosome 2 L (position 1), which is designated *MuDR(p1)* [17, 19]. This is likely due to the fact that the *MuDR* element at this position shows a strong position

Fig. 2 An ear segregating for *MuDR(p1)* and *Muk* generated by a cross between a plant carrying *MuDR(p1)* with one carrying *Muk*. All kernels contain the reporter at *a1-mum2*. Heavily spotted kernels contain *MuDR(p1)* by itself. Weakly spotted kernels contain *MuDR(p1)* with *Muk* (visible in F1 plants; nearly all somatic and germinal activity is lost in subsequent generations, even in the absence of *Muk*). Pale kernels lack *MuDR(p1)*

effect. Rather than duplicating at frequencies approaching 100 %, this element (as well as nonautonomous elements in the same background) duplicates at a frequency of only 10 % per generation, which is what made it possible to genetically isolate this minimal version of the system. Duplicate copies of *MuDR(p1)* at new positions show a restoration of high rates of duplication, suggesting that this position effect is reversible [37]. The cause of this effect is not known, but the recent sequencing of the maize genome reveals that *MuDR(p1)* is inserted into a heavily methylated helitron (Lisch, unpublished data). When combining the effects of the direction of cross and effects of position, duplication rates of single copy *MuDR* elements can range from 11 % for *MuDR(p1)* crossed as a female, to 70 % for a duplicate copy of that same element crossed as a male [37]. Interestingly, *MuDR* elements have also been shown to exhibit effects with respect to their capacity to condition somatic excision of a reporter element [37, 38] and the propensity to reactivate following epigenetic silencing [47], suggesting that *MuDR* elements may be particularly sensitive to their local chromatin environment.

3 Epigenetic Regulation

Like all TEs, *Mu* elements show evidence of epigenetic silencing. Typical high copy lines exhibit spontaneous silencing at roughly 10 % frequency [1, 48]. Silencing is associated with DNA methylation of *Mu* TIRs and a dramatic reduction in the quantity of polyadenylated cytoplasmic *mudrA* and *mudrB* transcripts [49–52]. The process is often progressive and clonal, resulting in increasingly large somatic sectors in which *MuDR* activity is lost over the course of development [50]. These somatic events lead to heritable silencing that is generally irreversible. Once initiated in one generation, silencing is transmitted to nearly all progeny [53]. Spontaneous silencing of this kind is poorly understood, although it is likely the result of the production of aberrant RNAs that trigger global epigenetic silencing. It is not, however, generally associated with the emergence of dominant factors that can act in trans, since fresh, active *MuDR* elements are not generally silenced when they are introduced into genetic backgrounds that have experienced spontaneous silencing [37, 54].

Minimal *Mutator* lines are exceptional in that they do not, as a rule, exhibit spontaneous epigenetic silencing. Lines carrying a single *MuDR* element have been propagated for many generations without evidence of silencing, a characteristic that has made genetic characterization of *MuDR* in these lines relatively easy. However, a dominant factor was isolated from a Minimal line that, when combined with *MuDR* can cause heritable silencing of one or many *MuDR* elements. This factor, which was designated *Mu killer* (*Muk*), is a rearranged deletion derivative of *MuDR* (*see* Fig. 1) [11, 12]. The element appears to have resulted from an aberrant transposition event that resulted in the deletion of one gene and portions of two others. In addition *MuDR* underwent a deletion/duplication/inversion event. The net effect of this rearrangement is a structure in which a portion of the 5′ end of *MuDR* is duplicated and inverted, resulting in a 2.4 kb perfect inverted repeat that includes portions of the *mudrA* gene as well the TIR adjacent to that gene. Although the *mudrA* promoter in TIRA is intact in *Muk*, it appears to be transcriptionally inactive. Instead, expression is initiated from a flanking promoter, proceeds through the duplicated and inverted *MuDR* sequence, and terminates in downstream flanking sequence. The resulting polyadenylated transcript contains a very long (2.4 kb) hairpin (*see* Fig. 1d). Processing of that hairpin produces small RNAs ranging in size from 21 to 24 nt that are competent to trigger silencing of *MuDR* elements.

Given the sequence identity between the hairpin *Muk* transcript and a portion of *mudrA* as well as its adjacent TIR (TIRA), the effects on this gene are what one would expect. In F1 plants that carry both *MuDR(p1)* and *Muk*, the TIRA in *MuDR* is heavily methylated and *mudrA* is transcriptionally inactive by the immature ear stage. DNA methylation is in all three sequence contexts (CG, CHG, and CHH, where H stands for any nucleotide other than guanine) and is associated with H3K9 and H3K27 dimethylation, classic marks associated with RNA-directed DNA methylation [55]. Interestingly, the initiation of silencing in F1 *MuDR;Muk* leaves requires the activity of *lbl1*, the maize homolog of SUPPRESSOR OF GENE SILENCING3 (SGS3) [55, 56]. For unknown reasons, expression of *lbl1* in maize is down regulated in leaves that represent a transition between juvenile and adult growth. This down regulation results in a transient loss of epigenetic silencing of *MuDR*. In the next generation, in progeny that carry *MuDR* but that lack *Muk*, levels of methylation remain high even in the absence of the trigger, consistent with stable and heritable silencing.

Surprisingly, *Muk* is also competent to silence *mudrB*, albeit via a distinct mechanism with very different results. In immature ears of F1 plants that carry both *MuDR(p1)* and *Muk*, *mudrB* transcript is present at levels roughly equivalent to those observed in the absence of *Muk*. However, the *mudrB* transcript in these F1 plants appears to lack polyadenylation. By the next generation *mudrB* transcript, like the *mudrA* transcript, is no longer detectable [11]. However, in marked contrast to what is observed at TIRA, there are no changes in DNA methylation between active and silenced TIRB at position 1; regardless of activity TIRB is heavily methylated in the CG and CHG sequence contexts but not in the CHH context. Neither are there differences in H3K9 or H3K27 dimethylation. Instead, the major changes associated with TIRB silencing are an increase in H3K27 trimethylation and a decrease in H3 acetylation (Lisch, in preparation). This would suggest that in addition to the direct effects on *mudrA* by small RNAs derived from *Muk*, events initiated within *mudrA* can trigger a spread of silencing information, in *trans* into the adjacent *mudrB* gene. It is not known what mediates this spread, but it does not appear to be small RNAs and it is not associated with a uniform spread of DNA methylation, as sequences between *mudrA* and *mudrB* are not methylated in silenced *MuDR* elements. H3K27me3 is not generally associated with TE silencing [57, 58]. However, a significant number of TEs and TE genes are in fact targeted by this modification, particularly in the endosperm, which has led to the suggesting that H3K27me3 may represent an alternative silencing pathway for a subset of TEs [59].

4 Maintenance of Silencing

Once a *MuDR* element is inactivated, that state is remarkably stable. We have propagated lines carrying silenced *MuDR* element for almost 20 generations, examined thousands of kernels and have yet to detect a spontaneous reactivation event (Lisch unpublished data). There are, however, mutants that can destabilize this silencing. The best studied of these is *mop1*, which encodes the maize homolog of RNA-dependent RNA polymerase 2 [60, 61]. When a *MuDR* element that has been stably silenced for multiple generations is put into a *mop1* mutant background and then test crossed to a plant that is heterozygous for *mop1* there is no obvious evidence of reactivation, despite the fact that *mop1* has clear and immediate effects on *Mu* TIR methylation. It is only after multiple generations in the mutant background that silenced *MuDR* elements become reactivated [27]. This gradual process of reactivation is associated with discrete changes in heritability of activity. In one experiment, detectable levels of somatic activity (spotted kernels) were not observed until four generations in a mutant background, and only after seven generations did the previously silenced *MuDR* element retain activity after a wild type copy of *mop1* was reintroduced.

5 Silencing in High Copy Lines

Despite active investigation over several years, we still know relatively little about spontaneous silencing of high copy lines. As with *Muk*-induced silencing, this process is associated with a loss of cytoplasmic *mudrA* and *mudrB* transcript and with the heritable loss of somatic excision activity and an absence of new insertions [49]. It is also associated with methylation of both *MuDR* and nonautonomous *Mu* element TIRs, consistent with epigenetic silencing targeting *MuDR* promoters, resulting in transcriptional gene silencing [50]. Unlike the case with *Muk*, however, the trigger for this process remains elusive. Presumably, this trigger (or triggers) is aberrant RNA that is processed into small RNAs that target the TIRs of *MuDR* elements. However, no small RNAs specifically associated with spontaneous silencing have been identified, nor have individual *MuDR* variants in these lines been associated with silencing. There is evidence for a relative increase in *MuDR* and/or *hMuDR* transcripts in the nuclei of plants that are undergoing silencing, but it is unclear as to the extent that these transcripts play a causal role in silencing of active elements [49]. Finally, as a rule, when inactivated high copy lines are crossed to a single *MuDR* element, or to an active high copy line, there is little evidence for a genetic factor present in silenced lines that is competent to silence active elements [37, 54]. Unfortunately, then, spontaneous silencing remains a poorly understood phenomenon. We do, however, have some clues. Inactivation tends to accumulate over developmental time, so that sectors of inactivation become larger and more frequent as plants develop [50]. This suggests that whatever is triggering silencing becomes more and more likely to occur in individual cells as the maize plant develops. Further, although stochastic, the process appears to be irreversible in any given clonal sector, suggesting a threshold of some kind that, once passed, leads to somatically heritable inactivation. If those sectors find their way into the germ line (an ear sector, for instance), the result is heritable silencing. This differs from instances in which *Muk* is crossed to *MuDR*, in which case silencing is initiated in the gamete [47].

Together with what we know about *Muk*-induced silencing, we can make some educated guesses as to the process of spontaneous silencing. Based on what is known about all known silencing pathways in plants, the triggers are almost certainly aberrant *MuDR* transcripts that are similar, at least functionally, to *Muk* transcript. Given that *Muk* is the result of a spontaneous rearrangement of a *MuDR* element, and somatic rearrangements of *MuDR* elements are common [37], it would seem likely that in a subset of cells, at least one of the multiple *MuDR* elements in a high copy line would become rearranged in such a way that it would be prone to produce a "killer" transcript that would be competent to silence multiple *MuDR* elements. According to this model, each somatic sector of lost activity would represent a clone of cells in which a single *MuDR* element had "gone bad" and triggered silencing. Sectors that include a germinal linage would transmit silenced *MuDR* elements. If, as the case with *Muk*, the silencing event triggers heritable silencing, all progeny, even if they did not inherit the trigger, would remain stably silenced. Why then do these germinally transmitted events not propagate a dominant silencing factor such as *Muk*? This may have to do with the somewhat idiosyncratic nature of *Muk*. Although this derivative carries two TIRs with sequence identity to TIRA in *MuDR*, these two TIRs appear to be transcriptionally silenced [12]. Instead, the major *Muk* transcript is initiated from a flanking ectopic promoter, which is not silenced. If that promoter were not there, it is likely that *Muk* would have simply silenced itself after its formation. If other, full-length *MuDR* elements were present in the same cell, it may also have silenced them, but in subsequent generations it may not have been competent to silenced *MuDR* elements in trans. It is possible that most "killers" arising in plants undergoing spontaneous silencing are similar; transiently competent to trigger heritable silencing, but destined for inactivation themselves.

This makes an attractive hypothesis, but there are observations that suggest that it is inaccurate. Southern blot analysis has failed to detect new *MuDR* polymorphisms in leaves undergoing spontaneous silencing, and heritably silenced lineages do not carry novel, germinally transmitted derivatives [49, 50, 62]. This suggests that the loss of *Mu* activity in these sectors is not due to rearrangements of *MuDR* within them.

A second possibility is that plants undergoing spontaneous silencing die the death of a thousand cuts. Rather than a few aberrant *MuDR* elements causing the process, perhaps this process is the result of thousands of such events that occur very late during development, when it is known that the vast majority of transpositions occur in *Mutator* active lines. Rearrangements in any individual small clone of cells would not be detectable, but they could have a cumulative effect if the small RNAs that are produced from them could be transported back to the meristem. According to this scenario, thousands of individual events, producing "microkillers" could collectively contribute to silencing in the whole plant. As development progressed, the odds that sufficient siRNAs would accumulate to trigger silencing in any given cell would increase, resulting in increasing frequent sectors in which all *MuDR* elements have been heritably silenced in a given clone of cells.

This hypothesis fits all of the observations, but it has not been tested. It does, however, make some predictions. First, it predicts that trans-acting, mobile small RNAs are produced in somatic cells. These small RNAs have not been detected as of yet, but resolution has been increased by orders of magnitude by the implementation of next generation sequencing. Analysis of small RNAs produced by *Muk* indicates that these small RNAs will be 21–22 nt and will be dependent on *lbl1* for their production. This could provide a key test of the hypothesis. *Muk* can heritably silence *MuDR* elements even in an *lbl1* mutant background, presumably because silencing in the shoot apical meristem does not require *lbl1* (Li and Lisch, unpublished). However, we have shown that methylation of *MuDR* TIRs in leaf tissue does require *lbl1* [55]. If silencing of high copy lines is the product of late somatic events in the leaves that are dependent on *lbl1*, then *lbl1* mutants would be expected to experience fewer spontaneous silencing events. Transport of microkiller siRNAs would be difficult to demonstrate, but it would not be that surprising given extensive evidence for transport of small RNAs that results in systemic silencing [63].

The reason that spontaneous silencing remains an important question is that it is likely to represent a far more typical process for TE silencing than does *Muk*. The vast majority of plant tissues are not germinal, but effective silencing of TEs must be germinal. Rather than waiting for a relatively rare "mistake" like *Muk* to be made by a given TE in the germinal lineage, plants may be competent to gather information in the form of microkiller small RNAs from all somatic tissue and transport it to the meristem, which could result in heritable silencing.

6 Uses of *Mu* Killer

6.1 Introduction

Like all mutagens, *Mutator* produces a significant mutational load, meaning that lines carrying mutations generated by this transposon should be out-crossed to one's favorite inbred background, in order to reduce overall copy number as rapidly as possible. However, unlike EMS backgrounds, *Mutator* can be the gift that keeps on giving, since the system can continue to amplify itself even after out crossing [32, 64]. This is not true of *Mu* lines produced by the McCarty group, since their protocol involves selection against activity prior to sequence indexing of *Mu* insertion and line propagation [65]. However, lines produced by Robert Meeley at Pioneer (which has been a wonderful and reliable resource) and those generated by the Barkan group often retain high levels of activity [6, 7, 66]. In addition to the mutational load carried by these lines, plants derived from these lines often exhibit a spectrum of phenotypes that has been referred to as "*Mu*-syndrome" [1] (*see* Fig. 3). These phenotypes include thin, somewhat necrotic leaves, short plants as well as small ears and tassels. Depending on variables such as levels of activity and overall *Mu* copy number, these phenotypes can manifest themselves to varying degrees. Work by the Walbot group has demonstrated that *Mu* activity is associated with changes in a wide variety of genes, particularly those involved in stress response, which would obviously have an impact on RNAseq analysis of mutants and wild type siblings [67]. These effects of activity have a considerable potential to complicate analysis of mutant phenotypes and should be avoided. In addition, many *Mu*-induced mutations have been identified that are *Mu-suppressible*, such that the phenotype is only visible in the presence of *Mu* activity [9]. Again, this can complicate phenotype analysis, particularly if one is unaware of whether or not the system is active in a given line. In this respect, it is important to point out that even a single active *MuDR* element can affect suppressible alleles, so the absence of *Mu*-syndrome phenotypes (which result from the activity of many *Mu* elements) is no guarantor that activity is in fact absent [19].

6.2 Materials

1. Available stocks. The best way to ensure that activity is lost is to cross *Mu*-active lines with insertions of interest into lines that are homozygous for *Muk*, which can efficiently silencing lines carrying multiple *MuDR* elements. Such lines are available at the Maize Genetics COOP Stock Center (http://maizecoop.cropsci.uiuc.edu/) in a number of inbred backgrounds, including B73, Mo17, W23, W22, and A188, as well as the original *a1-mum2* minimal line and a line carrying the *bz-mum9* reporter [68]. Flowering times for *Muk* in each of these backgrounds are roughly comparable to those of these

Fig. 3 The effects of *Mu* activity on plant phenotype and the loss of those effects upon crossing to plants carrying *Muk*. The *bottom panel* shows the effects of *Muk* on methylation of nonautonomous *Mu1* elements. Digestion with the methylation-sensitive *HinfI* restriction enzyme, which cuts in the ends of these elements, yields a 1.4 kb fragment if the *Mu1* is unmethylated (*arrow*) and larger fragments if the elements are methylated

backgrounds without *Muk*, although it should be noted that the *Muk* allele involves a deletion of three genes (*see* Fig. 2). When homozygous, this deletion causes a slight delay in flowering time, reduced germination rates and subtle morphological changes, including shortened stature and reduced tassel branch length (Lisch, unpublished). The *a1-mum2* and *MuDR(p1)* lines flower roughly 2 weeks earlier than B73 and produce an abundance of pollen and tiller extensively in the field (but not in pots). These lines are also quite sensitive to high temperatures (above 90 °C), which causes tassel "burning" and pollen inviability. *Mutator* active lines used will vary depending on the source. The most active lines are distinguishable because of their short stature, necrotic or yellowing and narrow leaves and small ears and tassels (*see* Fig. 3). Given that, an adequate number of seeds should be planted and lines to be crossed to them should be stagger planted at least 1 week before, during and after planting the *Mutator* lines to ensure nicking and a sufficient number of crosses.

2. PCR primers and conditions. Efficient use of *Muk* requires genotyping for the Muk allele and, if available, for the insertion allele.

   For genotyping *Muk*:

   1. 12-3R: CGGTATGGCGGCAGTGACA.
   2. TIRAR: AGGAGAGACGGTGACAAGAGGAGTA.
   3. PCR conditions:
      (a) 94 °C 5 min.
      (b) 94 °C 30 s.
      (c) 60 °C 45 s.
      (d) 72 °C 1 min.
      (e) Repeat **steps 2–4** 35 times.
      (f) 72 °C 10 min.
      (g) Soak at 4 °C.

   For genotyping *MuDR(p1)*:

   1. Ex1: ACATCCACGCTGTCTCAGCC.
   2. RLTIR2: ATGTCGACCCCTAGAGCA.
   3. PCR conditions:
      (a) 94 °C 5 min.
      (b) 94 °C 30 s.
      (c) 57 °C 45 s.
      (d) 72 °C 45 s.
      (e) Repeat **steps 2–4** 35 times.

(f) 72 °C 10 min.

(g) Soak at 4 °C.

For genotyping a *Mu* insertion of interest: This will depend on the particular insertion, but all protocols involve one primer in the gene of interest and one *Mu* TIR primer. Excellent (and detailed) advice for efficient primer design is available at Maize GDB.

6.3 Genetic Crossing Protocols

1. Cross plants that are homozygous for the insertion of interest into a *Muk* homozygous background prior to introgression (*see* Fig. 4 for a flow chart of recommended crosses). This should result in an immediate amelioration of *Mu*-syndrome symptoms (*see* Fig. 3). A control cross of each *Mu*-active plant should also be made to the same genetic background that lacks *Muk*. Each plant should be numbered and plant numbers should be recorded (*see* **Notes 1** and **2**). Plants can be crossed in either direction, but using *Muk* as a female may result in more efficient silencing and permit each *Mu*-active plant to be crossed to both the experimental and control lines.
2. For recessive mutations (which are the majority of insertion alleles), the resulting F1 progeny from both the *Muk* and the control crosses should be self-fertilized and examined for evidence of a co-segregating phenotype. Each plant that is self-fertilized should also be out-crossed to one's preferred introgression line. A sufficient number of F2 progeny from the self-fertilization (>100) should be examined to provide confidence in linkage between the phenotype and the insertion. This by itself cannot prove that the insertion is actually causing the phenotype, which requires the availability of at least one additional allele, but confirmation of linkage in these families provides reassurance and aids in subsequent introgressions if desired. If a reporter for *Mu* activity was present in the *Mu-active* parent, or if the *a1-mum2;Muk*, and *a1-mum2* tester lines were used, activity in this generation can be scored by examining kernels for the presence or absence of somatic excisions of the reporter element (*see* **Note 3**). The control cross makes it possible to determine whether or not a given allele is suppressible, which means that the mutant phenotype is only observed in a Mu active background. If the phenotype can be observed in the progeny of self-fertilized plants from the control cross, but is not observed in the progeny of those generated from the *Muk* cross, then the *Mu*-allele is likely suppressible. At this point many investigators may wish to discard the line and try for another allele or simply use the existence of the *Mu*-induced allele to confirm the identity of a reference allele.
3. Assuming a given insertion is not suppressible, recovery of low copy lines lacking *Mu* activity is relatively straightforward. Following the cross to homozygous *Muk* stocks, F1 progeny

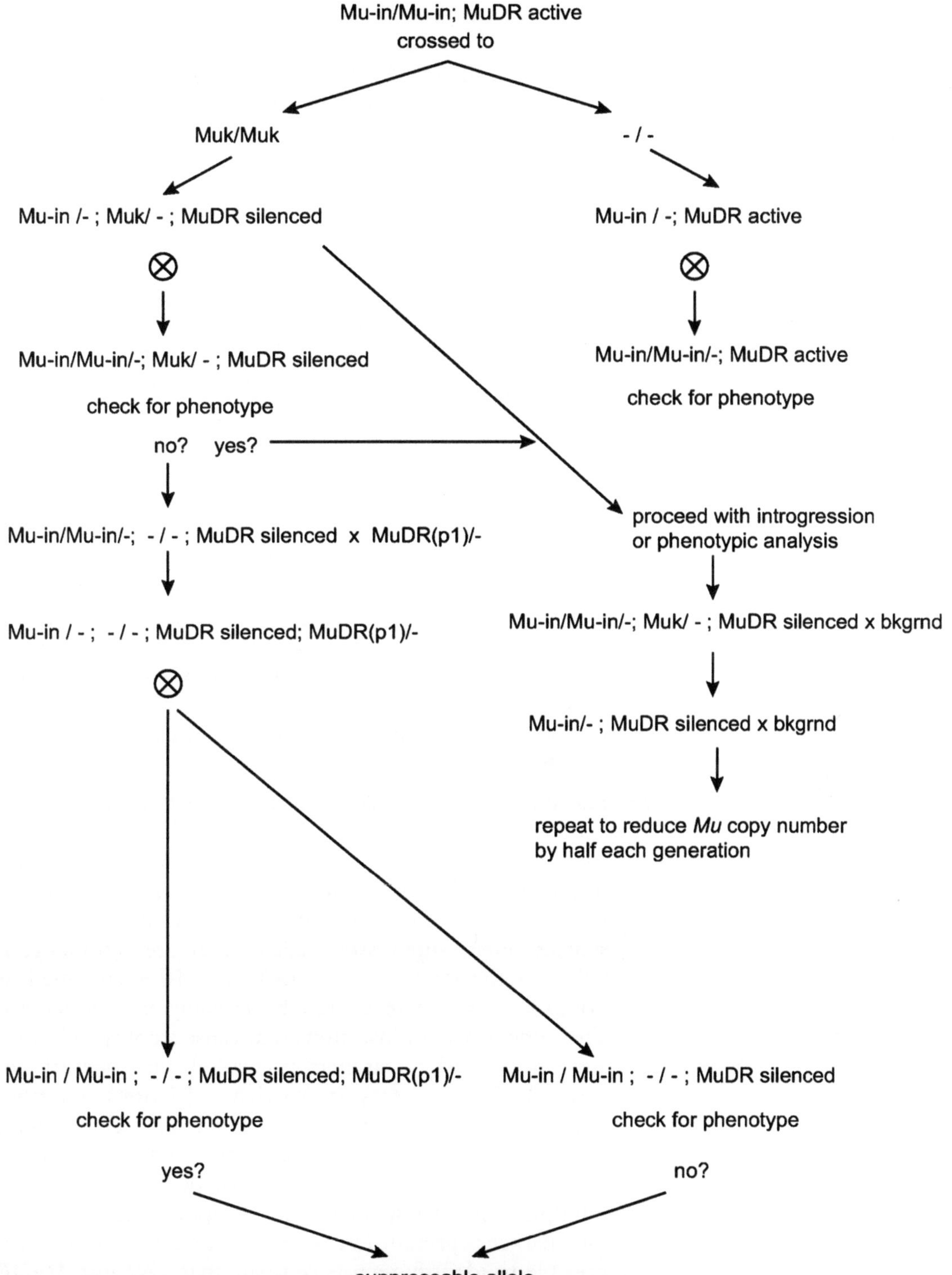

Fig. 4 A flow chart detailing crosses that should be used to silence *Mu* activity using *Muk*. "Mu-in" indicates a *Mu* insertion allele of interest. A *circle with an "x"* in it indicates self-fertilization

(that are heterozygous for the insertion of interest and for *Muk*) should be out-crossed to one's favorite inbred background(s). Standard introgression protocols can then be used, exploiting the availability of primers specific to the insertion to select the appropriate plants each generation. One caveat is that *Muk* should probably be eliminated at the first out-cross using *Muk*-specific primers to screen for plants that lack this element.

4. Using *Mu*-suppressible alleles.
   (a) Given that *Mu* activity can be unstable and unpredictable and can affect phenotype in its own right, analysis of the effects of these insertions are likely to be difficult to interpret. However, it is possible to exploit suppressibility as well, if one is willing to be patient. To confirm suppressibility, plants that are homozygous for the insertion (25 %) but that lack a phenotype due to the loss of activity induced by *Muk* can be crossed to plants that carry a single *MuDR* element, also available from the Maize Genetics COOP Stock Center (*see* Fig. 4). Ideally these plants should also be screened for a lack of *Muk* (25 %). Progeny of this cross that are heterozygous for the *Mu* insertion, lack *Muk* and that carry *MuDR* can then be self-fertilized and screened for an appearance of the phenotype. Progeny of self-fertilized siblings that lack *MuDR* can be used as a control; since they lack activity they would not be expected to show the phenotype.
   (b) Having established that a given insertion is suppressible and having introgressed the *Muk*-silenced line carrying the suppressible alleles into a minimal *Mutator* background, it becomes possible to use activity to regulate expression of the phenotype. This can be useful in a number of circumstances. First, suppressible alleles that are homozygous lethal in the presence of activity can be maintained as homozygotes and recovered by crossing in *Mu* activity. The same is true of insertions that cause sterility in homozygotes. Second, researchers can exploit the propensity for single *MuDR* elements to experience deletions at various points during development [35]. Previous work has demonstrated that these deletions are associated with a coordinate restoration of mutant phenotypes resulting from multiple suppressible alleles. If the appropriate suppressible marker is present (*a1-mum2* in the minimal line) it is possible to identify sectors of tissue that have lost *MuDR* activity and examine those tissues for the effects of a loss of the mutant phenotype of one's suppressible insertion of interest [69]. A similar (albeit less controlled) experiment

can be performed by crossing a line carrying a suppressible mutation of interest into a line carrying suppressible markers such as *a1-mum2* and examining plants undergoing spontaneous silencing, a process that can involve somatic sectors of lost activity that can be scored using the suppressible allele [50].

7 Notes

1. The genetic protocols described above represent best practices, but they are certainly not required to benefit from the availability of *Muk* lines. The simplest protocol is to simply cross one's *Mu* insertion line to *Muk* homozygotes and then self-fertilize a subset of F1s and begin introgression of lines that show the phenotype. If the expected phenotype (assuming there is one) is absent in the progeny of the self-fertilized F1 then subsequent introgression is probably a waste of time, since the phenotype is unlikely to come back. If *Mu*-syndrome is not eliminated in the F1 plants (*see* Fig. 3), then it is possible that the "*Muk*" line was not in fact *Muk*. Replacement stocks can be obtained from the Maize Genetics COOP stock center or from Damon Lisch directly.
2. Given that a limited number of crosses and plants are required for the initial steps of this protocol, it is advisable to collect leaf tissue from the initial *Mu*-active lines and the *Muk* lines for genotyping. The *Muk* lines should be genotyped to ensure that they carry *Muk* and the *Mu*-active lines should be genotyped for the insertion allele (assuming it is derived from a sequence-indexed resource).
3. Although not essential, the availability of a visible marker for *Mu* activity can be useful. These are available in some (but not all) *Mu*-active stocks. The most common of these is *bz-mum9*. In this case, the effects of *Muk* on *Mu* activity can easily be monitored by simply scoring for excisions of the reporter *Mu* element in the self-fertilized progeny of the F1 cross between *Muk* and the *Mu*-active line. For this purpose, the *bz-mum9 Muk* homozygous stock would be particularly useful, since all of the progeny seed would then be scorable for activity. Caution should be used, however, because the presence of the *r* allele of the *R* locus can result in mottling in the aleurone that is superficially similar to excision events. These can be distinguished from *Mu* reversions because unlike those events, which are quite uniform (*see* Fig. 1), *R* mottling results in nonuniform sectors of color.

References

1. Walbot V (1991) The Mutator transposable element family of maize. Genet Eng 13:1–37
2. Bennetzen JL (1996) The Mutator transposable element system of maize. Curr Top Microbiol Immunol 204:195–229
3. Chandler VL, Hardeman KJ (1992) The Mu elements of Zea mays. Adv Genet 30:77–122
4. Robertson DS (1978) Characterization of a mutator system in maize. Mutat Res 51:21–28
5. Candela H, Hake S (2008) The art and design of genetic screens: maize. Nat Rev Genetics 9:192–203
6. Williams-Carrier R et al (2010) Use of Illumina sequencing to identify transposon insertions underlying mutant phenotypes in high-copy Mutator lines of maize. Plant J 63:167–177
7. McCarty D, Meeley R (2009) Transposon resources for forward and reverse genetics in maize. In: Bennetzen J, Hake S (eds) Handbook of maize: genetics and genomics. Springer, Berlin, pp 561–584
8. Settles AM et al (2007) Sequence-indexed mutations in maize using the UniformMu transposon-tagging population. BMC Genomics 8:116
9. Lisch D, Jiang H (2009) Mutator and MULE transposons. In: Bennetzen J, Hake S (eds) Handbook of maize: genetics and genomics. Springer, Berlin, pp 277–306
10. Lisch D (2009) Epigenetic regulation of transposable elements in plants. Annu Rev Plant Biol 60:43–66
11. Slotkin RK, Freeling M, Lisch D (2003) Mu killer causes the heritable inactivation of the Mutator family of transposable elements in Zea mays. Genetics 165:781–797
12. Slotkin RK, Freeling M, Lisch D (2005) Heritable transposon silencing initiated by a naturally occurring transposon inverted duplication. Nat Genet 37:641–644
13. Lisch D (2002) Mutator transposons. Trends Plant Sci 7:498–504
14. Tan BC et al (2011) Identification of an active new mutator transposable element in maize. G3 (Bethesda) 1:293–302
15. Dietrich CR et al (2002) Maize Mu transposons are targeted to the 5′ untranslated region of the gl8 gene and sequences flanking Mu target-site duplications exhibit nonrandom nucleotide composition throughout the genome. Genetics 160:697–716
16. Hershberger RJ et al (1995) Characterization of the major transcripts encoded by the regulatory MuDR transposable element of maize. Genetics 140:1087–1098
17. Qin M, Robertson DS, Ellingboe AH (1991) Cloning of the mutator transposable element MuA2, a putative regulator of somatic mutability of the a1-Mum2 allele in maize. Genetics 129:845–854
18. James MG et al (1993) DNA sequence and transcript analysis of transposon MuA2, a regulator of Mutator transposable element activity in maize. Plant Mol Biol 21:1181–1185
19. Chomet P et al (1991) Identification of a regulatory transposon that controls the Mutator transposable element system in maize. Genetics 129:261–270
20. Hershberger RJ, Warren CA, Walbot V (1991) Mutator activity in maize correlates with the presence and expression of the Mu transposable element Mu9. Proc Natl Acad Sci USA 88:10198–10202
21. Benito M-I, Walbot V (1997) Characterization of the maize Mutator transposable element MURA transposase as a DNA-binding protein. Mol Cell Biol 17:5165–5175
22. Eisen JA, Benito MI, Walbot V (1994) Sequence similarity of putative transposases links the maize Mutator autonomous element and a group of bacterial insertion sequences. Nucleic Acids Res 22:2634–2636
23. Hua-Van A, Capy P (2008) Analysis of the DDE motif in the Mutator superfamily. J Mol Evol 67:670–681
24. Marquez CP, Pritham EJ (2010) Phantom, a new subclass of Mutator DNA transposons found in insect viruses and widely distributed in animals. Genetics 185:1507–1517
25. Pritham EJ, Feschotte C, Wessler SR (2005) Unexpected diversity and differential success of DNA transposons in four species of entamoeba protozoans. Mol Biol Evol 22:1751–1763
26. Lisch D, Girard L, Donlin M, Freeling M (1999) Functional analysis of deletion derivatives of the maize transposon MuDR delineates roles for the MURA and MURB proteins. Genetics 151:331–341
27. Woodhouse MR, Freeling M, Lisch D (2006) The mop1 (mediator of paramutation1) mutant progressively reactivates one of the two genes encoded by the MuDR transposon in maize. Genetics 172:579–592
28. Raizada MN, Walbot V (2000) The late developmental pattern of Mu transposon excision is conferred by a cauliflower mosaic virus 35S-driven MURA cDNA in transgenic maize. Plant Cell 12:5–21
29. Lisch D (2005) Pack-MULEs: theft on a massive scale. Bioessays 27:353–355

30. Jiang N et al (2004) Pack-MULE transposable elements mediate gene evolution in plants. Nature 431:569–573
31. Walbot V, Rudenko GN (2002) MuDR/Mu transposable elements of maize. In: Craig NL, Craigie R, Gellert M, Lambowitz AM (eds) Mobile DNA II. ASM Press, Washington, DC
32. Alleman M, Freeling M (1986) The Mu transposable elements of maize: evidence for transposition and copy number regulation during development. Genetics 112:107–119
33. Doseff A, Martienssen R, Sundaresan V (1991) Somatic excision of the Mu1 transposable element of maize. Nucleic Acids Res 19:579–584
34. Britt AB, Walbot V (1991) Products of Mu excision from the Bronze1 gene of Zea-mays. J Cell Biochem Suppl 99
35. Lisch D (1995) Genetic and molecular characterization of the Mutator system in maize. University of California at Berkeley, Berkeley, CA
36. Yu W et al (2007) Cytological visualization of DNA transposons and their transposition pattern in somatic cells of maize. Genetics 175:31–39
37. Lisch D, Chomet P, Freeling M (1995) Genetic characterization of the Mutator system in maize: behavior and regulation of Mu transposons in a minimal line. Genetics 139:1777–1796
38. Lisch D, Freeling M (1994) Loss of Mutator activity in a minimal line. Maydica 39:289–300
39. Donlin MJ, Lisch D, Freeling M (1995) Tissue-specific accumulation of MURB, a protein encoded by MuDR, the autonomous regulator of the Mutator transposable element family. Plant Cell 7:1989–2000
40. Li J, Wen TJ, Schnable PS (2008) Role of RAD51 in the repair of MuDR-induced double-strand breaks in maize (Zea mays L.). Genetics 178:57–66
41. Hsia A-P, Schnable PS (1996) DNA sequence analyses support the role of interrupted gap repair in the origin of internal deletions of the maize transposon, MuDR. Genetics 142:603–618
42. Liu S et al (2009) Mu transposon insertion sites and meiotic recombination events co-localize with epigenetic marks for open chromatin across the maize genome. PLoS Genet 5:e1000733
43. Fernandes J et al (2004) Genome-wide mutagenesis of Zea mays L. using RescueMu transposons. Genome Biol 5:R82
44. Robertson DS, Stinard PS (1989) Genetic analyses of putative two-element systems regulating somatic mutability in Mutator-induced aleurone mutants of maize. Dev Genet 10:482–506
45. Robertson DS, Stinard PS (1992) Genetic regulation of somatic mutability of two Mu-induced a1 mutants of maize. Theor Appl Genet 84:225–236
46. O'Reilly C et al (1985) Molecular cloning of the a1 locus of Zea mays using the transposable elements En and Mu1. EMBO J 4:877–882
47. Singh J, Freeling M, Lisch D (2008) A position effect on the heritability of epigenetic silencing. PLoS Genet 4:e1000216
48. Robertson DS (1986) Genetic studies on the loss of mu mutator activity in maize. Genetics 113:765–773
49. Rudenko GN, Ono A, Walbot V (2003) Initiation of silencing of maize MuDR/Mu transposable elements. Plant J 33:1013–1025
50. Martienssen R, Baron A (1994) Coordinate suppression of mutations caused by Robertson's mutator transposons in maize. Genetics 136:1157–1170
51. Bennetzen JL (1987) Covalent DNA modification and the regulation of Mutator element transposition in maize. Mol Gen Genet 208:45–51
52. Chandler VL, Walbot V (1986) DNA modification of a maize transposable element correlates with loss of activity. Proc Natl Acad Sci USA 83:1767–1771
53. Walbot V et al (1988) Regulation of mutator activities in maize. Basic Life Sci 47:121–135
54. Brown J, Sundaresan V (1992) Genetic study of the loss and restoration of mutator transposon activity in maize—evidence against dominant-negative regulator associated with loss of activity. Genetics 130:889–898
55. Li H, Freeling M, Lisch D (2010) Epigenetic reprogramming during vegetative phase change in maize. Proc Natl Acad Sci USA 107:22184–22189
56. Nogueira FT et al (2007) Two small regulatory RNAs establish opposing fates of a developmental axis. Genes Dev 21:750–755
57. Wang X et al (2009) Genome-wide and organ-specific landscapes of epigenetic modifications and their relationships to mRNA and small RNA transcriptomes in maize. Plant Cell 21:1053–1069
58. Zhang X et al (2007) Whole-genome analysis of histone H3 lysine 27 trimethylation in Arabidopsis. PLoS Biol 5:e129
59. Lafos M et al (2011) Dynamic regulation of H3K27 trimethylation during Arabidopsis differentiation. PLoS Genet 7:e1002040
60. Alleman M et al (2006) An RNA-dependent RNA polymerase is required for paramutation in maize. Nature 442:295–298

61. Woodhouse MR, Freeling M, Lisch D (2006) Initiation, establishment, and maintenance of heritable MuDR transposon silencing in maize are mediated by distinct factors. PLoS Biol 4:e339
62. Kim SH, Walbot V (2003) Deletion derivatives of the MuDR regulatory transposon of maize encode antisense transcripts but are not dominant-negative regulators of mutator activities. Plant Cell 15:2430–2447
63. Molnar A, Melnyk C, Baulcombe DC (2011) Silencing signals in plants: a long journey for small RNAs. Genome Biol 12:215
64. Walbot V, Warren C (1988) Regulation of Mu element copy number in maize lines with an active or inactive Mutator transposable element system. Mol Gen Genet 211:27–34
65. McCarty DR et al (2005) Steady-state transposon mutagenesis in inbred maize. Plant J 44:52–61
66. Bensen RJ et al (1995) Cloning and characterization of the maize An1 gene. Plant Cell 7:75–84
67. Skibbe DS et al (2009) Mutator transposon activity reprograms the transcriptomes and proteomes of developing maize anthers. Plant J 59:622–633
68. Slotkin RK, Freeling M, Lisch D (2007) Mu killer locus available in multiple inbred backgrounds. Maize Genetics Cooperation Newsletter 81
69. Fowler JE, Meuhlbauer GJ, Freeling M (1996) Mosaic analysis of the liguleless3 mutant phenotype in maize by coordinate suppression of mutator-insertion alleles. Genetics 143:489–503

Chapter 10

Using *MuDR/Mu* Transposons in Directed Tagging Strategies

Virginia Walbot and Julia Qüesta

Abstract

An introduction to *MuDR/Mu* transposons as mutagens is provided along with protocols for using these elements to tag maize genes. Selection for retention of *Mutator* activity is described as well as details for establishing and screening tagging populations efficiently.

Key words *Mutator*, Maize, Transposon silencing, Genetic screen

1 Introduction

Transposon-induced mutations are useful for generating sectors for clonal analysis of organ and tissue growth patterns, for generating diverse but functional revertant alleles, and for analyzing aspects of the regulation of gene expression. A major use of transposons in corn and other species that are difficult to transform is confirmation of gene cloning. In readily transformed species, a single mutant allele suffices to define the gene provided a transgenic stock expressing a functional copy of that gene results in phenotypic normalcy. Otherwise, definition of the gene rests on recovery and analysis of two independent mutants, each of which can be stringently associated with the mutant phenotype. If the defining allele of a locus arose spontaneously, or after chemical mutagenesis, or in lines bearing transposons, a directed tagging experiment with an active transposon line can generate a second allele quickly. Generation of mutant alleles can be carried out concurrently with efforts to map and clone the target gene.

MuDR/Mu are popular transposons, because they exist in high copy numbers and transpose to unlinked sites [1]. Consequently, they can be used to mutagenize most maize genes in reasonable population sizes without special stock preparation. An alternative strategy is to map a gene of interest and then select a stock with an

Thomas Peterson (ed.), *Plant Transposable Elements: Methods and Protocols*, Methods in Molecular Biology, vol. 1057, DOI 10.1007/978-1-62703-568-2_10, © Springer Science+Business Media New York 2013

Ac/Ds element nearby for high frequency local mutagenesis, achieving a similar forward mutation frequency for targets within a few centiMorgans of the transposon [2].

To learn more about *MuDR/Mu* transposons and the history of *Mutator* lines please consult review articles [1, 3, 4] as only a few key properties are mentioned here. Maize *MuDR/Mu* exist in 20–50 or more copies in active *Mutator* lines; they insert throughout the genome, favoring RNA PolymeraseII (protein coding) transcription units. Collectively, they increase the mutation frequency 50- to 100-fold per locus with the expectation of a new mutant in every 3,000–5,000 individuals screened. *MuDR* is the autonomous element and encodes the MURA transposase as well as MURB, a protein of unknown function. There are about a dozen types of *Mu* elements, all of which share highly similar terminal inverted repeats (TIRs) of ~215 bp; the intervening sequences are generally not shared with *MuDR*, and most *Mu* elements have little in common with other *Mu*. Because forward mutation is roughly proportional to the copy number of each element type, new mutants derived from individual *Mutator* lines will harbor mainly *Mu1,* if this is the highest copy number element, or *Mu2*, *Mu3*, *Mu7*, or *Mu8*, as these elements most frequently cause new mutants. The 4.9 kb autonomous element is mobile but at a lower frequency than the smaller 1.4 kb *Mu1* and the other small elements.

2 Materials

The original purple *Mutator* stock was supplied by Don Robertson of Iowa State University in 1982. This stock was used to recover the mutable *bz2-mu1::Mu1* allele [5, 6] and later other mutable *bz2* alleles and mutants of other steps in the anthocyanin pigment pathway. Active *Mutator* stocks are maintained by outcrossing to and by the *bz2* tester and scoring for heritability of the mutable phenotype as described in this chapter. Molecular protocols are standard methods.

3 Methods

3.1 Genetic Crossing Schemes

In the easiest cases of gene tagging, *MuDR/Mu* mutagenesis is conducted by crossing a *Mutator* plant as the pollen source onto recipient ears that are homozygous recessive for the target gene (*see* Fig. 1). In some cases a pure homozygous population is not possible. For male-sterile lines, for example, a stock segregating 1:1 sterile: fertile maintainer must be grown, and the fertile (heterozygous *Ms//ms*) individuals removed from the population. The tagging population is then as efficient as the homozygous viable case.

Cross to generate the tagging population: *bz2//bz2* x Mutator *Bz2//Bz2*

Screen

Nearly all purple kernels (*bz2//Bz2*) with rare spotted kernels (*bz2//bz2-mu*)

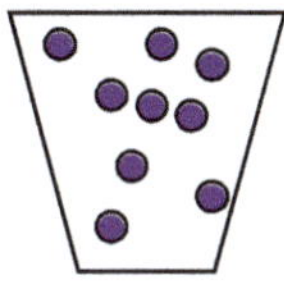

Fig. 1 Cross to perform to construct a tagging population with *bz2* as the target gene and the results of the screen

3.2 Key Steps in Successful Application of MuDR/Mu Tagging: Certify the Mutator Individuals

This process is important because 10–100 % of *Mutator* active individuals in a family will spontaneously silence in a single generation [7, 8]. At each step in the lifecycle a selection or enrichment should be made to identify the individuals most likely to produce new mutations in the pollen.

1. Use reporter alleles to select seeds. It is highly advantageous to use a *Mutator* line with a visible reporter gene to select seeds that are active. We use anthocyanin reporter alleles which are typically carried in a heterozygous condition, with a recessive, stable allele: *bz2-mu//bz2*. When *Mutator* is active, the kernels will be highly spotted from high frequency, late excision. *MuDR/Mu* generally excise in the terminal cell divisions, generating revertant sectors of 1–50 cells [8]. Mutable alleles in starch biosynthesis work equally well for selecting active seeds. Not only should individual seeds be spotted, but the ear should also have the expected ratio of spotted: colorless seed. In the cross [*bz2 x bz2-mu//bz2*] the expectation is 1:1 segregation; in the cross [*bz2 x bz2-mu*] the expectation is 100 % spotted kernels. Ears that deviate from these expectations by more than 5–10 % should not be used as the source of tagging individuals.
2. Use reporter alleles to visualize *Mutator* activity during somatic growth. In the correct regulatory background, anthocyanin pigmentation can be monitored in leaves, stems, glumes, and/or anthers. Alternatively, alleles that revert to dark green against a paler green leaf color can be used to visualize *Mutator* activity. This assay is less commonly available in tagging stocks than the seed phenotypic assessment.
3. Perform qRT-PCR of *mudrA* transcript in upper leaves. Abundant transposase transcript levels are a good indicator of *Mutator* activity. All maize lines examined contain some *hMuDR* (homolog of *MuDR*) elements that encode transcripts similar to *mudrA*, but at low levels [9], consequently

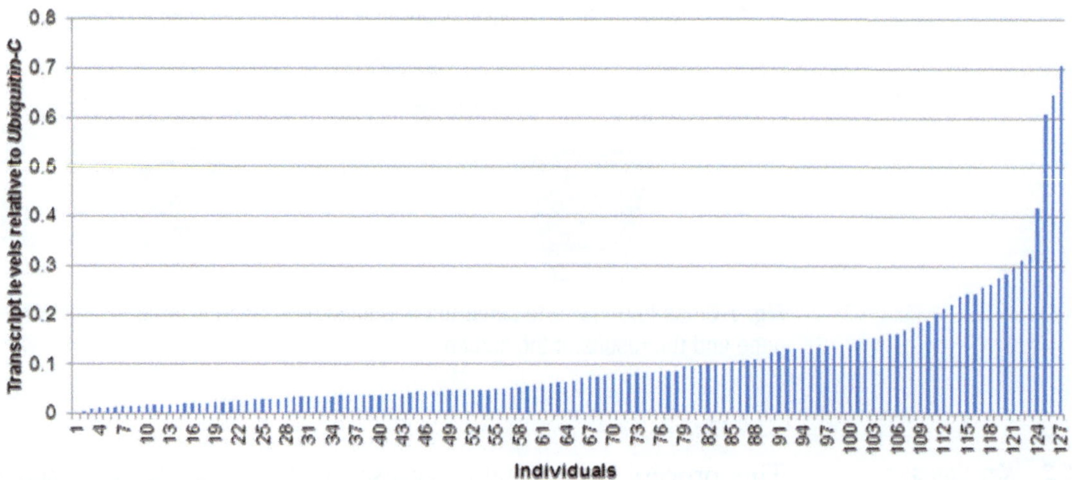

Fig. 2 *mudrA* transcript levels evaluated in the population of mutator plants used for 2010 tagging crosses. qRT-PCR analysis of *mudrA* transcript levels in 142 individuals belonging to 11 different families of mutator active plants. Amplification of Ubiquitin-C transcript was used for data normalization

mudrA-related transcripts are detectable by qRT-PCR in standard inbreds as well as in inactive (epigenetically silenced) *Mutator* individuals. Multi-copy *MuDR* active plants have substantially more transcript; however, a difference of at least tenfold above the level observed in the inbred line background should identify active *Mutator* individuals. Interestingly, *Mutator* individuals show an unexpectedly wide range of *mudrA* transcript abundance (*see* Fig. 2). In this population of 142 individuals belonging to 11 different families, all planted as certified active based on the spotted seed phenotype, the range is approximately 300-fold. Despite the wide variation in *mudrA* transcript levels, the amount of MURA protein is similar among active individuals, indicative of posttranscriptional regulation [9].

4. Use early-maturing plants as tagging individuals. Maturation of active *Mutator* individuals occurs 7–10 days before silenced siblings in the same family [5]. In our experience the *Mutator* active individuals shed pollen 7–10 days before the inbred line, suggesting that silencing restores the normal flowering date. The earliness of active *Mutator* individuals must be taken into account when planting the target populations, which typically must be planted 5–10 days before the *Mutator* line(s). In practice, it is a good idea to plant several *Mutator* families, spaced a few days apart, starting on the same date as the target populations and extending for about 2 weeks. As many target populations are in mixed or hybrid backgrounds, flowering date can be hard to predict for new materials. Repetitive planting of *Mutator* lines increases the chances of having active *Mutator*

individuals available at the correct time. *Mutator* families of 20–30 individuals are sufficient for crossing to several hundred to several thousand target ears.

5. Collect *Mutator* leaf samples. This step is optional; however, when new mutants are identified as progeny of a particular *Mutator* plant, the expectation is that the mutant was caused by a new *Mu* insertion that will be missing from the *Mutator* parent. This can be checked by a PCR test if a sample from each *Mutator* plant is available.
6. Testcross to assess reporter allele mutability. Near the *Mutator* source, plant sufficient tester individuals (twice the number of *Mutator* plants is a good rule of thumb) with the correct timing to be ear and pollen sources for testcrosses. We have derived *bz2* testers that can all be planted on the same day but are staggered in maturing by approximately 3 days over a 3 week period. Alternatively, the same tester can be planted in advance of the initial *Mutator* line and periodically thereafter; it is particularly important to have sufficient tester available when the first *Mutator* plants start shedding pollen so that crosses to tester ears can be performed from the same pollen collection that is used in the tagging crosses. It is better to have too many early tester plants than too few.

3.3 Organizing the Target Populations and Tagging Pollinations

Some care is required in developing and managing the target populations to optimize tagging efficiency and the ease of conducting the crosses (*see* **Notes 1–5**).

1. Temporal or spatial separation from exogenous sources of contaminating pollen is extremely important. The major source of frustration in transposon tagging is re-isolation of the original reference allele resulting from pollen contamination. At least 100 m distance or at least 1 week of temporal separation is recommended.
2. Within the target population eliminate all sources of contaminating pollen. For targets such as male-sterility loci, where a maintainer line is required, the *Ms//ms* fertile individuals should be eliminated daily before any pollinations are done onto the *ms//ms* target ears on sterile plants. A good practice is to detassel or entirely chop down the unwanted individuals once per day and to make a second check immediately before starting the tagging crosses. If unwanted individuals are spotted during the tagging crosses, stop and eliminate them rather than continuing to perform the tagging crosses.
3. Make careful pollinations to minimize contamination "as if there was contaminating pollen" in the air. Ears must be shoot bagged prior to silk emergence; possibly contaminated ears with a few emergent silks are best discarded. Do not pollinate an ear that has silks growing outside the shoot bag.

4. Keep careful records of the progress of the tagging crosses. Because spontaneous silencing will occur in virtually every *Mutator* line, it is important to spread the risk by using multiple individuals and multiple different *Mutator* families as pollen sources onto each target population. Generating a few thousand seed from a *Mutator* plant is easy from a single pollen collection, i.e., 10 ears of 250 kernels each = 2,500 progeny. To be reasonably sure of obtaining multiple alleles, a population of 20,000 individuals is a good goal; to ensure that you have this many, generate a population of 40,000 kernels so that only the post-pollination certified to be active *Mutator* progeny enter the screen.
5. Use marked bags to facilitate recovery of multiple ears with the same Mutator parent used on the same day. Use color stripe bags and further mark them with spray paint or written symbols on the bag, or attach a small piece of colored tape or paper; the first two bags in each run of pollinations from the same source are labeled with the date and the *Mutator* plant identification, but all subsequent bags have only the distinctive marking. Although the pollinated ears are not contiguous in the field, well marked bags are easy to spot and the final bag should be labeled “last” to make the harvesting process more efficient as the number of ears could range from just a few to ten or more.
6. Cut back ears and number the plants within a target population. Plant six or more target populations and in addition to stakes use plant tags to mark the borders between targets. The shoot bags within each target population should be labeled 1 to *N* during the daily step of cutting back ears for the next day’s pollinations. This practice makes it easier during pollination to know that you have found all of the cut back ears within each target. The target population for a specific gene could include several different families, but we do not keep track of the families if they share the same reference allele.
7. Bag the *Mutator* source plants the previous day or early in the morning and pollinate after maximum pollen shed. Because the goal is to generate a large progeny from each *Mutator* pollen collection, it is best practice to collect pollen from plants on their second and third day of pollen shed and to do so late in the morning when a large number of anthers have exerted. To facilitate pollination, the collected pollen can be transferred to a small glassine bag for ease of dispersal onto the silks.

3.4 Testcrosses with the Mutator Tagging Plants

1. Cross each *Mutator* source plant to an appropriate tester for the visible reporter allele using that day’s pollen collection. The purpose of this cross is to assess whether *Mutator* activity was transmitted through the pollen as expected; if somatic mutability is lost, there are unlikely to be any new mutations.

Table 1
Evaluating *bz2//bz2-mu* individuals for retention of mutator activity using *bz2* testcrosses

Individual	% Spotted kernels		Conclusion
	Mu as ear parent	*Mu* as pollen parent	
1	55	45	Active
2	41	No data	Risky to use
3	36	46	Probably OK as tassel is above 40 %
4	56	56	Active
5	49	45	Active
6	55	33	Going off in the tassel; discard
7	49	38	Going off in the tassel; discard
8	55	38	Going off in the tassel; discard
9	51	38	Going off in the tassel; discard
10	No data	47	Active
11	42	44	Active

Only a subset of family ZH555 is shown

The *Mutator* plants should each be crossed by tester pollen as a second test of the maintenance of activity and as a way of generating new seeds for future *Mutator* populations (*see* Table 1). In assessing 100 kernels for mutability in a population with a 1:1 high frequency spotted:colorless expectation, ears that show 40:60–60:40 segregation are in the 95 % probability of actually being 1:1. The important group to identify is *Mutator* individuals that show 39 % or fewer spotted kernels as pollen parent; *Mutator* activity is lost progressively during plant development, hence the tassel is more likely to have a silenced transposon family than the ear. In some cases, a *Mutator* plant will have been used as a pollen source on several days, and the transmission of *Mutator* activity must be checked on each day.

2. Discard all tagging progeny from silencing *Mutator* plants. Before screening for new mutants, it is a good idea to discard all of the inactive materials; we discard based on failing the 40:60 transmission ratio expectation (**step 1** above) for either test cross. Even for seed traits, which are readily observed, the forward mutation frequency is very low or zero, from silencing plants. For seedling or whole plant evaluation, the work involved in conducting the screen should be applied to the best progeny only. If progeny seeds are in short supply and the

Table 2
Timing of *Mu* insertions based on directed tagging experiments

		Number of individuals carrying new allele				
Target *ms* loci	**Average frequency**	**Unique**	**2**	**3**	**4**	**≥5 [max]**
13 loci	1/3,000	22	7	6	3	11 [32]

pollen test cross is acceptable but the ear fails, the progeny might be worth evaluating.

3. All the progeny of one *Mutator* plant on 1 day are treated as a single population and should be harvested together and then screened together. Do not pool seed from different *Mutator* plants, because multiple recovery of the same mutant is quite common in *MuDR/Mu* tagging (*see* Subheading 3.5). In practice, ears from the same pollen source and date are harvested into a single bag and treated as a single progeny. Screening for ear traits is quick (i.e., to identify colorless kernels with excision sectors on an otherwise purple seeded ear) (*see* **Note 6**).

3.5 Evaluating New Mutants

1. Evaluate yield of new mutants. Using the rigorous criteria for selecting appropriate *Mutator* plants as pollen donors, for careful target field management, and for evaluating progeny for activity before screening, we have had great success tagging targets with very few contaminants identified among new mutant plants. For 6 years, we have performed directed *Mu* tagging of male-sterile (*ms*) mutants. Overall, the frequency is approximately 1/3,000, with the highest frequency being 9.3×10^{-4} (13 mutants in 14,000 *tcl1* plants) and the lowest 10^{-4}. As shown in Table 2, 40 % (20/49) of newly tagged alleles were recovered in three or more progeny and therefore must be premeiotic insertions (*see* Table 2). Because pollen is typically in three- to tenfold excess, the true frequency of large groups with the same mutation must be higher. We conclude that the majority of heritable events occur during archesporial cell proliferation before meiosis. Confirmatory evidence for this premeiotic timing comes from high throughput sequencing of *RescueMu* insertions: >40 % were in multiple progeny [10]. The insertion mode persists through meiosis and in haploid gametophytes, because *MuDR/Mu* can transpose replicatively as late as the final mitotic division in haploid gametophytes [7]—the result is sperm cells that differ in transmission of mutations, recognized as non-concordance of the embryo and endosperm after double fertilization.

2. Putons (putatively *Mu*-tagged cases) are recognized in the progeny ear, sand bench seedlings or field whole plant screen as mutant individuals (*see* **Note 7**). To certify that the puton is not from contaminating pollen of the reference allele (or other source), a PCR test to detect the mutable reporter allele (*bz2-mu* in our case) is conducted; this mutable allele is transmitted to half the progeny. Twins (or more than two putons within a single family) should all carry the same mutant allele, and half of the group should be carriers for the reporter allele (*see* **Note 8**). For those progeny that lack the reporter allele, a qRT-PCR assay for *mudrA* transcripts should be performed. Only *Mutator* active individuals have a high level of transcript. Putons that pass one or both of these tests are considered to be *Mu*-tagged (*see* **Note 9**). The newly tagged individuals should be grown and crossed to perform an additional allelism test and crossed to at least one inbred line to start an introgression into a specific background and to generate progeny suitable for fine mapping.

4 Notes

1. Organize the field for ease of screening. It can be daunting to evaluate large numbers of plants for a mutant phenotype. Fields may range from 25,000 to 100,000 plants on 1.5–4.0 acres. If space is not limiting, start planting each family at the same edge of the field and leave space unused at the end of each family. Consider that you will be walking through the entire field daily, and layout a design of rows of corn and aisles that permits easy observations of all plants. If there is drip irrigation present, either bury the lines or place the lines between pairs of rows with corn with no drip line in the aisles. Each screening family consists of the pool of all progeny for your target gene from one *Mutator* plant pollen collection on a specific day. Families of a few hundred to several thousand kernels are typical. If space is tight, plant testers (pollen and ear sources for outcrossing with the identified mutants) into the unused field rows after all the tagging families are planted.
2. Make a field plan on paper, put the start of family stakes in first, then plant. Pre-label all of the stakes identifying families, including stakes labeled END with the family number to place after the last plant. Write using an indelible marker in large letters on both sides of the stake. We spray paint the stakes in light colors, a different color for each target gene to aid orientation in the field later. For families exceeding 1,000 individuals, additional stakes labeled CONTINUE with the family number are invaluable, and these continue stakes can be placed at the end of rows containing this family. If space permits, start each

family in a new row, with the unplanted area at the end the preceding row being a reminder that a family has finished. We are usually very pressed for space, and must plant all of the land. Therefore, the "start stake" for a family will be within a row, and the continue stakes at the ends of rows are critical for orientation. Both machine planting and hand planting (individual seed as well as push planters) may be used. Machine planting is quicker during the actual planting, but requires more staff (i.e., tractor driver, two people on the planter to keep seed stocks in order and to conduct the planting and someone on the ground to put in the END stakes). In our experience, more mistakes are made in machine than hand planting when people outside the project must be employed to do the machine planting. In hand planting one person familiar with the entire project can manage all of the tasks easily. With a push planter 30,000 seed per day is a reasonable goal. After planting, if rows are long (more than 1,000 plants), pick a common point mid-field and mark each planted row with a labeled stake.

3. Unusual distribution of mutants in the field. Strangely, many of the male-sterile mutants we have recovered after machine planting were within 10 % of the edge of the field, planted preferentially at the start of rows after the planter turned at the edge of the field. Such clustering can be explained by seed sorting within the planter. Clustering of mutants could occur if mutant seeds are a different shape or weight than the bulk of the population. Large items sort to the top of the pile during agitation if their density is similar to the smaller items, a phenomenon called the Brazil nut effect: these large nuts are at the top of the can, as are the intact crackers and cereal flakes in their boxes and large stones in a wheelbarrow filled with gravel. During hand planting, the lightest seeds are planted first; this causes a bias for male-sterile individuals to be clustered at the front of each family for many of the male-sterile lines. The bias is significant: in families segregating 1:1 for a number of male-sterility factors, two-thirds to three quarters of the sterile individuals are in the first half of the family.

4. Field maintenance. Because nearly all of the plants in the field will be discarded, the field should get minimal maintenance. Weed control must be sufficient to permit walking in the aisles safely. Because of high predation from birds, we must cover the entire field with bird netting, and we do hand weeding when the bird netting is removed. Low fertilizer and just sufficient water should be used during the initial growth phase. As there is no rain in the summer time at Stanford University, all of the weeds are near the drip lines and can be removed from damp soil with a hoe or by pulling. When putons are identified, we

give them and ~20 neighboring plants extra fertilizer and water to insure good seed set. Water is very expensive in California. Currently we refrain from irrigating the entire field, and instead use a large water barrel in the back of a truck plus a siphon and bucket to hand water the puton areas weekly for a month.

5. Minimize plant evaluation. For male-sterility screens, remove tillers when plants are young and focus on the main tassel on each plant. Fertile plants with exerted anthers are cut off at a comfortable level for each person (about ear height), and that plant is thus never evaluated again. For other phenotypes removing unwanted individuals might start earlier in the growth cycle. We put all of the plant debris into the furrows with drip irrigation to keep the aisles clear and to provide a mulch layer that helps reduce late season weeds. A 100,000 plant male-sterility screen will require about 25 person hours per day at the peak to check the entire field daily. Training sufficient helpers for these peak periods is essential. Otherwise the planting should be split into weekly groups if only one or two people are available to conduct the screen.
6. Conducting crosses with possibly contaminated ears. Open pollination often occurs in a male-sterility screen because the male-sterile individuals are identified by the absence of exerted anthers in a family in which most plants have been fertile and thus removed. Typically, the male-sterile individuals are found when 80 % or more of the family has been removed, and most, if not all, remaining plants have ears with some silks emerged. Knowledge of the pattern of silk outgrowth is helpful in recovering controlled pollination kernels from a partially open-pollinated ear. The first silks to emerge are from the center of the ear, then the base, then the tip. Silks attached to kernels that have been pollinated cease to grow. Thus an open-pollinated ear can be cutback, shoot bagged, and silks emerging the next day represent non-pollinated kernels, preferentially from the base + tip (if there are hundreds of silks), or tip (if there are just dozens of silks) of the ear.
7. Processing the putons. We sequentially number putons (*ms8-mu1, ms8-mu2*, etc.) We spray paint the putons with fluorescent paint along with about ten neighboring siblings to facilitate finding these plants easily; we also put bright tape on the puton tassel and attach a sign with the date, puton type (*ms8-mu1*) and the screening family name. The tassel is photographed with this sign for the permanent record. The rest of the family is searched thoroughly to determine if there are twins (additional individuals with the same phenotype), and these are then labeled sequentially as *ms8-mu1* twin 1, *ms8-mu1* twin 2, etc. We collect three leaf samples from each puton and twin and one sample from each of the ten nonmutant sib-

lings; samples are stored in a cooler or on dry ice in the field before transfer to −80 °C. Collecting into labeled shoot bags (no Dipel® treatment) or coin envelopes works well. These leaf samples are used to make DNA and RNA to test whether the puton had a *Mutator* parent. If possible, these tests should be done as quickly as possible.

8. Crosses with putons and siblings. The siblings and putons are shoot bagged with Dipel®-treated bags to suppress corn ear worms and later crossed with appropriate testers or self-pollinated if possible. We have tried crossing new male-steriles by *Mu killer* [11], but this line does not silence high copy number *Mutator* individuals in a single generation, especially through the pollen. Spontaneous silencing can be relied on to occur in sufficient progeny to acquire a silenced line. Furthermore, *Mu* elements do not excise germinally, consequently newly tagged alleles are permanently marked. Additional crosses can be done using split ear pollinations, even with partially open-pollinated ears. The silks that grow out after cutting back are separated by a card into two groups—one facing the stem and the second facing out. One group is pollinated by the male-sterile maintainer line (this is an allelism test and will generate a 1:1 sterile:fertile progeny with typically yellow kernels), and the second group by an inbred line with an anthocyanin color marker (*bz2* in our case, generating a progeny that is 1:1 purple:bronze + spotted from a puton that is *Bz2//bz2* or *Bz2//bz2-mu* from the tagging cross).

9. Use a *Mu* tag to clone a gene. Details of cloning strategies are beyond the scope of this chapter except for these brief comments. The nonmutant neighbors of the puton are crossed by a tester or inbred line as are the puton and its twins. The original siblings and their progeny will share segregating *Mu* element locations with the puton and its progeny, but none of the field siblings should have the *Mu* element insertion at the target gene; this insertion location should also be missing from the *Mutator* parent. *Mu* insertion locations can be assessed by DNA blot hybridization, inverse PCR and sequencing. As sequencing costs decline, high throughput sequencing may be employed to compare puton and sibling noncarriers to find the *Mu* element that is associated with the new mutation. These assessments are even more powerful if there are twins or higher multiple mutants within a family, as the individual mutant plants will carry different new *Mu* element insertions and the segregating parental elements, but all will share one *Mu* element responsible for the mutation.

Acknowledgments

Research support in the Walbot laboratory is provided from a grant from the National Science Foundation (PGRP 07–01880). J.Q. was supported by a Fulbright award for a 3 month visit to Stanford University during her Ph.D. program.

References

1. Walbot V, Rudenko GN (2002) MuDR/Mu transposons of maize. In: Craig NL, Craigie R, Gellert M, Lambowitz A (eds) Mobile DNA II. American Society of Microbiology, Washington, DC, pp 533–564
2. Athma P, Grotewold E, Peterson T (1992) Insertional mutagenesis of the maize P gene by intragenic transposition of *Ac*. Genetics 131:199–209
3. Lisch D (2002) *Mutator* transposons. Trends Plant Sci 7:498–504
4. Lisch D, Jiang N (2009) Mutator and MULE transposons. In: Bennetzen JL, Hake S (eds) Handbook of maize genetics and genomics. Springer, New York, pp 277–306
5. Walbot V (1986) Inheritance of mutator activity in *Zea mays* as assayed by somatic instability of the *bz2-mu1* allele. Genetics 114:1293–1312
6. McLaughlin M, Walbot V (1987) Cloning of a mutable *bz2* allele of maize by transposon tagging and differential hybridization. Genetics 117:771–776
7. Robertson DS (1986) Genetic studies on the loss of *Mu* mutator activity in maize. Genetics 113:765–773
8. Levy AA, Walbot V (1990) Regulation of the timing of transposable element excision during maize development. Science 248:1534–1537
9. Rudenko GN, Walbot V (2001) Expression and post-transcriptional regulation of maize transposable element *MuDR* and its derivatives. Plant Cell 13:553–570
10. Fernandes J et al (2004) Genome-wide mutagenesis of *Zea mays* L. using RescueMu transposons. Genome Biol 5:82
11. Slotkin RK, Freeling M, Lisch D (2003) Mu killer causes the heritable inactivation of the mutator family of transposable elements in *Zea mays*. Genetics 165:781–797

Chapter 11

Genetic and Molecular Analyses of UniformMu Transposon Insertion Lines

Donald R. McCarty, Masaharu Suzuki, Charles Hunter, Joseph Collins, Wayne T. Avigne, and Karen E. Koch

Abstract

The UniformMu transposon population is a large public resource for reverse genetics and functional genomics of maize. Users access the collection of UniformMu genetic stocks that are freely distributed by the Maize Cooperation Stock Center using online tools maintained at MaizeGDB.org. Genetic and molecular analyses of UniformMu stocks (UFMu insertion lines) typically require development of genotyping assays that use a gene-specific polymerase chain reaction (PCR) to follow segregation of transposon insertions in genes of interest. Here we describe methods for accessing the resource and recommended protocols for genotyping of transposon insertion alleles.

Key words *Robertsons mutator*, Transposon mutagenesis, Transposon genotype, Reverse genetics

1 Introduction

The UniformMu transposon population was developed as a functional genomics resource for the maize genetics community. The highly mutagenic, native *Robertson's Mutator* transposon system was harnessed to perform systematic insertional mutagenesis of the maize genome [1, 2]. Users can access the collection of sequence-indexed insertions using genome browser and search tools at MaizeGDB.org and PopCorn.org portals. The online tools enable users to request UniformMu seed stocks carrying germinal insertions directly from the Maize Cooperation Stock Center (maizecoop.cropsci.uiuc.edu). These seed stocks are distributed without charge. Genotyping of the *Mu* insertion alleles is typically performed by polymerase chain reaction (PCR) using a combination of *Mu*-specific and gene-specific primers. Due to the complexity of the maize genome, developing a successful PCR genotyping assay

Thomas Peterson (ed.), *Plant Transposable Elements: Methods and Protocols*, Methods in Molecular Biology, vol. 1057, DOI 10.1007/978-1-62703-568-2_11, © Springer Science+Business Media New York 2013

for transposon insertions in specific genes is sometimes challenging. This chapter outlines recommended methods and practices for genotyping and analysis of UniformMu insertion lines.

2 Materials

2.1 UniformMu (UFMu) Stocks

The UniformMu population was developed by backcross introgression of the active, autonomous, *Mu*DR transposon into a color-converted, W22 inbred [1]. All stocks in the public collection are bronze-colored and homozygous for the *bronze-1 mu-mutable-9* (*bz1-mum9)* mutation [3] used as a genetic marker for the presence of *MuDR*, the autonomous transposable element of the *Robertson's Mutator* system [1, 4]. The *bz1-mum9* allele contains a nonautonomous *Mu1* transposon insertion that disrupts the *Bz1* gene. *Bz1* encodes a UDP-glucose flavanol glucosyl transferase that catalyzes a key step in biosynthesis of purple anthocyanin pigment in the seed aleurone [5]. In the absence of *MuDR*, the *bz1-mum9* allele has a uniformly bronze-colored aleurone. When active *MuDR* is present in the genome, transposition of the *Mu1* element in *bz1-mum9* is induced in somatic tissues of the endosperm resulting in a spotted aleurone phenotype [3] (*see* Fig. 1). The spots are due to small, typically single-cell, revertant sectors that produce purple anthocyanin.

Viewed under a dissecting microscope, the pigmented sectors have non-cell autonomous halos due to diffusion of anthocyanin to adjacent aleurone cells. In UniformMu lines, genetic control of *Mutator* activity is achieved via Mendelian segregation of *MuDR* [1].

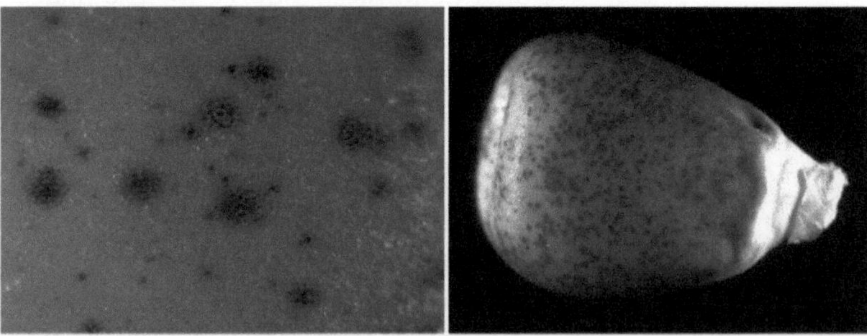

Fig. 1 Revertant purple sectors appear in the bronze-colored aleurone of kernels when an active MuDR transposon is present in the genome of UniformMu bz1-mum9 lines. The bronze color results from insertion of a nonautonomous Mu1 element in a key gene for anthocyanin biosynthesis (*see* text), and purple sectors appear when this gene is restored to normal function by excision of the Mu1 transposon; an event that indicates an active MuDR transposase in the line. The revertant sectors are typically small, often single-celled, and tend to show halos due to diffusion of the essential anthocyanin precursors from the now-normally functioning cells

Table 1
UniformMu genetic stocks

Stock	Genotype	Stock center catalog number
UFMuXXXXX	*bz1-mum9*, +Mu insertions	UFMu00001–UFMu08256
UF-W22 (ACR)	*A1, A2, C1, C2, Bz1, Bz2, R1-r*	X17EA
UFMu-MuDR	*bz1-mum9, MuDR(+)*	919JA

All UFMu insertion lines in the public resource are screened for loss of *Mutator* activity (absence of *MuDR*) using the *bz1-mum9* marker prior to sequence analysis in order to minimize occurrence of non-heritable, somatic insertions in DNA samples used for sequence-indexing and mapping of germinal insertions. Hence, UFMu seed stocks deposited (*see* Table 1) in the Maize Genetics Cooperation Stock Center have been carefully screened for absence of spotting as well as seed quality to ensure the insertion lines are genetically stable and no longer *Mu*-active. UFMu stocks are non-transgenic and require no special permits for propagation and distribution.

2.2 UF-W22 (ACR) Wild Type

The color-converted, W22 inbred strain, developed originally by R. A. Brink at the University of Wisconsin, carries dominant alleles for all anthocyanin pathway genes (*A1, A2, C1, C2, Bz1, Bz2, R1-r*) required for purple pigmentation of the seed aleurone [5]. Because this inbred strain has been maintained separately in several laboratories over the years, it is possible that lab-specific genetic differences have accumulated. For this reason, we recommend obtaining the University of Florida W22 strain used as the recurrent parent for UniformMu from the Maize Genetics Cooperation Stock Center (catalog # X17EA).

2.3 UF MuDR, bz1-mum9 Stock

This stock provides a source of active *MuDR* in the UniformMu background that can be used to reactivate transposition of *Mu-inactive* UFMu stocks. The seed aleurone phenotype of this stock is dense spotting on a bronze background. Reactivation of *Mu* is achieved by crossing the UFMu stock of interest to the *MuDR* stock and establishing a line that is homozygous for the insertion of interest and also carries an active *MuDR*. Reintroduction of *MuDR* may be useful for two reasons. (a) A significant fraction of *Mu*-induced mutations, so-called *Mu*-suppressible mutations, have mutant phenotypes only when *MuDR* is also present in the genome [1, 6, 7]. Typically *Mu*-suppressible mutations occur near the 5′-end of genes enabling transcription initiation within the *Mu-TIR*, whereas interaction of *MuDR* with the TIR can prevent transcription initiation [6]. In such cases, reintroduction of *MuDR* may result in a loss-of-function phenotype. (b) Transposon reactivation can be used to generate derivative deletion alleles from a *Mu* insertion allele [8]. Deletion derivatives may be especially useful in cases where the original insertion mutation does not cause a loss of gene expression. *Mu* insertions frequently occur near the 5'-end of genes in locations that may not strongly impact gene expression [7]. Although *Mu* insertion alleles rarely give rise to wild type germinal revertant progeny, deletions of genomic sequence adjacent to the insertion can be recovered at frequencies greater than 1 event per 1,000 gametes [8]. Such deletions will frequently cause complete loss of gene function. Das and Martienssen [8] describe an efficient PCR strategy for isolating deletion derivatives of a Mu insertion allele.

3 Methods

3.1 Searching the UniformMu Collection

Online tools for accessing the UniformMu collection are provided at MaizeGDB.org [9]. UniformMu insertions that have been mapped to the maize genome are displayed as a track in the MaizeGDB genome browser. Hence, to search for insertions in or near a specific gene of interest, a user can enter either a locus name or gene model identifier into the search window at the MaizeGDB website and open the genome browser view for that gene. Alternatively, UniformMu insertion sites may be searched by BLASTN using the PopCorn search tool [9].

The UniformMu data deposited in MaizeGDB conform to the locus/variation/stock schema of the MaizeGDB database; where each *Mu* insertion site in the genome is assigned a unique locus identifier (e.g., mu1016504). The locus page includes links to the genome browser view for that locus as well as a link to a variation page for that locus. In the MaizeGDB schema, each locus has at least one variation (in this example, mu1016504::*Mu*). The variation page displays a concise summary of information about the insertion locus including its physical map location in the genome and a list of available UFMu stocks that contain the insertion. Selecting a stock link will open the stock page. The stock page includes a list of all insertions that have been identified in the UFMu stock providing information about other mutations that may be segregating in the line. The side panel of the stock page includes a link to the Maize Genetics Cooperation Stock Center online seed request form. Seeds are distributed to the user at no cost, and by tradition, all Stock Center requests are confidential. If multiple stocks are listed on the variation page for an insertion, we recommend requesting at least two of the listed stocks.

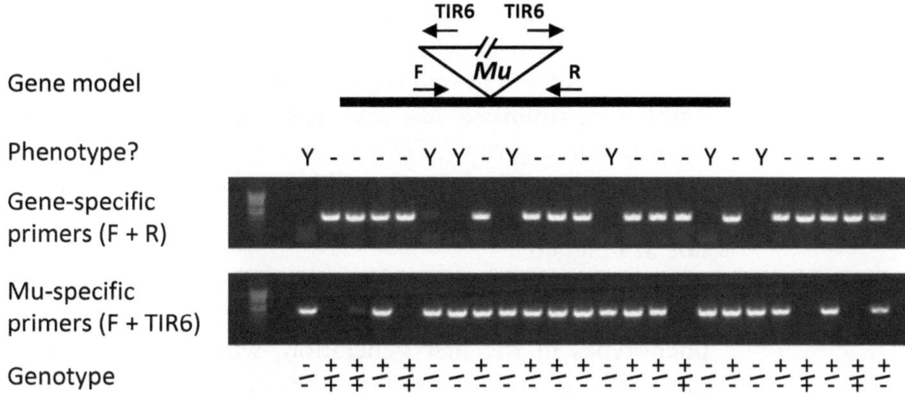

Fig. 2 Example showing co-segregation of a Mu insertion with a recessive phenotype. The PCR-based genotyping of a segregating F2 family carrying a Mu insertion in a gene of interest allows identification of individuals homozygous for the insertion. Note that normal Mendelian segregation is unlikely in first generation UniformMu lines because of the sib-mating strategy used to generate seed stocks (*see* text). To obtain a segregating family, backcrossing and selfing will often be required. Wild type allele present (+); Mu insertion allele present (–). Data are from a Mu insertion (mu1061500) in Cellulose Synthase-Like D1 (GRMZM2G015886), which co-segregates with a narrow-leaf warty phenotype [11]

3.2 Planting and Pollination of UFMu Stocks

At least 15 seeds will typically be provided for each request of a UFMu line from the Maize Genetics Cooperation Stock Center. Because the seed supply is limited, it is important to plan carefully before planting to ensure recovery of seeds carrying the insertion of interest from as many plants as possible.

The seeds provided are pooled from ears of between 2 and 5, sib-pollinated, F3 plants. The F3 lines are sib-pollinated to minimize selection against insertions that may have deleterious phenotypes (e.g., lethals and male steriles) thereby maximizing recovery of insertions. Many interesting, but deleterious mutations are thus preserved as heterozygous insertions, whereas roughly half of the *Mu* insertions in UFMu lines are homozygous and will be present in all plants. The sib-pollination strategy also enables long-term sustainability and maintenance of heterozygous mutations. For example, pooled seed of three or more sib-pollinated ears (six parent plants) will preserve heterozygous insertions with less than 2 % loss per generation. Even if the *Mu* insertion of interest is deleterious, it will likely still be carried by at least some heterozygous F3 seed in the sample provided.

A third advantage to maintaining heterozygosity where possible is that, in general, the most reliable approach for establishing linkage of an insertion to a phenotype is to compare homozygous mutant and nonmutant siblings within a segregating family (*see* Fig. 2). In cases where the insertion of interest is already homozygous in the UFMu line, the temptation to simply compare phenotypes of homozygous UFMu insertion line to the standard W22 wild type should be resisted for two reasons: (a) Although all lines share the same uniform inbred background, UFMu lines carry a variable number of unlinked *Mu* insertions. A list of mapped insertions identified in a given UFMu stock can be assessed by viewing the corresponding Stock page on MaizeGDB.org. (b) In addition to the known insertions, UFMu lines may carry mutations that are not *Mu* tagged.

For these reasons, UFMu lines that are segregating for the insertion of interest have the advantage of enabling evaluation of phenotypes in the first generation, whereas homozygous stocks must first be out-crossed to the standard wild type and self-pollinated in order to generate segregating materials. However, the heterozygosity/homozygosity status of UFMu insertions must be determined experimentally for each insertion. We recommend planting at least eight seeds in the field or greenhouse such that all plants can be sampled for DNA and pollinated either by selfing or outcrossing, preferably both.

We also recommend concurrently planting UF-W22 wild type seeds (available from the Stock Center—catalog # X17EA), so that plants are available as maternal parents or pollen donors that can be crossed to mutant plants. Additional W22 can also be planted approximately 10 days before and after the mutant seed to increase potential for silk and pollen availability at the time of mutant anthesis. This is important for three reasons. First, although UniformMu lines generally have good fertility, some may carry mutations that reduce vigor and fecundity (e.g., male sterility). Having wild type pollen and silks available helps ensure recovery of seed from the selected UFMu line. Second, backcrosses to wild type are often a necessary step towards developing a family of segregating progeny, where wild type and mutant siblings can be tested for co-segregation of any phenotypic features with presence of the *Mu*-insert of interest. This is especially important if the insertion is homozygous in the UFMu line, in which case generation of a suitable segregating family is necessary. Third, because UFMu lines typically carry 5–10 novel insertions (sometimes more), a backcross will aid separation and independent assortment of these in subsequent generations. An obvious phenotype observed in the first generation may or may not be due to the *Mu* insertion of interest.

UFMu stocks typically flower uniformly, with nearly concurrent silking and anthesis among individuals of a given line and also with those of the W22 wild type. However, in extreme hot, dry field environments, many homozygous *bz1* mutant stocks including UFMu lines are prone to poor pollen shed. We find that this problem can be ameliorated by regular application of over-head irrigation prior to- and during tassel emergence. In our experience, flooding or drip irrigation is not effective in improving pollen shed.

Detailed methods for hand pollination of maize plants are described by M. G. Neuffer in the Maize Handbook [10].

Table 2
Mu-TIR-specific primers

Primer	Sequence	Tm (°C)	Mu classes
TIR6	AGAGAAGCCAACGCCAWCGCCTCYATTTCGTC	71.7	All
TIR8.1	CGCCTCCATTTCGTCGAATCCCCTS	69.5	Mu7, Mu5
TIR8.2	CGCCTCCATTTCGTCGAATCCSCTT	67.9	Mu3, Mu2, Mul, Mu4
TIR8.3	SGCCTCCATTTCGTCGAATCCCCKT	68.7	MuA2, MuDR, Mu8
TIR8.2	CGCCTCCATTTCGTCGAATCACCTC	67.9	Mu7
TIR5	GCTCTTCKTCYATAATGRCAATT	58.3	All

3.3 PCR Confirmation and Genotyping of Mu Insertions

1. To confirm presence of the *Mu* insertions in a gene of interest, genomic DNA is extracted from individual plants and analyzed by PCR amplification of flanking sequences using a combination of *Mu*-TIR (*Mu*-Terminal Inverted Repeat) and gene-specific primers. An example of the primer-design strategy, the PCR results, and analysis of co-segregation for genotypes and phenotypes is shown in Fig. 2 for a *Mu*-induced mutation in the Cellulose Synthase-Like D1 gene (Csld1) [11]. Published mini-prep methods yield DNA from leaf tissue suitable for genotyping [12].
2. At least two gene-specific primers will need to be designed for each *Mu* insertion (one upstream and one downstream of the insertion site as shown in Fig. 2), and these will need to be compatible with a *Mu*-TIR-specific primer. *See* Table 2 for a list of *Mu*-TIR-specific primers [13]. We recommend TIR6 or TIR8 for routine confirmation and genotyping. Note that TIR8 is a mixture of four primers that are nested with respect to TIR6. Nesting enables their use in two stages for enhanced specificity. TIR8 amplicons contain sufficient TIR sequence downstream of the primer to authenticate the *Mu* insertion by DNA sequencing. For sequence validation of insertion sites, we typically do an initial PCR reaction with TIR6. Product is gel purified (*see* Fig. 2) and sequenced using the nested TIR8 primer and/or gene-specific primer. In our experience, the relative effectiveness of TIR6 and TIR8 vary widely depending on the *Mu* insertion and context, hence it is beneficial to use both primers.
3. Online tools such as Primer3 [14] may aid design of gene-specific primers. To expedite identification of suitable PCR primers, we typically design three pairs of 21–27 bp gene-specific primers that anneal to genomic sequence upstream and downstream of the insertion site. Ideally, all primers are compatible

Table 3
Two-step PCR protocol

1.	94 °C	1 min
2.	94 °C	25 s
3.	62 °C	30 s
4.	72 °C	1 min (depending on product size)
5.	8–10 cycles	Steps 2–4
6.	94 °C	25 s
7.	56 °C	30 s
8.	72 °C	1 min (depending on product size)
9.	27 cycles	Steps 6–8
10.	72 °C	5 min
11.	4 °C	Hold

so that they can be tested in alternate combinations (nine primer pairs). The upstream and downstream gene-specific primers should be separated by less than 1,000 bp in the wild type sequence to maximize amplification efficiency. BLASTN (blast.ncbi.nih.gov) searches of the maize genome should be used to avoid primer sequences that are present in multiple copies. Because there is currently limited genome sequence available for the W22 inbred background of UniformMu, we typically rely on the B73 reference genome [15] for primer design. Sequence polymorphisms in the target region can affect primer efficiency and predicted product sizes. Where possible, targeting exon sequences can reduce likelihood of sequence polymorphisms that will compromise optimal primer annealing. Table 3 contains our standard two-step, touchdown PCR protocol. We have found that including 5 % DMSO in the reaction or addition of commercial PCR "adjuvant" cocktails (e.g., PCRx Enhancer System, Invitrogen.com) can be effective for improving PCR success with GC-rich maize genomic sequences. We add fresh DMSO from aliquots stored at −20 °C.

4. Gene-specific primers that flank the insertion site are first PCR-tested in pairs with wild type W22 DNA (UF-W22 stock). This confirms capacity of the selected primers to amplify the expected wild type fragment.
5. The advantage of designing and ordering multiple pairs of gene-specific primers is that individual upstream primers can

be tested with alternative downstream primers to optimize amplification of the wild type sequence. All possible forward–reverse combinations can be tested with W22 DNA. W22-B73 sequence polymorphisms and multiple copy sequences are common causes of primer failure. The former are more likely to be encountered where one or both primers are located in the promoter or 5′-UTR of the gene. While designing primers in those regions is often unavoidable due to a strong 5′ bias for *Mu* insertions, testing multiple primer combinations will usually produce a successful pair.

6. Once suitable primer pairs have been identified that will amplify the region of interest in W22 control DNA, both forward and reverse primers are tested in combination with the TIR6 primer (*see* Table 2) using template DNA extracted from multiple individual UFMu plants. It is not uncommon that only one of the primers will produce a product with TIR6. A probable explanation for this is that *Mu* elements typically have non-identical TIR sequences that may differ in TIR6 priming efficiency. If no products are obtained from any plants using TIR6, the PCR analysis should be repeated using TIR8 and/or TIR5. If products of the expected size are obtained, the location of *Mu* insertion may be confirmed by direct sequencing of the PCR products. Parallel PCR reactions performed with the pair of upstream and downstream flanking gene-specific primers can be used to test for the presence of the wild type allele and heterozygosity. However, amplification of a wild-type-sized fragment may sometimes occur even in situations where an insertion is homozygous due to the presence of closely related paralogs in the maize genome.

7. Because duplicate genes are relatively common in the maize genome, it is sometimes necessary or expedient to use primers that are not strictly gene specific. In that case, care must be taken to ensure that plants which are homozygous for the insertion are not incorrectly genotyped as heterozygous because the flanking primer pairs generate PCR products of the expected size for the wild type allele. There are two independent approaches to resolve and confirm such insertions. First, the PCR products can be directly sequenced with the gene-specific primers that were used to amplify or with additional internal sequencing primers. If primers amplify one or more additional loci, polymorphisms may be detected within the PCR products that can be used to distinguish closely related paralogs. Second, a genetic test can be performed by backcrossing the lines to the W22 inbred to produce multiple F1 ears. If parental plants are homozygous for the *Mu* insertions, all the F1 plants will typically carry the insertion.

Acknowledgments

This work was supported by grants from the National Science Foundation to D.R. McCarty and K.E. Koch (IOS0703273 and IOS1116561).

References

1. McCarty DR et al (2005) Steady-state transposon mutagenesis in inbred maize. Plant J 44:52–61
2. Settles AM et al (2007) Sequence-indexed mutations in maize using the UniformMu transposon-tagging population. BMC Genomics 8:116
3. Brown JWS, Sundaresan V (1992) Genetic study of the loss and restoration of Mutator transposon activity in maize – evidence against dominant-negative regulator associated with loss of activity. Genetics 130:889–898
4. Lisch D, Chomet P, Freeling M (1995) Genetic characterization of the mutator system in maize: behavior and regulation of Mu transposons in a minimal line. Genetics 139:1777–1796
5. Dooner HK, Nelson OE (1979) Interaction among *C*, *R* and *Vp* in the control of the *Bz* glucosyltransferase during endosperm development in maize. Genetics 91:309–315
6. Barkan A, Martienssen RA (1991) Inactivation of maize transposon-Mu suppresses a mutant phenotype by activating an outward-reading promoter near the end of Mu1. Proc Natl Acad Sci USA 88:3502–3506
7. May BP et al (2003) Maize-targeted mutagenesis: a knockout resource for maize. Proc Natl Acad Sci USA 100:11541–11546
8. Das L, Martienssen RA (1995) Site-selected transposon mutagenesis at the *hcf106* locus in maize. Plant Cell 7:287–294
9. Cannon EK et al (2011) POPcorn: an online resource providing access to distributed and diverse maize project data. Int J Plant Genomics 2011:923035
10. Neuffer MG (1993) Growing maize for genetic studies. In: Freeling M, Walbot V (eds) The maize handbook. Springer, New York, pp 197–208
11. Hunter CT et al (2012) Cellulose Synthase-Like D1 is integral to normal cell division, expansion, and leaf development in maize. Plant Physiol 158:708–724
12. Suzuki M et al (2006) The maize *viviparous15* locus encodes the molybdopterin synthase small subunit. Plant J 45:264–274
13. Settles AM, Latshaw S, McCarty DR (2004) Molecular analysis of high-copy insertion sites in maize. Nucleic Acids Res 32:e54
14. Rozen S, Skaletsky HJ (2000) Primer3 on the WWW for general users and for biologist programmers. In: Krawetz S, Misener S (eds) Bioinformatics methods and protocols: methods in molecular biology. Humana Press, Totowa, NJ, pp 365–386
15. Schnable PS et al (2009) The B73 maize genome: complexity, diversity, and dynamics. Science 326:1112–1115

Chapter 12

Digestion–Ligation–Amplification (DLA): A Simple Genome Walking Method to Amplify Unknown Sequences Flanking *Mutator* (*Mu*) Transposons and Thereby Facilitate Gene Cloning

Sanzhen Liu, An-Ping Hsia, and Patrick S. Schnable

Abstract

Digestion–ligation–amplification (DLA), a novel PCR-based genome walking method, was developed to amplify unknown sequences flanking known sequences of interest. DLA specifically overcomes the problems associated with amplifying genomic sequences flanking high copy number transposons in large genomes. Two DLA-based strategies, *Mu*Clone and DLA-454, were developed to isolate *Mu*-tagged alleles. *Mu*Clone allows for the amplification of DNA flanking subsets of the numerous *Mu* transposons in the genome using unique three-nucleotide tags at the 3′-ends of primers, simplifying the identification of flanking sequences that co-segregate with mutant phenotypes caused by *Mu* insertions. DLA-454, which combines DLA with 454 pyrosequencing, permits the efficient amplification and sequencing of *Mu* flanking regions in a high-throughput manner.

Key words Genome walking, Ligation-mediated PCR, Next-gen sequencing, Gene cloning, Mu transposon, Mutator

1 Introduction

Insertional mutagenesis is widely used in functional genomics. The *Mutator* (*Mu*) transposons of maize have been effectively used for both forward and reverse genetics because of their high copy number and frequent transposition [1]. Both applications require a method to amplify DNA sequences that flank transposons responsible for mutant phenotypes. For example, after a mutant phenotype has been identified following a forward genetic screen in a *Mu*-tagging population, the challenge in cloning the affected

Thomas Peterson (ed.), *Plant Transposable Elements: Methods and Protocols*, Methods in Molecular Biology, vol. 1057, DOI 10.1007/978-1-62703-568-2_12, © Springer Science+Business Media New York 2013

gene is to identify the specific genic sequence that flanks the causative insertion. Alternatively, reverse genetic resources can consist of very large numbers of individuals, each of which contains multiple independent *Mu* insertions, all of which need to be amplified and sequenced.

Multiple versions of adaptor ligation-mediated PCR have been developed to amplify sequences flanking insertional mutagens [2–10]. Active *Mu* transposon lines contain 50–200 copies of *Mu* [11, 12]. Such high copy number improves efficiency of insertional mutagenesis but greatly complicates downstream analyses. To improve amplification of *Mu* flanking sequences, we developed a simple and efficient genome walking method, digestion–ligation–amplification (DLA) [13]. DLA uses a single-stranded oligonucleotide in place of the adaptor (which is by definition double stranded) used in adaptor ligation-mediated PCR methods. The single-stranded oligo anneals to the overhang created by restriction enzyme digestion. The nick between the single-stranded oligo and the digested end of the DNA molecule is repaired using ligase. After ligation, the *Mu*-specific primer is used in combination with the single-stranded oligonucleotide primer to specifically amplify *Mu* flanking sequences. Here, we provide the basic DLA protocol and two DLA-based approaches, *Mu*Clone and DLA-454 [13, 14]. *Mu*Clone is a simple approach for cloning *Mu*-tagged genes; DLA-454, combines DLA with a high-throughput next-generation sequencing technology to analyze *Mu* flanking sequences. For a detailed comparison of these two methods, please refer to Liu et al. [13].

2 Materials

2.1 Enzymes and Kits

1. *Nsp*I (NEB, 10,000 U/ml). For options of restriction enzyme selection, please *see* **Notes 1–3**.
2. T4 DNA ligase.
3. AmpliTaq Gold DNA Polymerase.

2.2 Oligo Sequences

1. MuTIR 32 mer: 5′ AGAGAAGCCAACGCCAWCGCC-TCYATTTCGTC.
2. Mu53s 19 mer: 5′ GCCTCYATTTCGTCGAATC.
3. NspI-5B 22 mer: 5′ CAGAACGTCACAGCATGTCATG.
4. NspI-5 20 mer: 5′ GAACGTCACAGCATGTCATG.
5. NspI-P 19 mer: 5′ AACGTCACAGCATGTCATG.

6. *Nsp*I tail primers (*N* = 32) for *Mu*Clone method:

Primer_name	Sequence (5′–3′)[a]
Nsp-16caa	GTCACAGCATGTCATGcaa
Nsp-15cac	TCACAGCATGTCATGcac
Nsp-15cag	TCACAGCATGTCATGcag
Nsp-16cat	GTCACAGCATGTCATGcat
Nsp-15cca	TCACAGCATGTCATGcca
Nsp-15ccc	TCACAGCATGTCATGccc
Nsp-15ccg	TCACAGCATGTCATGccg
Nsp-15cct	TCACAGCATGTCATGcct
Nsp-15cga	TCACAGCATGTCATGcga
Nsp-15cgc	TCACAGCATGTCATGcgc
Nsp-15cgg	TCACAGCATGTCATGcgg
Nsp-15cgt	TCACAGCATGTCATGcgt
Nsp-16cta	GTCACAGCATGTCATGcta
Nsp-15ctc	TCACAGCATGTCATGctc
Nsp-15ctg	TCACAGCATGTCATGctg
Nsp-16ctt	GTCACAGCATGTCATGctt
Nsp-16taa	GTCACAGCATGTCATGtaa
Nsp-16tac	GTCACAGCATGTCATGtac
Nsp-16tag	GTCACAGCATGTCATGtag
Nsp-16tat	GTCACAGCATGTCATGtat
Nsp-16tca	GTCACAGCATGTCATGtca
Nsp-15tcc	TCACAGCATGTCATGtcc
Nsp-15tcg	TCACAGCATGTCATGtcg
Nsp-16tct	GTCACAGCATGTCATGtct
Nsp-16tga	GTCACAGCATGTCATGtga
Nsp-15tgc	TCACAGCATGTCATGtgc
Nsp-15tgg	TCACAGCATGTCATGtgg
Nsp-16tgt	GTCACAGCATGTCATGtgt
Nsp-16tta	GTCACAGCATGTCATGtta
Nsp-16ttc	GTCACAGCATGTCATGttc
Nsp-16ttg	GTCACAGCATGTCATGttg
Nsp-16ttt	GTCACAGCATGTCATGttt

[a]The terminal bases (lower case) are the PCR selective bases

7. DLA-454 primers.

Primer	Sequence (5′–3′)
barcodeMu[a]	GCCTCCCTCGCGCCATCAGxxxxxxGCCTCYATT TCGTCGAATC
BnspI-P[b]	GCCTTGCCAGCCCGCTCAGAACGTCACAGCAT GTCATG

[a]Barcoded composite primers, containing 454 primer A (5′ GCCTCCCTCG-CGCCATCAG 3′), the barcode (xxxxxx, e.g., atgctg), and the *Mu*-specific primer (5′ GCCTCYATTTCGTCGAATC 3′)

[b]The barcoded composite primer, containing 454 primer B (5′ GCCTTGCCAGCC-CGCTCAG 3′) and the primer, NspI-P, which matches the single-stranded oligo

3 Methods

In this section, the basic DLA protocol is first described. Subsequently, the protocols for two DLA-related strategies, *Mu*Clone and DLA-454, are provided.

3.1 The Digestion–Ligation–Amplification (DLA) Method

1. Digest (300 ng) genomic DNA in the following mixture. Incubate for 1.5 h at 37 °C.

Component	Volume (μl)
H_2O	–
10× NEB buffer 2	4
100× Bovine serum albumin (BSA) (10 mg/ml)	0.4
Genomic DNA	300 ng
*Nsp*I (10,000 U/ml)	1
Total	40

CAUTION: RNase treatment to remove RNA during genomic DNA extraction is recommended. Otherwise, the apparent concentration of gDNA isolated may be overestimated.

2. After 1.5 h incubation, add the following mixture to the genomic DNA digestion reaction. Total volume should now be 60 μl. Incubate for 3 h at 16 °C.

Component	Volume (μl)
H_2O	14
oligo NspI-5B (100 μM)	2
10× NEB ligase buffer	2

(continued)

(continued)

Component	Volume (μl)
T4 DNA ligase	2
Total	20

3. Follow the protocol provided for PCR purification with the Qiaquick PCR purification kit to purify the ligated product. Dissolve purified DNA in 30 μl EB buffer. Measure the concentration of each purified PCR product using a Nanodrop instrument.
4. Conduct first PCR with the *Mu* primer and the NspI-5 primer. Add ~50 ng purified ligation product to the following PCR mixture. This PCR program consists of 94 °C for 10 min; 15 cycles of 94 °C for 30 s, 60 °C for 45 s, 72 °C for 2.5 min; and a final extension at 72 °C for 10 min in a 20 μl volume.

Component	Volume (μl)
Water	–
GeneAmp 10× PCR Buffer II	2
dNTP (2 mM)	2
$MgCl_2$ (25 mM)	1.2
MuTIR primer (5 μM)	1.2
NspI-5 primer (5 μM)	1.2
Ligation product	~50 ng
AmpliTaq Gold (5 U/μl)	0.2
Total	20

5. The first PCR product was diluted ten times and 2 μl of diluted product was used for second PCR with nested primers, a *Mu* primer (e.g., Mu53s) and NspI-P, using AmpliTaq Gold® DNA Polymerase. This PCR program consists of 94 °C for 10 min; suitable numbers of cycles (35 and 20 cycles were employed for *Mu*Clone and DLA-454, respectively) of 94 °C for 30 s, 60 °C for 45 s, 72 °C for 2.5 min; and a final extension at 72 °C for 10 min.

Component	Volume (μl)
Water	16
GeneAmp 10× PCR Buffer II	3
2 mM dNTP	4

(continued)

(continued)

Component	Volume (μl)
$MgCl_2$ (25 mM)	1.8
Mu53s (5 μM)	1.5
NspI-P (5 μM)	1.5
10× diluted first PCR product	2
AmpliTaq Gold (5 U/μl)	0.2
Total	30

3.2 MuClone Protocol for Co-segregation Analysis

DLA was adapted to facilitate co-segregation analysis that enables the cloning of *Mu*-tagged mutants. *Mu*Clone, a cost-efficient strategy, adds unique three-nucleotide tags to the 3′ ends of a common primer based on the single-stranded oligo so that subsets of high-copy *Mu* transposons can be separately amplified in a manner analogous to AFLP technology [15]. This reduces the complexity of the amplification products and facilitates the identification of the transposon that co-segregates with the mutant allele in a segregating family.

1. For the co-segregation analysis, DNA samples from multiple pairs of genetically related plants need to be compared. In each pair, one DNA sample carries the mutant allele; the other does not carry the mutant allele. Ideally, the genetic background of each pair should be highly similar as is the case for Bulk Segregant Analysis [16]. For example, DNAs collected from families carrying the mutant allele and from sibling families without the mutant allele would be suitable for the co-segregation analysis. Due to the high copy number of *Mu* elements in the genome, multiple "mutant-specific" amplicons may be identified from the initial analysis. Once the mutant-specific *Mu* insertion(s) is identified, another round of PCR screen should be conducted using a *Mu* primer and a candidate gene-specific primer targeting that *Mu* insertion with individuals in a larger population for confirmation. It is also highly recommend that the cloning of the candidate gene be confirmed with additional mutant alleles.
2. Follow Subheading 3.1 (**steps 1–4**) to obtain first PCR product for each DNA sample.
3. The first PCR product is diluted ten times and 2 μl of diluted product is used in the following second PCR with nested primers, a *Mu* primer and one of the 32 *Nsp*I tail primers, using

AmpliTaq Gold® DNA Polymerase. Thus there will be a total of 32 PCR reactions.

Component	Amount per well (μl)
Water	16
GeneAmp 10× PCR Buffer II	3
2 mM dNTP	4
$MgCl_2$ (25 mM)	1.8
MuTIR or Mu53s (5 μM)	1.5
*Nsp*I tail primer* (5 μM)	1.5
10× diluted first PCR product	2
AmpliTaq Gold (5 U/μl)	0.2
Total	30

*It is likely that these primers could be further optimized by alternative designs

This PCR program consists of 94°C for 10 min; 35 cycles of 94°C for 30 s, 60°C for 45 s, 72°C for 2.5 min; and a final extension at 72° for 10 min.

4. Because multiple species of amplicons will be produced, PCR reactions are subject to electrophoresis to resolve various amplicons. Analyze all or at least 20 μl of each PCR reaction via agarose gel (2 %) electrophoresis. Bands that are only present in mutant pool (mutant-specific) but not in control pool can be cut from the gel and separately purified with the Qiagen gel extraction kit (Qiagen, cat# 28704). Each resulting product can be sequenced directly, or sequenced following TOPO cloning per TOPO TA Cloning kit protocol (Invitrogen).

3.3 DLA-454

DLA-454 combines DLA with 454 pyrosequencing to amplify and sequence *Mu* flanking regions in a high-throughput manner. To enable subsequent 454 sequencing the *Mu*-specific primers and primers based on the single-stranded oligo are concatenated with the 454 sequencing primer to generate various composite primers. DNA barcodes [17] are inserted between the 454 sequencing primer and the *Mu*-specific primer to allow different input samples that will be pooled in the same 454 run to be distinguished after sequencing (*see* Subheading 2).

1. Collect DNA samples of interest and quantify each DNA sample.
2. Follow Subheading 3.1 (**steps 1–4**) to obtain first PCR product for each DNA sample.
3. Dilute the first PCR product ten times and use 2 μl of diluted product in the following second PCR with composite primers using AmpliTaq Gold® DNA polymerase.

Component	Amount per well (μl)
Water	16
GeneAmp 10× PCR Buffer II	3
2 mM dNTP	4
$MgCl_2$ (25 mM)	1.8
barcodeMu (5 μM)	1.5
BnspI-P (5 μM)	1.5
10× diluted first PCR product	2
AmpliTaq Gold (5 U/μl)	0.2
Total	30

This PCR program consisted of 94°C for 10 min; 20 cycles of 94°C for 30 s, 60°C for 45 s, 72°C for 2.5 min; and a final extension at 72°C for 10 min.

4. After PCR, purify the PCR product following the protocol provided for PCR purification with the Qiaquick PCR purification kit. Dissolve the purified product in 30 μl EB buffer. Measure each purified PCR product via Nanodrop.
5. Add the same amount of purified PCR product in a pool. This pool is the DNA library that includes the 454 sequencing primer for 454 sequencing. To ensure that the library was prepared correctly, the pooled DNA can be TOPO cloned per the protocol of TOPO TA Cloning kit and sequenced (~10 DNA molecules) with Sanger sequencing technology. If most of the sampled amplicons exhibit the expected sequence, the DNA library is ready for 454 pyrosequencing.

3.4 Other Adaptations

Next-generation sequencing (NGS) is a fast-evolving technology. Several other NGS platforms can also be used in conjunction with DLA. Compared with 454 pyrosequencing, Illumina sequencing promises higher amount of output data at a lower cost; Ion Torrent sequencing has a shorter running time at a relatively lower price. These two platforms can be easily integrated with DLA by modifying amplification primers. It is likely that future improvements in NGS technologies will broaden the applications of DLA and further reduce the cost of sequencing DLA products.

4 Notes

1. In the protocol described here, we use *Nsp*I to digest genomic DNA. This enzyme, which exhibits high efficiency and fidelity, has a 6-base degenerate recognition site (5′ RCATGY 3′) and

therefore generates DNA fragments with a 4-base 3′ overhang and an average size of ~500 bp. The single-stranded oligo was designed to match this 3′ overhang for the ligation. No modifications are required during oligo synthesis.

2. Other enzymes that generate 3′ overhangs could also be used, including those that generate overhangs smaller than the 4-base overhang employed here. In fact, the Guoying Wang lab at the Chinese Academy of Agricultural Science successfully adapted DLA for an enzyme, *Hha*I, that generates a 2-base 3′ overhang (Jiankun Li, Jun Zheng, and Guoying Wang, personal communication).

3. One concern with using enzymes that generate 3′ overhangs is that they could potentially lead to the generation of undesirable PCR products via the ligation of the single-stranded oligo to both ends of a genomic DNA fragment which could then be PCR amplified [13]. Although these amplified artifacts would not be sequenced in the current protocol, in an effort to eliminate them we have explored the use of enzymes that generate 5′ overhangs. A 5′ phosphorylated single-stranded oligo is used for the ligation. The 5′ digestion overhang cannot be extended, avoiding the introduction of DNA artifacts. We have tested several enzymes that generate 5′ overhangs, such as *Bfu*CI, sufficiently to demonstrate that this strategy has potential, but it still requires further testing.

References

1. Benito MI, Walbot V (1997) Characterization of the maize mutator transposable element MURA transposase as a DNA-binding protein. Mol Cell Biol 17:5165–5175
2. Devon RS, Porteous DJ, Brookes AJ (1995) Splinkerettes-improved vectorettes for greater efficiency in PCR walking. Nucleic Acids Res 23:1644–1645
3. Edwards D et al (2002) Amplification and detection of transposon insertion flanking sequences using fluorescent muAFLP. Biotechniques 32:1090–1092, 1094, 1096–1097
4. Frey M, Stettner C, Gierl A (1998) A general method for gene isolation in tagging approaches: amplification of insertion mutagenised sites (AIMS). Plant J 13:717–821
5. Hengen PN (1995) Vectorette, splinkerette and boomerang DNA amplification. Trends Biochem Sci 20:372–373
6. Jones DH, Winistorfer SC (1992) Sequence specific generation of a DNA panhandle permits PCR amplification of unknown flanking DNA. Nucleic Acids Res 20:595–600
7. Kilstrup M, Kristiansen KN (2000) Rapid genome walking: a simplified oligo-cassette mediated polymerase chain reaction using a single genome-specific primer. Nucleic Acids Res 28:E55
8. O'Malley R et al (2007) An adapter ligation-mediated PCR method for high-throughput mapping of T-DNA inserts in the Arabidopsis genome. Nat Protoc 2:2910–2917
9. Riley J et al (1990) A novel, rapid method for the isolation of terminal sequences from yeast artificial chromosome (YAC) clones. Nucleic Acids Res 18:2887–2890
10. Uren AG et al (2009) A high-throughput splinkerette-PCR method for the isolation and sequencing of retroviral insertion sites. Nat Protoc 4:789–798
11. Settles AM, Latshaw S, McCarty DR (2004) Molecular analysis of high-copy insertion sites in maize. Nucleic Acids Res 32:e54
12. Walbot V, Warren C (1988) Regulation of Mu element copy number in maize lines with an active or inactive mutator transposable element system. Mol Gen Genet 211:27–34

13. Liu S, Dietrich CR, Schnable PS (2009) DLA-based strategies for cloning insertion mutants: cloning the gl4 locus of maize using Mu transposon tagged alleles. Genetics 183:1215–1225
14. Liu S et al (2009) Mu transposon insertion sites and meiotic recombination events co-localize with epigenetic marks for open chromatin across the maize genome. PLoS Genet 5:e1000733
15. Yunis I, Salazar M, Yunis EJ (1991) HLA-DR generic typing by AFLP. Tissue Antigens 38: 78–88
16. Michelmore RW, Paran I, Kesseli RV (1991) Identification of markers linked to disease-resistance genes by bulked segregant analysis: a rapid method to detect markers in specific genomic regions by using segregating populations. Proc Natl Acad Sci USA 88:9828–9832
17. Qiu F et al (2003) DNA sequence-based "bar codes" for tracking the origins of expressed sequence tags from a maize cDNA library constructed using multiple mRNA sources. Plant Physiol 133:475–481

Chapter 13

Molecular Genetics and Epigenetics of CACTA Elements

Nina V. Fedoroff

Abstract

The CACTA transposons, so named for a highly conserved motif at element ends, comprise one of the most abundant superfamilies of Class 2 (cut-and-paste) plant transposons. CACTA transposons characteristically include subterminal sequences of several hundred nucleotides containing closely spaced direct and inverted repeats of a short, conserved sequence of 14-15 bp. The *Supressor-mutator* (*Spm*) transposon, identified and subjected to detailed genetic analysis by Barbara McClintock, remains the paradigmatic element of the CACTA family. The *Spm* transposon encodes two proteins required for transposition, the transposase (TnpD) and a regulatory protein (TnpA) that binds to the subterminal repeats. *Spm* expression is subject to both genetic and epigenetic regulation. The *Spm*-encoded TnpA serves as an activator of the epigenetically inactivated, methylated *Spm*, stimulating both transient and heritable activation of the transposon. TnpA also serves as a negative regulator of the demethylated active element promoter and is required, in addition to the TnpD, for transposition.

Key words CACTA elements, transposon, *Suppressor-mutator*, *Spm*, transposition, TnpA, TnpD, epigenetic regulation

1 Introduction

The CACTA superfamily of elements was so designated for the highly conserved 5-letter sequence that appears at the distal ends of the terminal inverted repeats of all superfamily members [1]. CACTA elements have been identified in a wide range of plants species, including maize, wheat, rice, sorghum, *Arabidopsis*, soybean, carrot, pea, snapdragon, and morning glory, as well as fungi and some animal species [2–4]. CACTA elements are Class 2 DNA transposons, employing a direct cut-and-paste transposition mechanism in contrast to Class 1 retrotransposons, whose transposition involves an RNA intermediate [5]. The paradigmatic and first-described element of the CACTA family is the *Suppressor-mutator* (*Spm*) transposon. Mutations caused by *Spm* transposons were first described by McClintock in 1951, although she did not name

Thomas Peterson (ed.), *Plant Transposable Elements: Methods and Protocols*, Methods in Molecular Biology, vol. 1057, DOI 10.1007/978-1-62703-568-2_13, © Springer Science+Business Media New York 2013

this transposon family until several years later [6, 7]. Peterson identified the same element family, naming the autonomously transposing element *Enhancer* (*En*) and the transposition-defective nonautonomous elements *Inhibitor* (*I*) [8, 9]. McClintock did not assign a separate name to the category of transposition-defective *Spm* elements; I gave them the categorical designation *defective Spm* (*dSpm*) [10–12].

Recent decades have witnessed a veritable explosion of genomic data, facilitating the identification, cataloging, and categorization of transposons in increasing detail. There are many excellent reviews that describe the numbers and types of transposons that comprise the bulk of the genomic DNA of plants [3–5, 13, 14]. Indeed, Bennetzen has made the point that most of the plant DNA on the planet comprises transposons [15]. CACTA elements, once thought to be relatively low-copy-number DNA transposons, are emerging as among the most abundant DNA transposons, comprising as much as 10 % of the DNA in some grass genomes [16–19]. The expanding appreciation of their abundance is likely due to the improving ability to identify CACTA elements based on the conservation of subterminal structural features, which characteristically include a sequence of several hundred nucleotides at each end containing closely spaced direct and inverted repeats of a short motif of 10–20 bp [19].

Despite the rapidly expanding catalog of transposon sequences, our understanding of the function and regulation of DNA transposons remains limited. The genetic and molecular studies on the maize *Spm* transposon family remain unique in providing insight into both the genetic and epigenetic regulatory circuitry controlling expression and transposition of a CACTA DNA transposon. McClintock's genetic studies on the *Spm* transposon family were possible because she was able to identify and study a single autonomous *Spm* element and its influence on the expression of alleles caused by insertion of nonautonomous *dSpm* family members [10, 20, 21]. As well, her pioneering work on the epigenetic regulation of *Spm* activity was predicated on the ability to track a small number of autonomous elements and their interactions using a reporter gene. The results of these genetic studies and the alleles they generated made it possible, in turn, to explicate the genetic and epigenetic regulatory interactions within the *Spm* family at the molecular level once the elements were isolated [22, 23]. In what follows, McClintock's purely genetic observations are integrated with the results of more recent molecular studies to sketch a picture of the simple yet intricate regulatory circuitry of the *Spm* transposon family.

2 Molecular Genetics of the *Spm* Transposon Family

2.1 Spm-Suppressible Alleles

McClintock named the *Spm* transposon for its effect on certain mutant alleles of the *A1* locus, designated the *a1-m1* alleles [7, 24]. Expression of the *A1* gene is attenuated by what is now known to be an insertion of a transposition-defective *Spm* at an exon–intron boundary within the *A1* gene, one of several genes required for anthocyanin pigment production in the maize kernel's aleurone layer [25, 26]. Kernels carrying an *a1-m1* allele either homozygous or heterozygous with *a1*, a colorless null allele, but no *Spm*, are much less intensely pigmented than kernels carrying the wild-type *A1* allele (*see* Fig. 1a–c). Pigmentation is completely suppressed in the presence of an *Spm* element elsewhere in the genome in response to what McClintock designated as the *suppressor* component of *Spm*, giving a colorless kernel aleurone (*see* Fig. 1d). We later called this type of allele *Spm-suppressible* (*see* Fig. 2a) to distinguish it from insertions that render expression of the gene dependent on the presence of *Spm*, described below. The insertion is excised in some cells during development in response to the *mutator* component of *Spm*, restoring full *A1* gene expression in tissue sectors derived from such cells (*see* Fig. 1d).

Members of the *a1-m1* allelic series were derived from a single initial insertion, but differed from each other in both the basal level of expression of the *A1* gene in the absence of *Spm*, as well as in the timing and frequency of reversion to the wild-type level of gene expression. The various alleles, which McClintock designated "changes of state," were stably heritable mutations, like those she

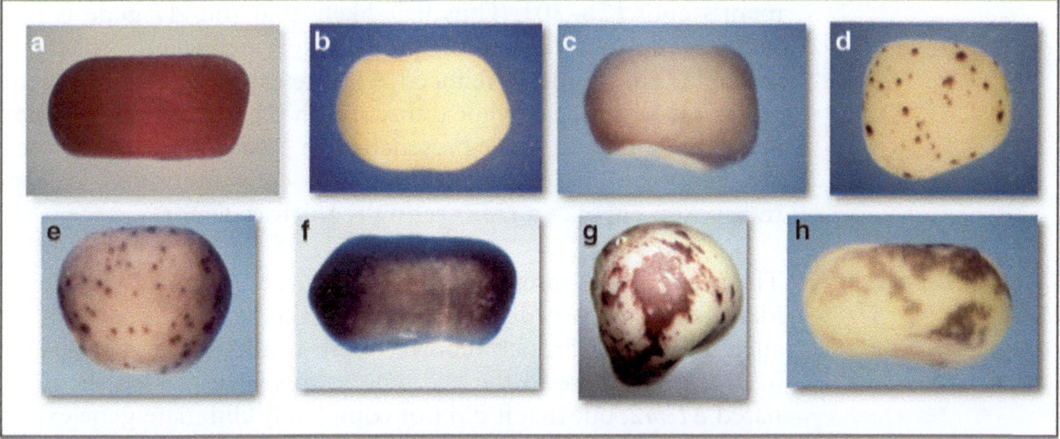

Fig. 1 Kernel phenotypes. (**a**) *A1* (*red color* due to homozygous *pr* allele); (**b**) homozygous recessive *a1* allele; (**c**) *a1-m1/a1*, no Spm; (**d**) *a1-m1/a1*, with *Spm*; (**e**) *a1-m2 7995/a1*, with *Spm*; (**f**) *a2-m1* (class II)/*a2* without *Spm*; (**g**) *a2-m1* (class II)/*a2* with an *Spm-c*; (**h**) the "preset" pattern exhibited by *a1-m2 7995* just after meiotic segregation of the *Spm* element

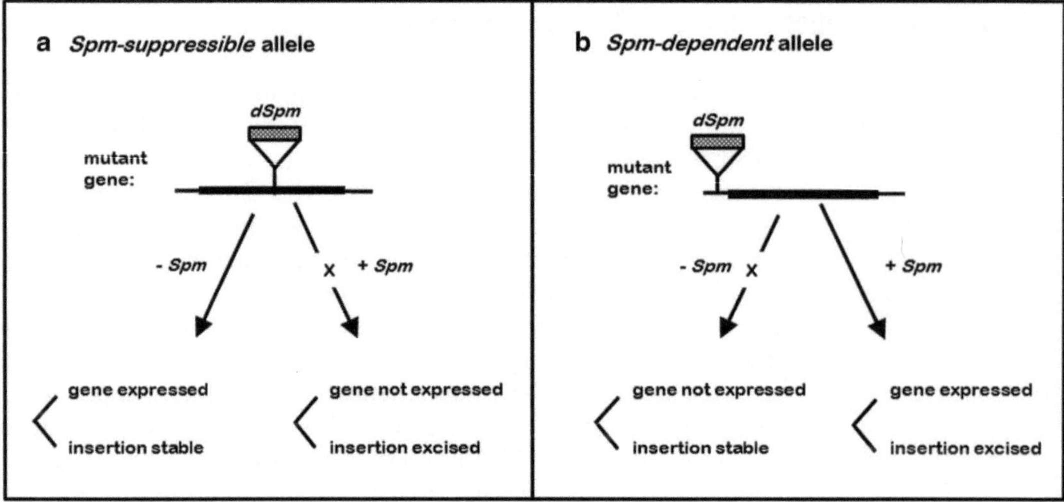

Fig. 2 Effect of the *Spm* element on genes with *dSpm* insertion mutations. Two different types of interactions between the *trans*-acting *Spm* element and a gene with a *dSpm* insertion mutation are illustrated. (**a**) The response of an *Spm-suppressible* allele. In the absence of an *Spm* element, the mutant gene (*filled box*) is expressed at an intermediate level and the insertion is not excised. In the presence of an *Spm*, the gene is not expressed, and the insertion is excised in some cell lineages to give revertant sectors. (**b**) The response of an *Spm-dependent* allele. The mutant gene is not expressed in the absence of an *Spm* and the insertion is stable. In the presence of an *Spm*, the mutant gene is expressed and the insertion is excised to give revertant sectors (figure and legend reproduced from Masson et al. [11])

reported in the *Ac–Ds* element family [27]. These are now known to constitute a series of internal deletions in the *dSpm*, which is an internally deleted *Spm* transposon [25, 26]. Subsequent studies on *Spm* revealed that the transposon-encoded TnpA protein binds to a short sequence that is repeated in close juxtaposition at the element's ends [28–30]. Thus it is likely that the *A1* gene is transcribed but that the transcript is either inefficiently spliced or unstable. This results in reduced *A1* gene expression in the absence of a *trans*-acting *Spm*, while transcription is blocked by TnpA binding to the insertion in the presence of an *Spm* element elsewhere in the genome. These observations were early hints that the *Spm* transposon encodes a *trans*-acting regulatory factor.

2.2 Spm-Dependent Alleles

The second *Spm* insertion mutation at the *A1* locus McClintock identified was markedly different from the *Spm-suppressible a1-m1* allele [31]. McClintock reported that a fully functional *Spm* was inserted very close to the *A1* gene in this allele, which she designated *a1-m2*, but that it did not completely eliminate gene expression. McClintock then identified a number of derivatives that exhibited what she called a "two-element system of control of gene action," in which a transposition-defective element capable of being mobilized in *trans* remained at the locus [32]. By contrast to

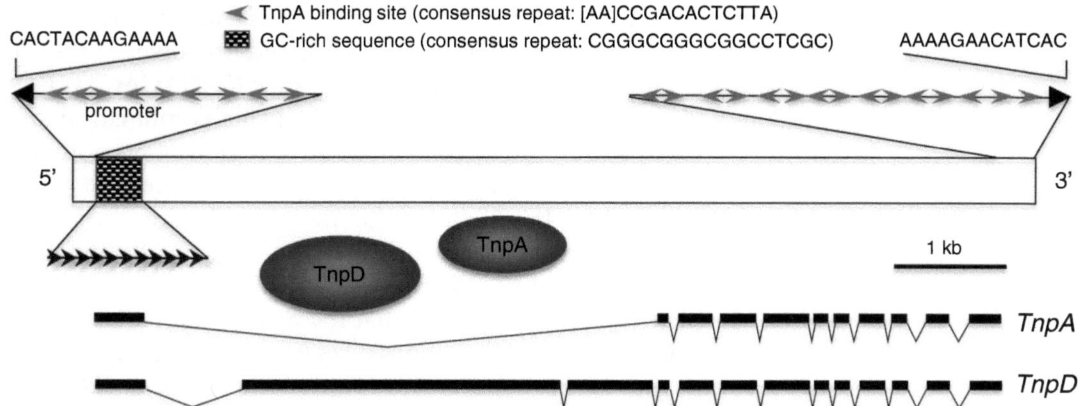

Fig. 3 Structure of the *Spm* transposon, the *TnpA* and *TnpD* transcripts encoding the TnpA and TnpD proteins required for transposition. TnpA is a 68-kDa protein with a dimerization domain and TnpD is a 131-kDa protein with a DDE transposase motif. TnpD is encoded within the first intron of the TnpA transcript

the behavior of the *Spm-suppressible* alleles, pigmentation resulting from expression of the *A* gene is not observed in the absence (*see* Fig. 1b) but is observed in the presence of an active *Spm* elsewhere in the genome (*see* Fig. 1e). Thus expression of the *A* gene in the *a1-m2* mutant series is dependent on a gene product supplied in *trans* by an *Spm*, hence the designation *Spm-dependent* alleles (*see* Fig. 2b). The observation that the gene bearing the insertion mutation is expressed only in the presence of a fully functional *Spm* suggested that the gene had come under the control of a positive auto-regulatory system encoded by the transposon.

2.3 Spm Structure and Auto-Regulation

Cloning and molecular analysis of the *Spm* revealed it to be an 8.3-kb sequence with a single transcription unit that gives rise to a number of alternatively spliced transcripts (*see* Fig. 3) [11, 33, 34]. Two proteins, a 68-kDa protein encoded by the *TnpA* transcript and a 131-kDa protein encoded by the *TnpD* transcript, were found to be required for transposition based on cDNA-mediated excision of the insertion from a *dSpm*-disrupted β-glucuronidase gene in tobacco cells [33, 35]. The two transcripts are derived by alternative splicing from the same primary transcript and the larger protein, TnpD, is encoded entirely within the long first intron of the smaller and more abundant *TnpA* transcript (*see* Fig. 3). As described below, TnpD appears to be the transposase, as judged by the presence of a characteristic transposase DDE motif [36], while TnpA is a multifunctional regulatory protein and participates in transposition, as well.

In addition to its 13-bp terminal inverted repeats, CACTACAAGAAAA, the *Spm* has subterminal sequences of several hundred nucleotides at each end, designated the "subterminal repetitive regions," characterized by the presence of multiple copies of closely spaced, highly homologous direct and inverted 12-bp repeats, 9 copies at the element's 5′ end, and 15 copies at the element's 3′ end (*see* Fig. 3; consensus sequence CCGACACTCTTA, half of the copies are preceded by AA). This combination of a CACTA motif and a subterminal repetitive region of homologous 10- to 20-bp direct and inverted repeats is characteristic of the superfamily and has been exploited in the identification and molecular characterization of family members, which vary substantially in length and sequence [17, 19, 37].

Subsequent studies revealed that the *Spm* promoter is coextensive with the subterminal repetitive region at the transposon's 5′ end and the 12-bp repeats are TnpA-binding sites [28, 38]. Additionally, TnpA is both a transcriptional activator of its own methylated promoter and a repressor of its unmethylated promoter [28, 39]. Because the TnpA-binding sites are present in both orientations and in multiple copies at both ends of the transposon, *Spm* ends are bidirectional, TnpA-dependent promoters [28]. Thus the ability of a *trans*-acting element to activate expression of the *A1* gene in the *a1-m2* allele series reflects the operation of the transposon's positive auto-regulatory system.

McClintock collected a number of *dSpm* derivatives of the original *a1-m2* allele that had arisen by insertion of a fully functional *Spm* [40]. All but one of the *Spm-dependent* alleles analyzed contained an internally deleted *Spm* element at the same insertion site in the promoter of the *A1* gene [26]. These differed in the timing and frequency of *dSpm* excision promoted by a *trans*-acting *Spm*, comparable to the "states" of the *a1-m1* allelic series [11, 12, 25–27]. Internal deletions that encroach on the subterminal repetitive regions decrease the excision frequency [11]. Deletion derivatives in the *a1-m2* series retaining at least 1 kb of the element's 5′ end, but lacking just four or five copies of the 15 subterminal repeats at the element's 3′ end, transpose at a lower frequency and later in development than elements with the full complement of 3′ terminal repeats [11]. Thus the multiple subterminal repeats appear to be important determinants of transposition timing and frequency. But in addition, the very low transposition frequency of an element retaining all of the subterminal repeats, but lacking a 0.4-kb sequence containing 11 direct repeats of an extremely GC-rich sequence (consensus sequence CGGGCGGGC-GGCCTCGC), suggests that additional internal element sequences play a role in transposition [11, 28].

The presence of multiple copies of the TnpA-binding sites at both ends of the transposon suggests that interactions between *Spm*-encoded TnpA proteins bound at transposon ends may serve to bring the ends together in the correct register for transposition. Direct biochemical evidence that TnpA serves in such a capacity and in proportion to the redundancy of its binding sites was obtained by examining the ability of TnpA to accelerate the ligation of DNA fragments containing 3, 6, and 9 TnpA-binding sites [30]. Thus TnpA's primary role in transposition is to bring together the element ends in a transposition complex [30], while DNA cleavage is likely to be carried out by TnpD.

McClintock reported that mutations that affect the genetic properties of *dSpm* transposons occur in the presence, but not in the absence, of a fully functional *Spm* in the genome [41]. Subsequent molecular studies established that almost all such mutations are intra-element deletions [11, 12, 25, 26]. Interestingly, six of the eight sequenced deletion endpoints in the *a1-m2* allelic series are within or at the end of one of the subterminal, TnpA-binding repeats. Although the DNA-binding properties of TnpD have not been investigated, the protein contains the characteristic transposase DDE motif [42]. Since both TnpA and TnpD are required for transposition, the observation that deletion endpoints co-localize with TnpA-binding sites suggests that TnpD ordinarily cleaves at the transposon's termini upon binding to the terminal inverted repeat and an adjacent TnpA, but occasionally mis-cleaves the transposon internally upon binding solely to a TnpA bound elsewhere within the element's sequence to give rise to an internal deletion [11].

3 Epigenetics of the *Spm* Transposon Family

McClintock is widely acknowledged for her discovery of transposition, but her extraordinarily prescient work on epigenetics remains virtually unrecognized. Her work on transposon "changes of phase" [21, 43, 44], together with that of R. A. Brink on paramutation at the maize *R* locus [45–47] and Mary Lyon on mammalian X chromosome inactivation [48–50], presaged the now-thriving field of epigenetic regulation.

3.1 Changes of Phase

McClintock identified a series of mutant alleles of the *A2* locus with genetic properties similar to those of the *Spm-suppressible a1-m1* allelic series [43]. Her analysis of the *a1-m2* alleles was initially confused by the occurrence of an alteration in the transacting *Spm* element in these cultures that changed its "phase of activity," as well as a "change in state" of the transposition-defective element at the locus that rendered it unable to excise [44]. Both of these alterations proved important in understanding the epigenetic inactivation of the *Spm* element, as well as the regulatory interactions between elements [44]. In the Carnegie Institution of Washington Year Book 57, McClintock wrote:

> By means of various kinds of experiment with this state, it was first learned that the *Spm* element in the *a2-m1* cultures may undergo frequent changes in activity during the development of a plant, each such

> change affecting its capacity to serve as a suppressor-mutator. Clearly, some regulatory mechanism controls the time of occurrence of such changes, although it is not yet understood.... The class II state of *a2-m1* readily reveals these changes in action capacity of *Spm*, for, with this state, the *Spm* element in its active phase serves only to inhibit expression of gene action at *a2-m1*.

The kernels in Fig. 1 illustrate McClintock's observations. Kernels having the genetic constitution *a2-m1* (class II)/*a2* (*a2* is a stable recessive allele that gives a colorless kernel when homozygous) are intensely pigmented in the absence of an *Spm* element (*see* Fig. 1f) and colorless in the presence of a fully active *Spm* (*see* Fig. 1b). Hence this allele does not exhibit the pigmented sectors observed with transposition-competent *dSpm* elements that arise when the transposon is excised during kernel development, restoring wild-type gene expression. Kernels of this genetic constitution vividly reveal changes in expression of the *Spm* transposon itself. McClintock wrote:

> Changes in *Spm* action phase may alternate, and both the times and the types of change are revealed in the kernel phenotypes. In kernels having one *Spm* element, these alternating changes may be observed readily. For example, a large pigmented area may be seen in an otherwise colorless region of a kernel. Within this large pigmented area, smaller colorless areas may be observed and within these, in turn, specks of deep pigmentation. In this illustration, the sequence of changes of phase of *Spm* activity during development of the kernel was from active to inactive to active, and again to inactive.

The variegation pattern McClintock observed with a single *Spm* undergoing changes in its phase of activity during development is shown in Fig. 1g. Colorless sectors correspond to tissue in which the *Spm* is active and colored sectors to tissue in which it is inactive. The occurrence of colorless sectors within colored sectors reveals that the *Spm* can cycle between active and inactive states during kernel development. I designated this type of element with rapidly alternating phases of activity and inactivity, a *cycling Spm* (*Spm-c*) [10].

McClintock found that the kernel pattern of variegation observed depends on the number of *Spm-c* elements present. By introducing the *Spm-c* transposon through the male, the female, or both, she was able to observe the patterns produced with 1, 2, or 3 elements in the triploid endosperm [21]. She found that the deeply pigmented areas corresponding to the element in an inactive phase were progressively smaller as the number of *Spm-c* copies increased. This implies that the deeply pigmented inactive-*Spm* phenotype requires all of the resident elements to be inactive, while the colorless active-*Spm* phenotype will result if any one of the elements is active. This observation, in turn, implies that an active element produces a "suppressor" that acts in *trans* on the transposition-defective element inserted at the *A2* locus to suppress expression of the *A2* gene. As described above, it is now known that it is the TnpA protein encoded by the *Spm* transposon that binds to the element termini and can repress the unmethylated *Spm* promoter at the element's 5′ end [28, 30, 39].

3.2 Heritable, Reversible Silencing of Spm

McClintock observed that some plants that apparently lacked an *Spm* could occasionally produce a few kernels on tiller ears that showed an active-*Spm* phenotype [21, 43, 44]. The inactive *Spm*, today more commonly referred to as "silenced," could be inherited in the inactive phase through many generations, showing only occasional activation. She then discovered that when an *Spm-c* was introduced into a plant carrying such an inactive *Spm*, the kernels that received both the inactive *Spm* and the *Spm-c* exhibited the variegation pattern expected for multiple copies of an *Spm-c* [21]. She inferred that the active element could reactivate the inactive element and that the *Spm-c* and the reactivated *Spm* then underwent inactivation during development independently. The implication of these observations is that a transposon-encoded protein can act in *trans* to activate the inactive one. McClintock's analysis of the *a1-m2* alleles, described earlier, was also consistent with the existence of such a positive, transposon-encoded regulator of transposon gene expression. But an important additional implication was that the inactive element was maintained in the inactive phase by some mechanism that could be overcome by a factor that could be supplied in *trans* by an active transposon.

3.3 Spm Promotes Heritable Activation of a Cryptic Spm

Molecular analysis of the several *a1-m2* alleles in McClintock's collection identified one that appeared to have an intact element at the locus, designated *a1-m2 8167B* [11]. As noted, the results of McClintock's studies on the interaction between active and inactive elements at the *A2* locus, described above, implied that an active *Spm* could *trans*-activate an inactive one. However, she reported that the activation was transient and that the inactive element reverted to the inactive form when segregated away from the active *Spm* [51]. Our subsequent studies showed that an active element could heritably activate an inactive one [20] and we were able to reactivate the resident *Spm* at the *A* locus in the *a1-m2 8167B* allele by maintaining it in the presence of an active *Spm* for several generations [52]. No spontaneous activation of this element was observed among several hundred thousand control kernels carrying the *a1-m2 8167B* allele produced on plants lacking an active *Spm*. We therefore designated an *Spm* in the very stably inactive state, such as the one in the *a1-m2 8167B* allele, a *cryptic Spm*. The clear implication of these studies is that an *Spm*-encoded gene product can promote the heritable activation of an epigenetically inactive element.

3.4 Machinery of Epigenetic Regulation

Progress has been made in recent years toward understanding the molecular basis of epigenetic silencing. It is now known that although histone modification and DNA methylation stabilize silencing and inactivation of genes and other genetic elements in many organisms, a genetic feedback mechanism based on small, noncoding RNAs is at the heart of epigenetic regulation [53–55]. Transposon silencing in particular is the result of RNA-directed DNA methylation (RdDM), a complex process that involves two unique plant RNA polymerases, pol IV and pol V, is mediated by 24-nt siRNAs, and results in both DNA methylation and histone modifications that suppress transcription [55, 56]. In brief, RdDM is initiated by pol IV-directed transcripts that are converted to RNA duplexes by RNA-DEPENDENT RNA POLYMERASE 2 (RDR2); the duplexes are then cleaved into 24-nt siRNAs by the RNAse III-family enzyme DICER-LIKE 3 (DCL3). These are stabilized by 3′-terminal methylation by HUA-ENHANCER 1 (HEN1) and the appropriate strand associates with ARGONAUTE 4 (AGO4) protein and is incorporated into an RNA-induced silencing complex (RISC). Pol V transcripts of the target locus bind to AGO4 through base-pairing with the siRNA in a large silencing complex, within which the REQUIRED FOR DNA METHYLATION 1 (RDM1) protein interacts with both AGO4 and the de novo cytosine methyltransferase DOMAINS REARRANGED METHYLTRANSFERASE 2 (DRM2), which in turn triggers local DNA methylation [55, 56]. As well, activation of RdDM promotes histone deacetylation and methylation changes that lead to the establishment of chromatin structures that repress transcription [55, 56].

Demethylation is mediated by 5-methylcytosine (5-meC) DNA glycosylase/lyases, encoded by four genes in *Arabidopsis*: REPRESSOR OF SILENCING 1 (ROS1), DEMETER (DME), DEMETER-LIKE 2 (DML2), and DEMETER-LIKE 3 (DML3). The DNA glycosylase activity removes the 5-meC, following which the DNA backbone is cleaved at the abasic site, followed by DNA repair [57].

Understanding of the molecular infrastructure involved in epigenetic regulation of transposons is growing rapidly, as is awareness that epigenetic regulation is at the heart of developmental processes [58–62]. The direct involvement of small RNAs in both transcriptional and posttranscriptional gene silencing explains how these gene regulatory processes are targeted to specific sequences [54, 55, 63]. But it does not explain how transposons, genes, and other genetic elements are selected for epigenetic inactivation or how the process is initiated.

Conversely, although major epigenetic reprogramming occurs in both animal and plant development, including widespread DNA demethylation [59, 60, 64], little is known about the mechanisms by which specific genes are targeted. Nor is it known how specific sequences are targeted for demethylation at specific developmental stages or in a subset of cells. Recent reports suggest that DNA-demethylating enzymes may themselves target subsets of genes and that there may be RNA-binding proteins that mediate targeting of demethylation [65, 66].

3.5 Targeting Spm Epigenetic Regulation

Insight into the epigenetic regulation of CACTA transposons derives from the analysis of *Spm* inactivation and reactivation. Molecular analysis of active, inactive, and cycling *Spm* elements revealed that the transposon's phase of activity is well correlated with both the abundance of TnpA transcripts and the extent of methylation of the promoter and the GC-rich repetitive sequence in the first exon [52, 67]. At one extreme, the *cryptic* element in the *a1-m2 8167B* allele is extensively methylated and transcriptionally inactive, while at the other extreme, a fully active element is unmethylated in this region and transcribed, although the remainder of the transposon is methylated irrespective of its activity phase [52].

Elements exhibiting an intermediate level of methylation in the GC-rich first exon sequence can also show heritable regularities in the timing or spatial pattern of activity and were therefore designated *programmable* [67]. Although heritable, such spatial or temporal patterns (programs) of activity are relatively labile, readily giving rise to different patterns. Active *programmable Spm* elements have unmethylated promoters, while inactive *programmable* elements have methylated promoters. Introduction of a *trans-activating Spm* into a plant carrying an inactive *programmable* element results in transcriptional activation of the element and demethylation of the promoter [67]. Thus it appears that the genetic stability of the inactive state is determined by the extent (and perhaps pattern) of methylation within the GC-rich first exon sequence. Transcriptional activity is determined by whether or not the promoter is methylated, but only if the GC-rich sequence is not extensively methylated. Moreover, *trans*-activation of an inactive element by an active one is associated with promoter demethylation.

A *cryptic Spm* does not undergo spontaneous activation, but can be reactivated by maintaining it in the presence of an active *Spm* over several generations, in each generation selecting for kernels exhibiting the phenotype of an active *Spm* [20, 67, 68]. The active state is also highly stable, although less so than the *cryptic* state and heritably inactive derivatives can be selected over several generations, exhibiting increasing methylation of the GC-rich sequence in each generation [67]. *Spm* transposons with intermediate levels of methylation can also be inactive in the absence of a transacting element, but fully active in the presence of a transacting active *Spm*. Strikingly, such elements can remain transiently active immediately after segregation of the *trans*-activating *Spm*, suggesting an explanation for the curious phenomenon of presetting, described below [20]. All of these observations indicate that the development of the *cryptic Spm* state is a multistep process that involves some measure of stochasticity.

The *Spm* transposon becomes reversibly inactivated in transgenic tobacco plants [69]. Further dissection of the sequence requirements for inactivation showed that they are confined to a short sequence at the transposon's 5′ end. A reporter gene expressed from the *Spm* promoter is inactivated and methylated if the promoter sequence includes the GC-rich sequence in the first exon, but not if it lacks the sequence [39]. It follows that the first exon sequence, with its 11 closely spaced repeats of a highly GC-rich sequence (consensus sequence: CGGGCGGG-CGGCCTCGC), targets *Spm* for epigenetic silencing. Thus the transposon carries its own internal inactivation signal.

Of the two transposon-encoded proteins necessary for transposition, TnpA and TnpD [33], only TnpA is necessary to reactivate an inactive *Spm* in transgenic tobacco. Moreover, the methylated and inactive promoter can be reactivated and demethylated in the presence of TnpA [39]. Subsequent studies showed that TnpA promotes active demethylation of the promoter [70]. It should be noted that the sequences that undergo methylation and demethylation are both within the promoter, as each TnpA-binding site contains three to seven C residues, and downstream from it is the GC-rich repetitive region. It is therefore a reasonable conjecture that TnpA itself recruits demethylating enzymes to the promoter.

3.6 Presetting

Some of the *a1-m2* alleles had an odd property that McClintock called "presetting" [40, 71]. Such alleles showed no *A1* gene expression when consistently maintained in the absence of an active *Spm* element. But in some kernels on ears in which the *Spm* is newly separated from the insertion allele by meiotic segregation, some kernels exhibit continued, albeit irregular expression of the *A1* gene (*see* Fig. 1h). McClintock inferred that the *Spm* element in some way "presets" the gene containing the transposition-defective element to continue expressing in its absence. The likely explanation for presetting is that the transposon sequence inserted just upstream from the *A1* gene in the *a1-m2* alleles does not itself prevent expression of the gene, but does readily attract inactivating methylation to the promoter. The presence of a *trans*-acting *Spm* supplies TnpA, which recruits the demethylation machinery to the gene, allowing its expression. However, upon removal of the *Spm* by genetic segregation, the remaining transposition-defective element in the *A1* gene's promoter is not instantly remethylated, allowing its transient "preset" expression in some of the aleurone cells (*see* Fig. 1h).

As noted earlier, an *Spm* transposon with intermediate level of methylation in the repetitive GC-rich first exon sequence can be inactive in the absence of a *trans*-acting element, but fully active in the presence of an active *Spm*. The observation that such elements can remain transiently active immediately after segregation of the *trans*-activating *Spm* is consistent with the foregoing interpretations of both presetting and element inactivation [20]. That is, TnpA promotes active demethylation of the *Spm* promoter and GC-rich sequence, while inactivating methylation is a somewhat stochastic process; hence the activated, demethylated element can continue to be expressed and transpositionally active after the *trans*-activating element is gone and before it is sufficiently methylated to render it transcriptionally inactive.

4 Summary

McClintock's perceptive analysis of the "changes in the phase" of *Spm* activity constitutes one of the earliest genetic investigations of an epigenetic regulatory system, which is only now beginning to be understood at the molecular level. An active negative feedback regulation mechanism, now known to be triggered by silencing siRNAs derived from transposon transcripts and maintained by DNA methylation, is fundamental to keeping transposon activity in check and maintaining chromosome stability. The *Spm* transposon carries an internal sequence in its first exon that targets the element for methylation and inactivation. Extensive methylation of the *Spm* promoter and first exon sequence gives rise to a *cryptic Spm*, a deeply silenced element that neither undergoes spontaneous reactivation nor can be readily reactivated in *trans* by an active *Spm*, and which can only be excised by an active element at a low frequency.

The *Spm* transposon encodes two proteins required for transposition, one of which, TnpA, is a regulatory protein with both positive and negative regulatory functions. TnpA is a transcriptional activator of a methylated *Spm* promoter capable of both overriding and actively reversing the DNA methylation that maintains the transposon in a silent state. It is also a repressor of the unmethylated *Spm* promoter. This property, in turn, may reflect its transpositional role, likely to bind to the multiple TnpA-binding sites at both element ends to bring them into close juxtaposition for transposition.

Thus *Spm* transposons tend toward one of the two highly heritable states, active and cryptic. Cryptic elements are extremely difficult to extricate from inactivity and even difficult to mobilize by providing a source of transposase in the form of an active *Spm* elsewhere in the genome. It is likely that the stability of the cryptic state is enforced by the high density of methylatable C residues in the promoter and downstream GC-rich repeats. By contrast, once activated and demethylated, *Spm* transposons stay active by virtue of a positive auto-regulatory loop in which the transposon-encoded TnpA protein promotes the demethylation of its own promoter and adjacent GC-rich first exon. What remains enigmatic is how and why cryptic transposons can be activated by a variety of abiotic stresses, conditions that disrupt development, such as tissue culture, and disruptions of chromosome structure and mechanics [72].

References

1. Kunze R, Weil CF (2002) The hAT and CACTA superfamilies of plant transposons. In: Craig NL, Craigie R, Gellert M, Lambowitz AM (eds) Mobile DNA II. ASM Press, Washington DC, pp 565–610
2. DeMarco R, Venancio TM, Verjovski-Almeida S (2006) SmTRC1, a novel *Schistosoma mansoni* DNA transposon, discloses new families of animal and fungi transposons belonging to the CACTA superfamily. BMC Evol Biol 6:89
3. Feschotte C, Pritham EJ (2007) DNA transposons and the evolution of eukaryotic genomes. Annu Rev Genet 41:331–368
4. Wicker T et al (2007) A unified classification system for eukaryotic transposable elements. Nat Rev Genet 8:973–982
5. Feschotte C, Jiang N, Wessler SR (2002) Plant transposable elements: where genetics meets genomics. Nat Rev Genet 3:329–341
6. McClintock B (1951) Mutable loci in maize. Carnegie Inst Wash Yr Bk 50:174–181
7. McClintock B (1954) Mutations in maize and chromosomal aberrations in *Neurospora*. Carnegie Inst Wash Yr Bk 53:254–260
8. Peterson PA (1953) A mutable pale green locus in maize. Genetics 38:682–683
9. Peterson PA (1965) A relationship between the *Spm* and *En* control systems in maize. Am Nat 99:391–398
10. Fedoroff NV (1983) Controlling elements in maize. In: Shapiro J (ed) Mobile genetic elements. Academic, New York, pp 1–63
11. Masson P et al (1987) Genetic and molecular analysis of the *Spm*-dependent *a-m2* alleles of the maize *a* locus. Genetics 117:117–137
12. Schiefelbein JW et al (1985) Deletions within a defective suppressor-mutator element in maize affect the frequency and developmental timing of its excision from the bronze locus. Proc Natl Acad Sci USA 82:4783–4787
13. Bennetzen JL (2000) Transposable element contributions to plant gene and genome evolution. Plant Mol Biol 42:251–269
14. Kazazian HH Jr (2004) Mobile elements: drivers of genome evolution. Science 303:1626–1632
15. Bennetzen JL (2005) Transposable elements, gene creation and genome rearrangement in flowering plants. Curr Opin Genet Dev 15:621–627
16. Kwon S-J et al (2006) CACTA and MITE transposon distributions on a genetic map of rice using F15 RILs derived from Milyang 23 and Gihobyeo hybrids. Mol Cells 21:360–366
17. Langdon T et al (2003) A high-copy-number CACTA family transposon in temperate grasses and cereals. Genetics 163:1097–1108
18. Sergeeva EM et al (2010) Evolutionary analysis of the CACTA DNA-transposon Caspar across wheat species using sequence comparison and in situ hybridization. Mol Gen Genet 284:11–23
19. Wicker T et al (2003) CACTA transposons in Triticeae. A diverse family of high-copy repetitive elements. Plant Physiol 132:52–63
20. Fedoroff NV (1989) The heritable activation of cryptic *Suppressor-mutator* elements by an active element. Genetics 121:591–608
21. McClintock B (1971) The contribution of one component of a control system to versatility of gene expression. Carnegie Inst Wash Yr Bk 70:5–17
22. Fedoroff N, Schlappi M, Raina R (1995) Epigenetic regulation of the maize Spm transposon. Bioessays 17:291–297
23. Fedoroff N (1989) Maize transposable elements. In: Howe M, Berg D (eds) Mobile DNA. American Society for Microbiology, Washington, pp 375–411
24. McClintock B (1953) Mutation in maize. Carnegie Inst Wash Yr Bk 52:227–237
25. Schwarz-Sommer Z et al (1985) Sequence comparison of 'states' of *a1-m1* suggest a model of Spm (En) action. EMBO J 4:2439–2443
26. Schwarz-Sommer Z et al (1987) Influence of transposable elements on the structure and

function of the A1 gene of *Zea mays*. EMBO J 6:287–294
27. McClintock B (1956) Intranuclear systems controlling gene action and mutation. Brookhaven Symp Biol 8:58–74
28. Raina R, Cook D, Fedoroff N (1993) Maize *Spm* transposable element has an enhancer-insensitive promoter. Proc Natl Acad Sci USA 90:6355–6359
29. Raina R, Fedoroff N (1995) The role of TnpA and TnpD in transposition of *Spm*. Maize Genet Coop Newsl 69:13–15
30. Raina R et al (1998) Concerted formation of macromolecular *Suppressor-mutator* transposition complexes. Proc Natl Acad Sci USA 95:8526–8531
31. McClintock B (1961) Further studies of the suppressor-mutator system of control of gene action in maize. Carnegie Inst Wash Yr Bk 60:469–476
32. McClintock B (1962) Topographical relations between elements of control systems in maize. Carnegie Inst Wash Yr Bk 61:448–461
33. Masson P, Strem M, Fedoroff N (1991) The tnpA and tnpD gene products of the Spm element are required for transposition in tobacco. Plant Cell 3:73–85
34. Masson P et al (1989) Essential large transcripts of the maize Spm transposable element are generated by alternative splicing. Cell 58:755–765
35. Masson P, Toohey K, Fedoroff N (1988) Excision of *Spm* in tobacco. Maize Genet Coop Newsl 62:26–27
36. Yuan YW, Wessler SR (2011) The catalytic domain of all eukaryotic cut-and-paste transposase superfamilies. Proc Natl Acad Sci USA 108:7884–7889
37. Tian P-F (2006) Progress in plant CACTA elements. Yi Chuan Xue Bao 33:765–774
38. Gierl A, Lutticke S, Saedler H (1988) TnpA product encoded by the transposable element En-1 of *Zea mays* is a DNA binding protein. EMBO J 7:4045–4053
39. Schlappi M, Raina R, Fedoroff N (1994) Epigenetic regulation of the maize *Spm* transposable element: novel activation of a methylated promoter by TnpA. Cell 77:427–437
40. McClintock B (1963) Further studies of gene-control systems in maize. Carnegie Inst Wash Yr Bk 62:486–493
41. McClintock B (1955) Controlled mutation in maize. Carnegie Inst Wash Yr Bk 54:245–255
42. Hickman AB, Chandler M, Dyda F (2010) Integrating prokaryotes and eukaryotes: DNA transposases in light of structure. Crit Rev Biochem Mol Biol 45:50–69
43. McClintock B (1957) Genetic and cytological studies of maize. Carnegie Inst Wash Yr Bk 56:393–401
44. McClintock B (1958) The suppressor-mutator system of control of gene action in maize. Carnegie Inst Wash Yr Bk 57:415–429
45. Brink RA (1956) A genetic change associated with the R locus in maize which is directed and potentially reversible. Genetics 41:872–889
46. Brink RA (1956) A regularly reversible change in determinative action at the R locus in maize. Genetics 41:636
47. Brink RA (1958) Paramutation at the R locus in maize. Cold Spring Harb Sym 23:379–391
48. Lyon M (1961) Gene action in the X-chromosome of the mouse. Nature 190:372–373
49. Lyon MF (1971) Possible mechanisms of X chromosome inactivation. Nat New Biol 232:229–232
50. Lyon MF (1993) Epigenetic inheritance in mammals. Trends Genet 9:123–128
51. McClintock B (1959) Genetic and cytological studies of maize. Carnegie Inst Wash Yr Bk 58:452–456
52. Banks JA, Masson P, Fedoroff N (1988) Molecular mechanisms in the developmental regulation of the maize *Suppressor-mutator* transposable element. Genes Dev 2:1364–1380
53. Meyer P (2011) DNA methylation systems and targets in plants. FEBS Lett 585: 2008–2015
54. Simon SA, Meyers BC (2011) Small RNA-mediated epigenetic modifications in plants. Curr Opin Plant Biol 14:148–155
55. Zhang H, Zhu JK (2011) RNA-directed DNA methylation. Curr Opin Plant Biol 14:142–147
56. Haag JR, Pikaard CS (2011) Multisubunit RNA polymerases IV and V: purveyors of non-coding RNA for plant gene silencing. Nat Rev Mol Cell Biol 12:483–492
57. Zhu JK (2009) Active DNA demethylation mediated by DNA glycosylases. Annu Rev Genet 43:143–166
58. Bourc'his D, Voinnet O (2010) A small-RNA perspective on gametogenesis, fertilization, and early zygotic development. Science 330:617–622
59. Feng S, Jacobsen SE, Reik W (2010) Epigenetic reprogramming in plant and animal development. Science 330:622–627
60. He G, Elling AA, Deng XW (2011) The epigenome and plant development. Annu Rev Plant Biol 62:411–435
61. Henderson IR, Jacobsen SE (2007) Epigenetic inheritance in plants. Nature 447:418–424
62. Mattick JS (2011) The central role of RNA in human development and cognition. FEBS Lett 585:1600–1616
63. Almeida R, Allshire RC (2005) RNA silencing and genome regulation. Trends Cell Biol 15:251–258

64. Johnson MA, Bender J (2009) Reprogramming the epigenome during germline and seed development. Genome Biol 10:232
65. La H et al (2011) A 5-methylcytosine DNA glycosylase/lyase demethylates the retrotransposon Tos17 and promotes its transposition in rice. Proc Natl Acad Sci USA 108:15498–15503
66. Zheng X et al (2008) ROS3 is an RNA-binding protein required for DNA demethylation in Arabidopsis. Nature 455:1259–1262
67. Banks JA, Fedoroff N (1989) Patterns of developmental and heritable change in methylation of the Suppressor-mutator transposable element. Dev Genet 10:425–437
68. Fedoroff NV, Banks JA (1988) Is the *Suppressor-mutator* element controlled by a basic developmental regulatory mechanism? Genetics 120:559–577
69. Schlappi M, Smith D, Fedoroff N (1993) TnpA trans-activates methylated maize *Suppressor-mutator* transposable elements in transgenic tobacco. Genetics 133: 1009–1021
70. Cui H, Fedoroff NV (2002) Inducible DNA demethylation mediated by the maize Suppressor-mutator transposon-encoded TnpA protein. Plant Cell 14:2883–2899
71. McClintock B (1964) Aspects of gene regulation in maize. Carnegie Inst Wash Yr Bk 63:592–602
72. Lisch D (2009) Epigenetic regulation of transposable elements in plants. Annu Rev Plant Biol 60:43–66

Chapter 14

Activation Tagging Using the Maize *En-I* Transposon System for the Identification of Abiotic Stress Resistance Genes in Arabidopsis

Amal Harb and Andy Pereira

Abstract

Activation tagging is a high-throughput method of overexpressing genes by using an enhancer present in insertion sequences that are randomly inserted in the genome to enhance the expression of adjacent genes. Gain-of-function approaches are advantageous to identify the functions of redundant genes that are not identifiable by knockout (KO) mutations, and for identification of phenotypes with small effects, which are enhanced by activation. An activation tag (ATag) library of 800 lines was generated in Arabidopsis ecotype Columbia using the *En-I* (*Spm*) transposon system. The ATag lines were used in a forward genetics strategy to identify novel genes that confer resistance/tolerance to abiotic stresses. The ATag lines were screened for altered drought and salt stress response phenotypes using quantitative assays for biomass accumulation under stress, revealing a number of resistant and sensitive ATag mutants.

Key words Activation tagging, *En-I* transposon, Abiotic stress, Arabidopsis, Forward genetics

1 Introduction

Forward and reverse genetics techniques have been used to understand plant responses to different stresses [1, 2]. Insertion mutagenesis using T-DNA and transposon knockout (KO) mutants as well as RNAi knockdown lines are useful methods to dissect the pathways that are involved in plant responses to stress [3]. Although loss-of-function mutagenesis is a direct way to reveal gene function, gene redundancy and lethality in Arabidopsis make the loss-of-function mutations of less practical use [1]. Approximately two-thirds of the Arabidopsis genome is duplicated, and many genes have redundant functions [4]. Thus, a gain-of-function strategy using activation tagging was developed that could identify the phenotypic functions of redundant genes [5–7]. An attractive tool to investigate genes with redundant function or minor mutant phenotypes with small genetic effects is activation tagging, a gain-of-function approach.

Thomas Peterson (ed.), *Plant Transposable Elements: Methods and Protocols*, Methods in Molecular Biology, vol. 1057,
DOI 10.1007/978-1-62703-568-2_14,

In activation tagging, a DNA insertion sequence such as transposon or T-DNA containing an enhancer sequence is used. When this activation tag (ATag) element inserts near genes in the genome, it can increase the expression of adjacent genes (to a distance of ~10 kb) and often displays an overexpression phenotype of the tagged gene. Activation tagging systems using T-DNA insertions bearing a 35S CaMV enhancer [8, 9] generated over 30 dominant mutants out of more than 30,000 transformed plants [8]. However, the low frequency of activation-tagged mutants in Arabidopsis due to multiple inserts and methylation makes a T-DNA approach less attractive for saturation of the Arabidopsis genome [10].

Transposon-based activation tagging is an efficient mutagenesis method in many plant species. The *En-I* (*Spm-dSpm*) system has been developed with an *I/dSpm* element modified as an ATag and used for activation tagging in Arabidopsis [5, 11]. The system uses the herbicide selectable markers, the BAR and SU1 genes, for greenhouse-based selection of stable transposon inserts [5, 11, 12]. The *En-I* ATag system generated a high frequency of dominant mutations (31 dominant mutants, approximately 1 %) from a population of 2,900 insertions [5]. Later unpublished work with more precise screening of progeny lines revealed roughly 200 independent mutations in 8,000 ATag lines, giving a frequency of 2.5 %.

Using the technique of activation tagging, many novel mutants have been isolated [5, 13–15]. These include mutations for flowering time [13], parthenocarpy [16], reduced apical dominance [7], multiple leaf phenotypes [5], as well as other developmental [11, 17] and biochemical mutants [11, 18, 19] including abiotic and biotic stress resistance mutants [15, 18–22].

In this study, an ATag population of about 800 lines was generated in Arabidopsis ecotype Columbia (Col) as has been described previously for ecotype Ws [5]. Activation tagging mutant lines were screened for drought and salt stress phenotype. Drought- and salt-resistant lines were analyzed molecularly and the tagged genes were identified.

2 Materials

2.1 Generation of Activation Tagging Population in Arabidopsis Columbia Ecotype

1. Arabidopsis seeds of activation tagging mutants in Columbia ecotype.
2. Flats: 1,020 per box Flat no holes.
3. Humidity domes.
4. Peat moss.
5. Perlite.
6. Vermiculite.
7. Needle for sowing.

8. Smooth surface paper for sowing.
9. Cold room set at 4 °C.
10. Controlled plant growth room.
11. Liberty herbicide.
12. R7402.

2.2 Drought Screen

1. Peat pellets.
2. Trays: TOP press fill tray of 2.5′ SVD pots.
3. Flats: 1,020 per box Flat no holes.
4. Humidity domes.
5. Red marker pen.
6. Labels.
7. Seeds of knockout mutants.
8. Seeds of activation tagging mutants.
9. Wild-type Arabidopsis seeds.
10. Needle for sowing.
11. Smooth surface paper for sowing.
12. Cold room set at 4 °C.
13. Controlled plant growth room.
14. Balance.
15. Syringes (5 and 10 mL) to adjust the water content.
16. Beakers.
17. Blades.
18. Analytical balance.
19. Glassine bags (1 oz. flat glassine bags (2¾ × 3¼).
20. Card boxes.
21. Oven set at 75 °C.

2.3 Salinity Screen

1. Peat pellets.
2. Trays: TOP press fill tray of 2.5′ SVD pots (32 pots/tray).
3. Flats: 1,020 per box Flat no holes.
4. Humidity domes.
5. Red marker.
6. Labels.
7. Seeds of knockout mutants.
8. Seeds of activation tagging mutants (generated during this work).
9. Wild-type Arabidopsis seed.
10. Needle for sowing.

11. Smooth surface paper for sowing.
12. Cold room set at 4 °C.
13. Controlled plant growth room.
14. Sodium chloride (NaCl).
15. Analytical balance.
16. Glassine bags: 1 oz. Flat glassine bag (2¾ × 3¼).
17. Card boxes.
18. Oven set at 75 °C.

2.4 DNA Isolation from *Arabidopsis* Leaves

1. Plant material from activation tagging mutants for DNA isolation.
2. Eppendorf tubes.
3. Pestle that fits Eppendorf tube.
4. Liquid nitrogen.
5. DNA extraction buffer: 0.3 NaCl, 50 mM Tris–HCl pH 7.5, 20 mM ethylenediamine tetra-acetic acid (EDTA), 2 % sarkosyl, 0.5 % sodium dodecylsulfate (SDS), 5 M urea, 5 % phenol (equilibrated in 0.5 M Tris–HCl pH 8.0). Mix the first five ingredients in 2× stock solution, and add urea and phenol before use.
6. Phenol:chloroform (1:1).
7. Isopropanol.
8. 70 % ethanol.
9. Tris–EDTA (TE) buffer.
10. RNase.
11. Microcentrifuge.
12. Agarose gel electrophoresis equipment.
13. Power supply.
14. Agarose.
15. Ethidium bromide.
16. Gel documentation machine.
17. Nanodrop for DNA quantification.

2.5 Identification of Candidate Genes in Activation Tag Mutants by TAIL PCR

1. Plant material from activation tagging mutants for DNA isolation.
2. PCR machine.
3. Nucleotides (dNTPs).
4. Taq polymerase.
5. Primers (*see* Table 1).
6. PCR tubes.

Table 1
List of primers used in TAIL PCR analysis

Primer name	Primer sequence (5′–3′)
Int2	CAGGGTAGCTTACTGATGTGCG
Irj-201	CATAAGAGTGTCGGTTGCTTGTTG
DSpm1	CTTATTTCAGTAAGAGTGTGGGGTTTTGG
AD1	TG(A/T)G(A/T/G/C)AG(A/T)A(A/T/G/C)CA(G/C)AGA
AD2	(G/C)TTG(A/T/G/C)TA(G/C)T(A/T/G/C)CT(A/T/G/C)TGC
AD3	CA(A/T)CGIC(A/T/G/C)GAIA(G/C)GAA
AD4	TC(G/C)TICG(A/T/G/C)ACIT(A/T)GGA
AD5	(A/T)CAG(A/T/G/C)TG(A/T)T(A/T/G/C)GT(A/T/G/C)CTG
AD6	AG(A/T)G(A/T/G/C)AG(A/T)A(A/T/G/C)CA(A/T)AGG

7. Agarose gel electrophoresis equipment.
8. Power supply.
9. Agarose.
10. Ethidium bromide.
11. Gel documentation machine.
12. Blades.
13. Eppendorf tubes.
14. Kit for DNA extraction from agarose gel.

2.6 Expression Analysis of Candidate Genes by RT-PCR

1. Plant material from activation tagging mutants and wild-type plants for RNA isolation.
2. RNAeasy kit.
3. RNAase-free DNAse kit.
4. iScript cDNA synthesis kit.
5. Primers specific to the candidate genes.
6. Primers for ubiquitin 10 (UBQ10, AT4G05320) as a reference gene.
7. PCR machine.
8. Nucleotides (dNTPs).
9. Taq polymerase.
10. PCR tubes.

11. Agarose electrophoresis set.
12. Power supply.
13. Agarose.
14. Ethidium bromide.
15. Gel documentation machine.

3 Methods

3.1 Generation of Activation Tagging Population in Arabidopsis Columbia Ecotype

1. Seeds of T2 transformants were generated in Arabidopsis ecotype Columbia (Col-0) [5].
2. Sow T_2 seeds in peat moss:perlite:vermiculite mix in plastic flats and incubate in the cold room at 4 °C for 2 days.
3. Grow plants in the growth room under 16-h (100 μmol m/ s^{-1}) light at 22 °C.
4. After germination, spray seedlings twice with (0.5 μL/L) Liberty herbicide (18.19 % glufosinate ammonium) [23].
5. Transfer plants with SU1 phenotype (dwarf, dark, reduced apical dominance plants) to fresh peat moss-filled pots, and grow to maturity and seed formation under the same conditions as above.
6. Collect seeds from SU1 phenotype plants.
7. Sow seeds collected from SU1 plants at high density in peat moss:perlite:vermiculite mix in flats (*see* **Note 1**).
8. After germination, spray the seedlings with (0.5 μL/L) Liberty herbicide (contains 18.19 % glufosinate ammonium) three times, and with 100 g/L R7402 daily for 8 days [24].
9. Count the double-resistant seedlings to estimate the stable transposition frequency (STF) (*see* **Note 2**).
10. Transfer only double-resistant seedlings with STF less than 3 % (stable and unique insertions) to plastic pots to grow to maturity for morphological and phenotypic characterization (*see* Fig. 1) (*see* **Note 3**).
11. Sow seeds from stable insertion lines to study inheritance of mutation, segregation, and dominance.

3.2 Drought Screen

1. Grow plants from individual stable activation-tagged lines and wild-type (WT) plants in Jiffy peat pellets in growth chamber/ room under 10-h light (100 μmol m/s^{-1}) at 22 °C (*see* **Note 4**).
2. Using a balance that is connected to the computer, weigh the labeled pellets and print the weights directly on an Excel sheet (*see* **Note 5**).
3. Conduct controlled gravimetric drought screen as described in ref. 25.

Fig. 1 Columbia activation tag (ColATag) mutants showing morphological phenotypes. Each ColATag morphological mutant is indicated by *C* followed by a number; a wild-type Columbia plant (*WT Col*) is shown for comparison

3.3 Salinity Screen

1. Grow activation tagging and WT plants in Jiffy peat pellets (*see* Subheadings 3.1 and 3.2).
2. At 30 days after sowing (DAS), treat a set of plants with 150 mM NaCl by placing the pots in a tray of NaCl solution, and place another set of pots in a tray of water as a control group.
3. Harvest plants after 7 days of treatment.
4. After 2 days of drying plant samples, measure the dry weight (biomass).
5. Calculate the relative reduction in biomass as shown in the following equation:

$$\text{Relative reduction in biomass (RB)} = \frac{(\text{Biomass of water} - \text{treated control}) - (\text{biomass of salt} - \text{treated})}{(\text{Biomass of water} - \text{treated control})}.$$

Figure 2 shows reduction in biomass under salt treatment in some activation tagging mutants compared to the WT.

3.4 DNA Isolation from Arabidopsis Leaves

1. Put about 200 mg of Arabidopsis in Eppendorf tube and freeze in liquid nitrogen [26].
2. Grind the frozen leaves with a suitable pestle to a fine powder.
3. Add 250 μL of the DNA extraction buffer and continue grinding of the sample.
4. Add another 250 μL of the DNA extraction buffer and 500 μL phenol/chloroform. Hand vortex and keep sample at room temperature for 20 min.
5. Centrifuge the samples for 10 min at 10,600 × *g*.

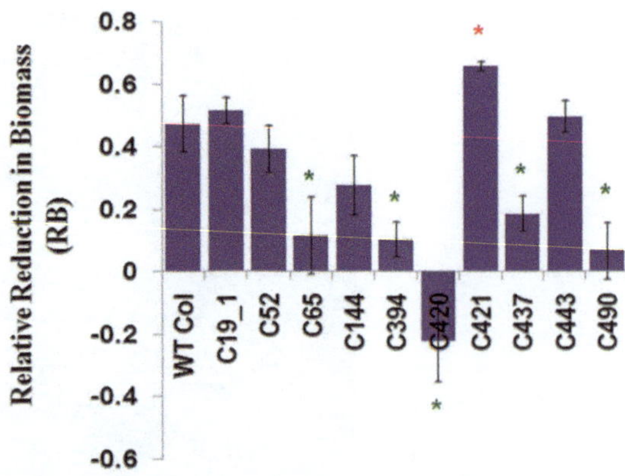

Fig. 2 Salt stress screen of Col ATag mutants. Relative reduction in biomass (RB) of ten ATag mutants compared to wild-type Columbia (WT Col). Bars represent ±SE, *N*= 4. *Significance at *p*-value < 0.05. *Red asterisk* indicates lines sensitive to salt stress compared to the WT Col, *green asterisk* indicates salt-resistant lines compared to WT Col

6. Transfer the supernatant to a clean Eppendorf tube.
7. Add 0.7 volume isopropanol, mix, and keep the tube at room temperature for 5 min.
8. Centrifuge for 5 min at 16,000 rpm (full speed).
9. Discard the supernatant and wash the pellet with 500–1,000 μL 70 % ethanol.
10. Air-dry the DNA pellet at room temperature for 10 min.
11. Dissolve the DNA pellet in 100 μL TE buffer containing 10 μg/mL RNase.
12. Run 5 μL of the DNA solution on 0.8 % agarose gel to check the quality of the isolated DNA.
13. Take 2 μL to measure the purity and the quantity of the DNA solution using a Nanodrop spectrophotometer.

3.5 Identification of Candidate Genes in Activation Tag Mutants by TAIL PCR

1. Use 20 ng of genomic DNA of mutant genotypes to carry out TAIL PCR as described below based on published methods [27–29] using the primers listed (*see* Table 1).
2. To set up primary PCR reaction, mix the following components: 20 ng of the DNA, 1× PCR buffer, 1.5 mM $MgCl_2$, 0.1 mM dNTPs, 0.15 μM Int2 (the furthest right border primer of the transposon), 2 μM of each one of the degenerate primers (AD1–AD6) in six reaction mixes, 1 U Taq polymerase, and sterile distilled water to complete the desired volume of the reaction mix (*see* **Note 6**).
3. Put the primary PCR reaction mixes in a thermal cycler following the protocol shown below:

1 Cycle denaturation: 94 °C 2 min.

5 Cycles: 94 °C 1 min, 62 °C 1 min, 72 °C 2 min.

1 Cycle: 94 °C 1 min, 25°C 3 min, 72 °C 2 min.

15 Cycles:

- 2 Cycles: 94 °C 30 s, 65 °C 1 min, 72 °C 2 min.
- 1 Cycle: 94 °C 30 s, 45 °C 1 min, 72 °C 2 min.

4. Dilute the primary PCR product with sterile distilled water by 1:40.
5. Set up secondary PCR reaction by mixing the following components: 1 μL of the diluted primary PCR product, 1× PCR buffer, 0.1 mM dNTPs, 0.2 μM Irj-201 (the middle right border primer of the transposon), 2 μM of each one of the degenerate primers (AD1–AD6) in six reaction mixes, 1 U Taq polymerase, and sterile distilled water to complete the desired volume of the reaction mix.
6. Put the secondary PCR reaction mixes in a thermal cycler following the protocol shown below:

   1 Cycle: 93 °C 1 min.

   13 Cycles:

   - 2 Cycles: 94 °C 30 s, 62 °C 1 min, 72 °C 2 min.
   - 1 Cycle: 94 °C 30 s, 45 °C 1 min, 72 °C 2 min.
7. Dilute the secondary PCR product with sterile distilled water by 1:10.
8. Set up tertiary PCR reaction by mixing the following components: 1 μL of the diluted secondary PCR product, 1× PCR buffer, 1.5 mM $MgCl_2$, 0.1 mM dNTPs, 0.2 μM DSpml (the closest right border primer of the transposon), 2 μM of each one of the degenerate primers (AD1–AD6) in six reaction mixes, 1 U Taq polymerase, and sterile distilled water to complete the desired volume of the reaction mix.
9. Put the tertiary PCR reaction mixes in a thermal cycler following the protocol shown below:

   1 Cycle: 93 °C 1 min.

   20 Cycle: 94 °C 30 s, 45 °C 1 min, 72 °C 2 min.
10. Run the PCR products of the three TAIL PCR reactions on 1 % agarose gel.
11. Purify amplicons of the third PCR reaction using a suitable gel extraction kit.
12. Sequence the purified DNA fragment using the closest right border primer of the transposon (DSpml, 5′-CTTATTTCAGTAAGAGTGTGGGGTTTTGG-3′).

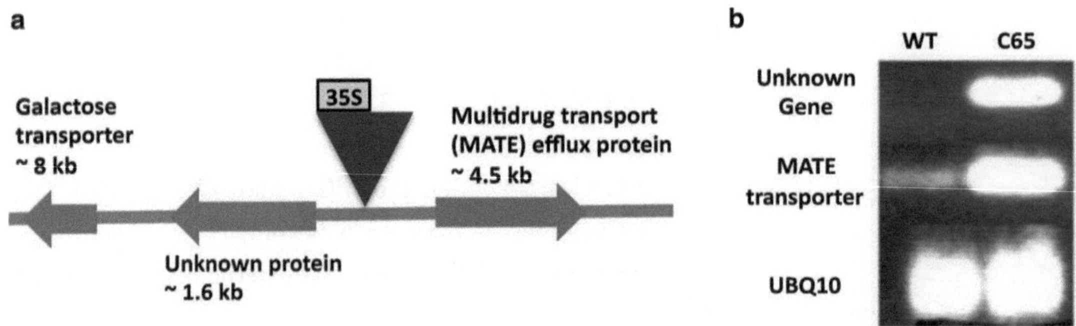

Fig. 3 Structure and expression analysis of a representative activation-tagged mutant. (**a**) Schematic illustration depicting the position of stable transposon (AIE) insertion in Arabidopsis genome in Col ATag mutant C65. *Blue line* indicates a portion of chromosome 4; *thick red arrows* show the positions and transcriptional orientations of nearby candidate genes. *Numbers* of kb indicate the distances from each gene promoter to the CaMv 35S enhancer in the AIE. (**b**) RT-PCR expression analysis of candidate genes in ColATag line mutant C65. Semiquantitative RT-PCR results indicate overexpression of two candidate genes in mutant C65 compared to wild type (WT); ubiquitin gene UBQ10 is used as a control

13. Blast the resulted sequence against the public gene databases to identify the tagged gene(s); an example is shown for mutant C65 (*see* Fig. 3a).

3.6 Expression Analysis of Candidate Genes by RT-PCR

1. Isolate RNA from activation tagging mutants and the WT using RNAeasy kit.
2. Eliminate DNA by RNAase-free DNAse kit.
3. Synthesize cDNA using iScript cDNA synthesis kit.
4. Use gene-specific primers to study the expression of candidate genes.
5. Set the PCR machine on the following protocol:
   (a) Initial denaturation cycle at 95 °C for 5 min.
   (b) 30 Cycles of denaturation at 95 °C for 1 min, annealing at 56 °C for 30 s, and extension at 72 °C for 1 min.
   (c) Final extension at 72 °C for 5 min.
6. Use UBQ10 (AT4G05320) as a reference gene in analysis.
7. Run PCR products on agarose gel and study the results using gel documentation machine. The results of mutant C65 are shown (*see* Fig. 3b).

4 Notes

1. This step is done to generate the stable activation tagging population.
2. STF is the ratio of surviving plants to the total number of sown seeds.
3. Based on Southern analysis, activation tagging mutants with different insertions and independent transpositions were obtained from plants with STF less than 3 %.
4. The peat pellets are wrapped with a biodegradable net, but each box of Jiffy pellets contains some inappropriately wrapped pellets; these should not be used for stress testing.
5. The software supplied with the balance is activated during the weighing process; it will transfer the data from the balance to a computer Excel sheet when the "print" button on the balance is pressed.
6. Each one of the activation tagging mutants has different degenerate primers that give good products in the three PCR reactions. So, the six degenerate primers must be tested to find combinations of these primers with the specific primer that gives the best result for each mutant.

References

1. Pereira A (2001) Genetic dissection of plant stress responses. In: Bucner P, Hawkesford MJ (eds) Molecular analysis of plant adaptation to the environment. Kluwer Academic, New York, pp 17–42
2. Papdi C et al (2010) Genetic screens to identify plant stress genes. Methods Mol Biol 639: 121–139
3. Xiong L, Zhu JK (2002) Molecular and genetic aspects of plant responses to osmotic stress. Plant Cell Environ 25:131–139
4. Bounchè N, Bounchez D (2001) Arabidopsis gene knockout: phenotypes wanted. Curr Opin Plant Biol 4:111–117
5. Marsch-Martinez N et al (2002) Activation tagging using the En-I maize transposon system in Arabidopsis. Plant Physiol 129: 1544–1556
6. Hirschi K (2003) Insertional mutants: a foundation for assessing gene function. Trends Plant Sci 8:205–207
7. Nakazawa M et al (2003) Activation tagging, a novel tool to dissect the functions of a gene family. Plant J 34:741–750
8. Weigel D et al (2000) Activation tagging in Arabidopsis. Plant Physiol 122:1003–1013
9. Tani H et al (2004) Activation tagging in plants: a tool for gene discovery. Funct Integr Genomics 4:258–266
10. Chalfun-Junior A et al (2003) Low frequency of T-DNA based activation tagging in Arabidopsis is correlated with methylation of CaMV 35S enhancer sequences. FEBS Lett 555:459–463
11. Scheinder A et al (2005) A transposon- based activation – tagging population in Arabidopsis thaliana (TAMARA) and its application in the identification of dominant developmental and metabolic mutations. FEBS Lett 579: 4622–4628
12. Tissier A et al (1999) Multiple independent defective suppressor – mutator transposon insertions in Arabidopsis: a tool for functional genomics. Plant Cell 11:1841–1852
13. Kardailsky I et al (1999) Activation tagging of the floral inducer FT. Science 286:1962–1965
14. Borevitz JO et al (2000) Activation tagging identifies a conserved MYB regulator of phenylpropanoid biosynthesis. Plant Cell 12: 2383–2394
15. Karaba A et al (2007) Improvement of water use efficiency in rice by expression of HARDY, an Arabidopsis drought and salt tolerance gene. Proc Natl Acad Sci U S A 104:15270–15275
16. Ito T, Meyerowitz EM (2000) Overexpression of a gene encoding a cytochrome P450 CYP78A9, induces large and seedless fruit in Arabidopsis. Plant Cell 12:1541–1550

17. Marsch-Martinez N et al (2006) BOLITA, an Arabidopsis AP2/ERF-like transcription factor that affects cell expansion and proliferation/differentiation pathways. Plant Mol Biol 62: 825–843
18. Aharoni A et al (2004) The SHINE clade of AP2 domain transcription factors activates wax biosynthesis, alters cuticle properties, and confers drought tolerance when overexpressed in Arabidopsis. Plant Cell 16:2463–2480
19. Seo PJ, Park CM (2010) MYB96-mediated abscisic acid signals induce pathogen resistance response by promoting salicylic acid biosynthesis in Arabidopsis. FEBS Lett 579: 4622–4628
20. Chini A et al (2004) Drought tolerance established by enhanced expression of the CC-NBS-LRR gene, ADR1, requires salicylic acid, EDS1 and ABI1. Plant J 38:810–822
21. Yu H et al (2008) Activated expression of an Arabidopsis HD-START protein confers drought tolerance with improved root system and reduced stomatal density. Plant Cell 20: 1134–1151
22. Aboul-Soud MA et al (2009) Activation tagging of ADR2 conveys a spreading lesion phenotype and resistance to biotrophic pathogens. New Phytol 183:1163–1175
23. Thompson CJ et al (1987) Characterization of the herbicide-resistance gene bar from Streptomyces hygroscopicus. EMBO J 6: 2519–2523
24. O'Keefe DP et al (1994) Plant expression of a bacterial cytochrome P450 that catalyzes activation of a sulfonylurea pro-herbicide. Plant Physiol 105:473–482
25. Harb A, Pereira A (2011) Screening Arabidopsis genotypes for drought stress resistance. Methods Mol Biol 678:191–198
26. Pereira A, Aarts M (1998) Transposon tagging with the En-I system. In: Martínez-Zapater JM, Salinas J (eds) Arabidopsis protocols. Humana Press, Totowa, NJ, pp 329–338
27. Liu YG et al (1995) Efficient isolation and mapping of *Arabidopsis thaliana* T-DNA insert junctions by thermal asymmetric interlaced PCR. Plant J 8:457–463
28. Liu YG, Whittier RF (1995) Thermal asymmetric interlaced PCR: automatable amplification and sequencing of insert end fragments from P1 and YAC clones for chromosome walking. Genomics 25:674–681
29. Tsugeki R, Kochieva EZ, Fedoroff NV (1996) A transposon insertion in the Arabidopsis SSR16 gene causes an embryo-defective lethal mutation. Plant J 10:479–489

Chapter 15

Reverse Genetics in Rice Using *Tos17*

Delphine Mieulet, Anne Diévart, Gaëtan Droc, Nadège Lanau, and Emmanuel Guiderdoni

Abstract

Transposon of *Oryza sativa* 17 (*Tos17*), a *Ty*1-Copia Class I retroelement, is one of the few active retroelements identified in rice, the main cereal crop of human consumption and the model genome for cereals. *Tos17* exists in two copies in the standard Nipponbare japonica genome ($n=12$ and 379 Mb). *Tos17* copies are inactive in the plant grown under normal conditions. However, the copy located on chromosome 7 can be activated upon tissue culture. Plants regenerated from 3- and 5-month-old tissue cultures harbor, respectively, an average of 3.5 and 8 newly transposed copies that are stably inserted at new positions in the genome. Due to its favorable features, *Tos17* has been extensively used for insertion mutagenesis of the model genome and 31,403 sequence indexed inserts harbored by regenerants/T-DNA plants are available in the databases. The corresponding seed stocks can be ordered from the laboratories which generated them. Both forward genetics and reverse genetics approaches using these lines have allowed the deciphering of gene function in rice. We report here two protocols for ascertaining the presence of a *Tos17* insertion in a gene of interest among R2/T2 seeds received from *Tos17* mutant stock centers: The first protocol is PCR-based and allows the identification of azygous, heterozygous and homozygous plants among progenies segregating the insertion. The second protocol is based on DNA blot analysis and can be used to identify homozygous plants carrying the *Tos17* copy responsible for gene disruption while cleaning the mutant background from other unwitting mutagen inserts.

Key words Insertion mutagenesis, Retroelement, Reverse genetics, Rice, *Tos17*

1 Introduction

Class I of mobile elements creates stable insertions because they transpose through an RNA intermediate and the reverse transcriptase. As a consequence of this "copy and paste" mode of transposition, they tend to accumulate in the host genome and may reach a high copy number. However, retroelements are generally inactive and not transcribed. Active retroelements have been isolated by reverse transcriptase-polymerase chain reaction (RT-PCR)

Delphine Mieulet and Anne Diévart contributed equally to this work.

Thomas Peterson (ed.), *Plant Transposable Elements: Methods and Protocols*, Methods in Molecular Biology, vol. 1057, DOI 10.1007/978-1-62703-568-2_15, © Springer Science+Business Media New York 2013

based on the conservation of their reverse transcriptase gene. The transposon of *Oryza sativa* (rice) *Tos17*, a Tyl copia long terminal repeat (LTR) retroelement, is one of six active transposons identified in rice. It has been isolated and characterized from a cell culture [1]. This 4,114 bp element is a member of a larger family of largely inactive *Tos* elements and, contrasting with other LTR transposons, exist in low [1–10] copy number in accessions of cultivated rice [1, 2]. Its transposition is regulated through DNA methylation at a transcriptional level in plants grown under normal conditions [3]. However, the stress accompanying cell dedifferentiation in tissue culture induces the demethylation and transposition of some copies of the element. A glycosylase lyase protein responsible for the active demethylation of *Tos17* has recently been characterized [4].

The standard cultivar of japonica rice, Nipponbare, contains two copies of *Tos17*, the first, active and located on chromosome 7 and the second, inactive and located on chromosome 10. The latter copy is actually transcribed as a fusion gene, a read through of an endogenous ABC transporter gene in which it is inserted [5]. Plants regenerated following 5 months of tissue culture contain on average 8 and a range of 5–30 newly transposed copies. As the *Agrobacterium*-mediated transformation procedure in rice involves tissue culture steps, T-DNA insertion lines also contain new copies of the element. In a given cultivar, accumulation of copies is a function of the activity of the resident copies and of the duration of tissue culture. An average of 0.2, 3, 3.5, and 4 newly transposed copies of *Tos17* is found in T-DNA plants of cv Tainung67, Zhonghua 11 Nipponbare and Dong Jin [6–8] (G. An, unpublished), respectively. As new copies accumulate in a cell lineage during tissue culture, some can be shared by the several plants regenerated from a same callus.

Due to its interesting features, *Tos17* has been extensively used as a mutagen through the generation of insertion libraries [9]. *Tos17* inserts in T-DNA insertion lines have also been characterized [8]. These lines have been used for both forward and reverse genetics. Though the success of establishing linkage between a phenotypic alteration observed in a mutant line and the mutagen remains rather low (3–5 %) in rice [5, 10], the functions of a number of important genes have been unveiled through *Tos17* tagging.

Tos17 has also been extensively used in reverse genetics: Insertions in a target gene have been first identified in DNA pools of regenerated lines using four combinations of two gene-specific primers and two insertion-specific primers in forward and reverse orientation [5], followed by the large-scale sequencing of *Tos17* insertion sites [9]. Large-scale sequencing has revealed (a) a strong insertion preference likely guided by the GC content of the host DNA region, (b) the presence of hot spots for insertions, and (c) frequent insertion within genes (80 %) with a higher and equal frequency in intron and exon sequences (ca. 45 % each) than in promoter (7 %) and 3′ UTR regions (4 %).

In a representative T-DNA insertion line library of cv Nipponbare, it has been shown that the genes tagged by a *Tos17* insert have an average additional 2.3 other discrete *Tos17* insertion alleles present in the library, whereas genes tagged by the T-DNA have only 0.14 other T-DNA insertion allele [8]. 274 genes were found to have from 5 to >100 allelic *Tos17* insertions in the library. A consequence of that finding is that informative allelic series of *Tos17* inserts are often available for a given gene. On the other hand, from a genome saturation perspective, *Tos17* mutagenesis has to be combined with another mutagen with less or different insertion bias.

The function of an increasing number of genes has been deciphered due to the use of knockouts created by *Tos17* insertions. Identification of *Tos17* inserts in a favorite gene can be browsed through Web-accessible genome navigator databases such as OryGenesDB [11, 12]. As *Tos17* is frequently spliced out when inserted into introns, insertions in exons or promoter sequences will be preferred. *Tos17* inserts in your favorite gene(s) can be ordered from the National Institute of Agrobiological Resources (cv Nipponbare) (NIAS Tsukuba, Japan), the Oryza Tag Line Génoplante library (cv Nipponbare) (CIRAD, Montpellier, France), and the Rice Mutant Database (cv Zhong Hua 11) (HZAU, Wuhan, China). The two latter collections also contain T-DNA insertions and seeds should therefore be handled as genetically modified organisms (GMOs). Generally 20–30 seeds (R2 or T2 generation) are received from these laboratories. Insertions are therefore largely segregating among the plants raised from these seeds. The first step upon seed receipt is the identification of plants homozygous, heterozygous, and azygous for the insertion in a gene of interest (GOI). Once identified, homozygous and azygous (control) T3/R3 progenies can then be evaluated for specific alterations with confidence.

We detail hereafter several protocols to order *Tos17* mutant lines, grow the plants and genotype them. We will describe two genotyping protocols, based on either PCR or DNA blot analysis, to identify homozygous and azygous plants among progenies segregating for *Tos17* inserts. DNA blot has the advantage of determining the total number of *Tos17* copies present in the line under study, in order to retain, if possible, progeny plants containing only the insert of interest. Through re-hybridization of the filter with different specific probes, other possible mutagens (T-DNA or other rice transposons known to transpose upon tissue culture) can be also detected and segregated out.

2 Materials

2.1 Seed Order and Bioanalysis

1. Accession number or Flanking Sequence Tag (FST) number of the GOI with *Tos17* insertion.
2. Web access to OryGenesDB (http://orygenesdb.cirad.fr/) and seed repository databases.

2.2 Plant Growth

1. Laminar flow hood.
2. Growth chamber at 28 °C and 60 % humidity (photoperiod 16:8).
3. Greenhouse at 28 °C and 60 % humidity (photoperiod 12:12).
4. Seed sterilization solutions: (1) 40 mL of 70 % ethanol; (2) Mix 30 mL of sterile distilled water in a 50 mL sterile Falcon tube, add 20 mL of a commercial hypochlorite solution (9.6 % of active chlorite) and 20 μL of Tween 20.
5. Petri dishes, tweezers, and sterile, round filter papers.

2.3 Quick DNA Extraction

1. Collection microtubes in racks (96 tubes of 1.2 mL) and caps.
2. Freeze-drying machine (Thermo Fischer, Heto Lyolab PowerDry PL 3000).
3. Mixer mill (Retsch, Mixer Mill MM 300, 20.746.0001), adapter set for microtube racks grinding balls of 3 mm diameter.
4. Extraction buffer: In 325 mL of distilled water, add 100 mL of Tris–HCl 1 M (pH 7.5), 25 mL of EDTA 0.5 M (pH 8.0), 25 mL of 10 % SDS, and 25 mL of NaCl 5 M. Mix and keep this solution in an incubator at 65 °C.
5. Isopropanol.
6. Centrifuge, rotor and adapter for racks.
7. Chemical hood.
8. Incubator at 65 °C.

2.4 Multiplex Polymerase Chain Reaction

1. Thermocycler, Taq DNA polymerase with standard Taq buffer, dNTPs Mix, primers for PCR amplification of the *Tos17* and GOI specific loci.
2. Electrophoresis tank, tray, and comb.
3. Agarose gel preparation: 1 % agarose in Tris–acetate–EDTA (TAE) (1×) buffer. Weigh 3.5 g of agarose powder and transfer it to an Erlenmeyer flask containing 350 mL of TAE (1×). Boil several minutes in a microwave to melt the powder and wait until the gel has cooled to ~60 °C before pouring. Prepare the gel tray by taping or blocking ends depending on the manufacturer instructions. Place the comb in the gel tray and pour the

agarose solution. Allow the gel to cool and solidify at room temperature. Remove the comb when ready to load samples.

4. Loading dye buffer for DNA electrophoresis.
5. DNA ladder for agarose gel.
6. UV transilluminator apparatus.

2.5 DNA Extraction for Southern Blotting

1. 2 mL microtubes.
2. Freeze-drying machine (Thermo Fischer, Heto Lyolab PowerDry PL 3000).
3. Mixer mill (Retsch, Mixer Mill MM 300, 20.746.0001), adapter set for microtube racks grinding balls of 3 mm diameter.
4. MATAB DNA extraction buffer. In a 500 mL cylinder, pour 50 mL of 1 M Tris–HCl (pH 8), 140 mL of 5 M NaCl (pH 8), 20 mL of 0.5 M EDTA (pH 8), 10 g of mixed alkyltrimethyl ammonium bromide (MATAB), 2.5 g of Na_2SO_3, and 5 g of PEG 6000. Then add sufficient water to get 500 mL final volume. Heat this solution in microwave and agitate to completely dissolve powders. Protect the bottle containing the solution from light with aluminum foil. Place in a water bath at 72 °C and left for 1 h prior to use. This solution can be kept at room temperature for 10 days.
5. CIAA (chloroform isoamyl alcohol) preparation. In a 500 mL cylinder, mix 480 mL of chloroform with 20 mL of isoamyl alcohol. Keep at room temperature.
6. Rnase A at 0.5 U/μL.
7. Flame-sealed Pasteur pipets, one per sample.
8. TE (Tris–EDTA) preparation. In a 200 mL cylinder, mix 2 mL of Tris–HCl (1 M pH = 8) with 400 μL of EDTA (0.5 M pH = 8). Then add sufficient water to get 200 mL final volume. Autoclave and keep at room temperature.

2.6 Southern Blotting

1. At least 8 μg of plant leaf genomic DNA extracted from ~1-month-old fresh plant leaves.
2. Enzymatic digestion of DNA: *Xba*I enzyme with buffer supplied by manufacturer.
3. Electrophoresis tank, tray, and comb.
4. Agarose gel preparation: 0.8 % agarose in Tris–acetate–EDTA (TAE) (1×) buffer. Weigh 2.8 g of agarose and transfer to an Erlenmeyer flask containing 350 mL of TAE (1×). Boil several minutes in a microwave to melt the powder and wait until the gel has cooled to ~60 °C before pouring. Prepare the gel tray by taping or blocking ends depending on the manufacturer instructions. Place the comb in the gel tray and pour the

agarose solution. Allow the gel to cool and solidify at room temperature. Remove the comb when ready to load samples.

5. Loading dye buffer for DNA electrophoresis.
6. DNA ladder for agarose gel.
7. UV transilluminator apparatus.
8. Depurination of DNA: 0.25 N HCl. To 800 mL distilled water in a graduated cylinder, add the appropriate volume of HCl stock solution for 1 L of 0.25 N HCl final concentration, then add sufficient water to get 1 L final volume (*see* **Note 1**).
9. Alkaline solution: 0.4 N NaOH. To 3.5 L distilled water in a graduated cylinder, add the appropriate volume of NaOH stock solution for 4 L of 0.4 N NaOH final concentration, then add sufficient distilled water to get 4 L final volume.
10. Southern blotting system: tank, towel, and Whatman papers, nylon membrane.
11. Stock solutions for hybridization: (a) saline-sodium citrate (SSC) buffers. The stock solution (25×) consists of 65.6 g of sodium chloride and 33 g of trisodium citrate (adjusted to pH 7.0 with HCl) in the appropriate volume of distilled water to get 1 L. This solution is autoclaved. To prepare 1 L of the (20×) SSC stock buffer, add 200 mL of distilled water to 800 mL of 25× SSC. (b) (50×) Denhardt's solution. Add 10 g of bovine serum albumin (BSA), 10 g of Ficoll 400, and 10 g of polyvinylpyrrolidone to the appropriate volume of distilled water to get 1 L of solution. Store at −20 °C.
12. Prehybridization buffer: In 14 mL of distilled water, add 1 mL of Tris–HCl 1 M (pH 8.0), 400 μL of 0.5 M EDTA (pH 8.0), and 4 mL of (25×) SSC. Heat this solution in a microwave up to 80 °C, do not boil it. Then add 200 μL of 20 % SDS and 400 μL of (50×) Denhardt's solution. Denature 200 μL of salmon sperm DNA at 100 °C for 10 min and add it to the prehybridization buffer. Heat again in a microwave up to 80 °C. Keep this solution in an incubator at 65 °C (*see* **Note 2**).
13. A random-prime labelling kit including primer, labelling buffer, and Klenow enzyme. Store at −20 °C.
14. [α^{32}P]-dCTP.
15. Thermal cycler, Taq polymerase and primers for PCR amplification of the *Tos17* and GOI-specific probes (*see* **Note 3**).
16. Hybridization buffer: In 7 mL of distilled water, add 500 μL of 1 M Tris–HCl (pH 8.0), 200 μL of 0.5 M EDTA (pH 8.0), and 2 mL of (25×) SSC. Heat this solution in a microwave up to 80 °C, do not boil it. Then add 100 μL of 20 % SDS, 200 μL of (50×) Denhardt's solution, 100 μL of the denatured salmon

sperm DNA, and 3 mL of Dextran sulfate 50 % solution. Heat in a microwave up to 80 °C. Keep this solution in an incubator at 65 °C (*see* **Note 2**).

17. S1 and S2 washing solutions. S1: in 970 mL of distilled water, add 25 mL of (20×) SSC and 5 mL of 20 % SDS. S2: in 990 mL of distilled water, add 5 mL of (20×) SSC and 5 mL of 20 % SDS. Heat solutions to 80 °C if used immediately or keep them at 65 °C in incubator until use.
18. Phosphorimager system and screen or alternatively, X-ray films, cassettes, and processors.
19. De-hybridization solution. 0.1 % SDS: pour 5 mL of 20 % SDS in 995 mL of distilled water in a 2 L glass beaker. Heat in microwave up to boiling.

3 Methods

3.1 Seed Order

1. Go to the following Website pages (*see* **Note 4**) to order seeds of *Tos17* insertion lines:
   (a) NIAS: http://www.rgrc.dna.affrc.go.jp/order.html.
   (b) OTL: http://oryzatagline.cirad.fr/order.htm.
   (c) RMD: http://rmd.ncpgr.cn/contact.cgi?nickname=.
2. Download and sign the Material Transfer Agreement (MTA) in 2–3 copies to be sent to the indicated address (*see* **Note 5**).
3. Upon confirmation of seed availability, pay handling fee and secure and send an import permit (*see* **Note 6**).
4. Receive the seeds by courier mail. They are accompanied by a copy of the signed MTA, the original import permit and a phytosanitary certificate. Generally, 10–30 seeds of T2 generation are provided (*see* **Note 7**).

3.2 Plant Growth and Material

1. Use 10–30 seeds of the *Tos17* insertion line of interest.
2. Remove the glumes and transfer the dehulled seeds into a 50 mL falcon tube.
3. Sterilization: Wash the seeds with 50 mL of seed sterilization solution 1 and shake by inverting 2 min. Remove the ethanol and add 50 mL of the sterilization solution 2. Mix gently during 30 min. Rinse five times with 40 mL of sterile distilled water. Discard the liquid and spread the seeds in a Petri dish containing a sterile round Whatman filter.
4. Germination: Add 10 mL of distilled water, seal the Petri dish, and incubate it under light at 28 °C (photoperiod 16:8).
5. After 7 days of growth, transfer the plantlets into soil in a greenhouse (*see* **Note 8**).

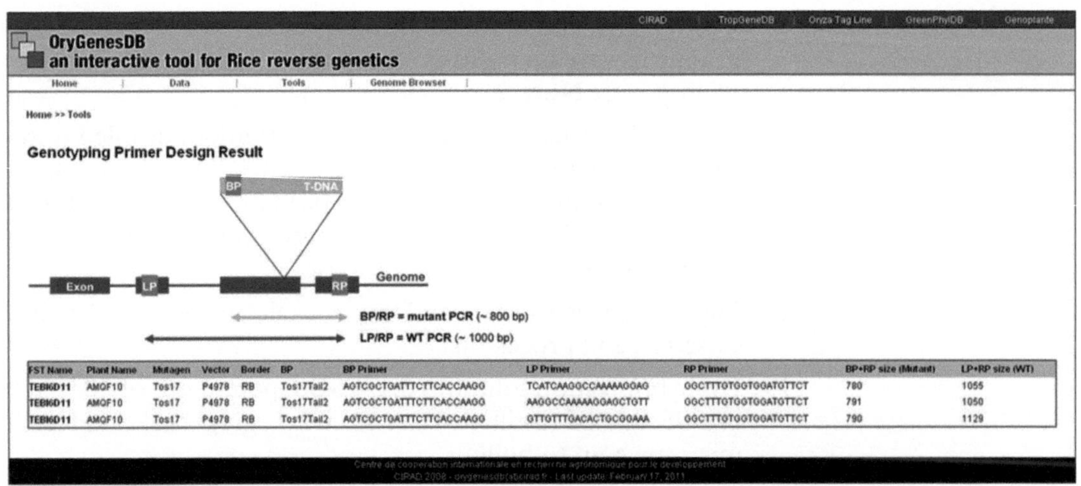

Fig. 1 Screen shot of the genotyping primer design tool of OryGenesDB (http://orygenesdb.cirad.fr/)

3.3 Protocol: PCR-Based Genotyping of Tos17 Insertions

3.3.1 Primer Design for Multiplex PCR

1. Go to the OryGenesDB Website (http://orygenesdb.cirad.fr/).
2. Select "Genotyping Primer Designer" in the "tools" tab and enter the accession number of the *Tos17* FST in the "Input" box.
3. Press "submit" (default parameters).

The three primers: Left Primer (LP), Right Primer (RP), and Backbone Primer (BP) are designed automatically (*see* Fig. 1). The design is based on Primer3Plus (www.bioinformatics.nl/cgi-bin/primer3plus/primer3plus.cgi) and the bioinformatics pipeline uses the Salk Institute parameters (http://signal.salk.edu/tdnaprimers.2.html). In these standard conditions, the PCR product is ~800 bp for a mutant plant and ~1,000 bp for a wild type plant.

3.3.2 Quick DNA Extraction

1. After 2 weeks of growth, for each plant, 2 cm of the second plant leaf is cut and collected separately in 1.2 mL 8-strip collection microtubes of a 96 racked plate.
2. Freeze-dry overnight.
3. Add one grinding ball per microtube and seal the tubes with microtube caps. Make sure tubes are balanced properly, and then disrupt tissue for 1.5 min at maximum speed (30 C/s) with a mixer mill. Turn racks over and disrupt for another 1.5 min.
4. Centrifuge briefly to make the powder fall to the bottom of the tube.
5. Add 180 μL of extraction Buffer and seal the tubes with caps.
6. Place at 65 °C, 15 min.

7. Centrifuge 15 min at max speed (~1,240 × *g*).
8. Transfer 120 μL of the supernatant to a 96-well assay plate containing 100 μL of isopropanol. Mix by pipetting. Incubate samples at room temperature for at least 5 min.
9. Cover samples with tape sheet and centrifuge for 10 min at max speed.
10. Decant the supernatant by turning plate upside down carefully, drain plate frame on a towel paper, and then add 100 μL of 75 % ethanol.
11. Cover samples with tape sheet and centrifuge for 5 min at max speed.
12. Decant the supernatant, drain gently on a towel paper, and allow to dry completely (from several hours to overnight).
13. Add 80 μL of sterile water and let DNA resuspend at room temperature for at least 30 min.

3.3.3 Multiplex PCR

1. For one multiplex PCR reaction, mix 39.75 μL of sterile distilled water, 5 μL of (10×) standard *Taq* reaction buffer, 1 μL of 10 mM dNTP mix, 1 μL of LP primer (10 μM), 1 μL of RP primer (10 μM), 1 μL of BP primer (10 μM), 1 μL of genomic DNA extracted by Quick DNA extraction (*see* Subheading 3.3.2) and 0.25 μL of Taq DNA Polymerase in a 0.5 mL eppendorf tube.
2. PCR conditions: 1 cycle: 95 °C for 2 min (hot start), 35 cycles: 95 °C for 20 s, 56 °C for 30 s, and 72 °C for 1 min, 1 cycle: 72 °C for 5 min (final extension). Hold at 15 °C.

3.3.4 Agarose Gel Migration

1. Prepare a 1 % agarose gel with the appropriate comb depending on the number of samples. Put the appropriate volume of TAE (1×) buffer in the electrophoresis tank to cover the gel.
2. Load first the DNA ladder in the first well and then all the samples.
3. Electrophorese at 90 V for 1 h.
4. To visualize DNA, stain the gel in 1 μg/mL ethidium bromide solution for 10 min and rinse in water for 10 min (*see* **Note 9**). Put the gel on a UV transilluminator and take a picture (*see* Fig. 2).

3.4 Protocol: Southern-Based Genotyping of *Tos17* Insertions

3.4.1 Primer Design for Probes

1. On the Gbrowse tab of OryGenesDB, enter the accession number of the GOI in the "Search" box. You can activate or remove tracks with the "Select Tracks" box at the bottom of the Web page. Activate the "Flanking Sequence Tag" box and check on the browser that your FST is actually there. By selecting "Restriction site annotation" on the right of the Web page, you will be able to visualize the size of the genomic fragments you

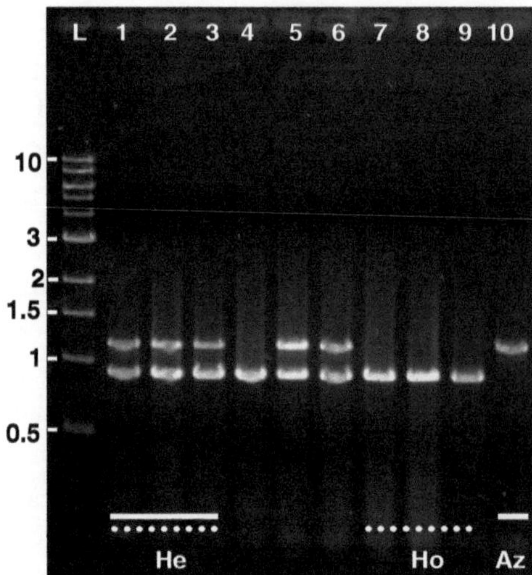

Fig. 2 Agarose gel electrophoresis of multiplex PCR products identifying individual plants either heterozygous (*He*), Homozygous (*Ho*), or Azygous (*Az*) for the *Tos17* insert disrupting the gene of interest among ten segregating progeny plants. *L*, DNA ladder with size in kb; 1–10

expect by cutting the wild type and rearranged DNA of the analyzed plants with the *Xba*I restriction enzyme (*see* **Note 10**).

2. Design primers for PCR amplification of a 300–500 bp DNA fragment located between an *Xba*I restriction site and the *Tos17* insertion. Use the *Tos17* primers to generate the *Tos17* probe to quantify the number of *Tos17* copies in your plants (*see* **Note 11**). These probes will be used, respectively, as gene-specific and *Tos17* probes in the Southern blot experiment.

3.4.2 Genomic DNA Extraction

1. After approximately 1 month of growth, for each plant, four pieces of 2 cm of a fresh leaf is cut and collected separately in 2 mL microtubes.
2. Freeze-dry overnight. Add two grinding ball per microtube and seal the tubes with microtube caps. Make sure tubes are balanced properly, and then disrupt tissue for 1.5 min at maximum speed (30 C/s) with a mixer mill. Turn racks over and disrupt for another 1.5 min (*see* **Note 12**).
3. Centrifuge briefly to make the powder fall to the bottom of the tube.
4. Add 900 μL of MATAB DNA extraction buffer pre-heated in 72 °C water bath.
5. Homogenize for 10 s with a vortex and incubate 45–60 min at 72 °C. Vortex every 15–20 min.

6. Add 900 μL of CIAA in each tube and mix gently by twisting approximately 50 times (*see* **Note 13**).
7. Centrifuge 10 min at max speed (~16,000×*g*) at room temperature.
8. Transfer the supernatant (aqueous phase) to a new 2 mL microtube and add 4 μL of RNAse A (2 U) per tube. Mix by pipetting. Incubate samples at 37 °C for 30 min.
9. Add 900 μL of CIAA in each tube and mix gently by twisting.
10. Centrifuge 10 min at max speed (~16,000×*g*) at room temperature.
11. Transfer the supernatant (aqueous phase) to a new 2 mL microtube and add 720 μL of isopropanol (0.8× volume). Mix gently by twisting until DNA precipitates forming a white pellet.
12. Using a sealed Pasteur pipet, spool out the DNA pellet, dry it on the microtube side and pour it into a new microtube containing 50 μL of TE.
13. Let DNA resuspend overnight at room temperature or 48 h at 4 °C. Keep at −20 °C for storage.

3.4.3 Genomic DNA Digestion

1. Mix 170 μL of distilled water, 8 μL of genomic DNA (1 μg/μL) (*see* **Note 14**), 20 μL of (10×) restriction enzymes buffer (*see* **Note 15**), 1 μL of *Xba*I (20 U), 2 μL of BSA (10 μg/mL) in a 1.5 mL eppendorf tube (*see* **Note 16**).
2. Centrifuge briefly to collect all the components at the bottom of the tube.
3. Incubate overnight at 37 °C.
4. Centrifuge briefly the next day.

3.4.4 DNA Precipitation

1. Add 10 μL (1/20 volume) of 5 M NaCl (pH 8.0) in each tube. Then add 500 μL (2.5 volumes) of absolute ethanol. Mix vigorously by vortexing and centrifuge 15 min at 4 °C at ~16,000×*g*.
2. Drain each tube by pouring carefully to keep the DNA pellet at the bottom of the tube. Add 500 μL of 70 % ethanol and wait 5 min. Centrifuge at 4 °C 20 min at ~16,000×*g*. Drain each tube again by pouring carefully and dry the top of the tube on a paper towel. Remove the ethanol at the bottom of the tube by pipetting carefully.
3. Resuspend the DNA pellet into 20 μL of distilled sterile water. Mix gently by pipetting and let resuspend overnight at room temperature.
4. Add ~3 μL of (10×) loading dye buffer in each tube.

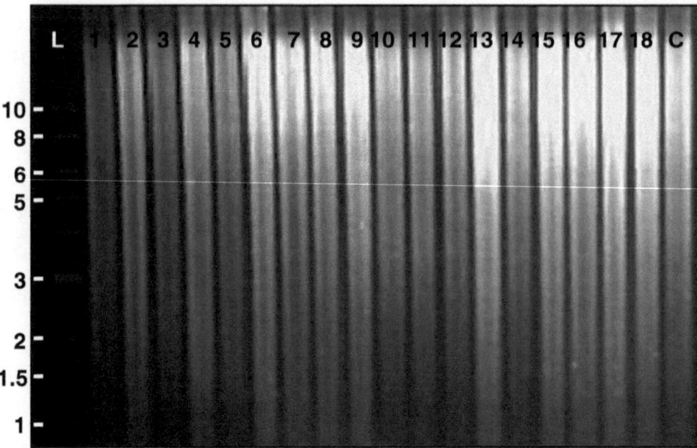

Fig. 3 Agarose gel electrophoresis of genomic DNA digested with *Xba*I restriction enzyme stained with ethidium bromide under UV illumination. *L*, DNA ladder with size in kb; 1–18, genomic DNA samples; *C*, genomic DNA of the control plant of the background cultivar harboring only native copies of *Tos17*

3.4.5 Agarose Gel Migration

1. Prepare a 0.8 % agarose gel with the appropriate comb depending on the number of samples. Add the appropriate volume of TAE (1×) buffer to the electrophoresis tank to cover the gel.
2. Load first the DNA ladder in the first well and then all the samples.
3. Electrophorese at 25 V for ~17–18 h.
4. To visualize DNA, stain the gel in 1 μg/mL ethidium bromide solution for 5 min and rinse in water for 5 min (*see* **Note 9**). Put the gel on a UV transilluminator and take a picture (*see* Fig. 3).

3.4.6 Southern Blotting

1. Lay the agarose gel on a plexiglass tray in a tank.
2. Pour 1 L of the depurination solution (0.25 N HCl) in the tank and agitate gently for 15 min.
3. Rinse the gel and the tank with distilled water.
4. Pour 1 L of the alkaline solution (0.4 N NaOH) in the tank and agitate gently for 30 min.
5. Remove the gel from the tank and insert a sponge (*see* **Note 17**). Pour 2.5–3 L of 0.4 N NaOH up to 2/3 of the sponge height. Cut six rectangles of Whatman paper and one nylon membrane of gel size. On the sponge, stack in order, three pieces of Whatman papers, the agarose gel, and the nylon membrane (*see* **Note 18**). Put a mark on the nylon membrane with an ink pen to locate both the wells and the face on which DNA will be transferred. Stack on the nylon membrane, three pieces of

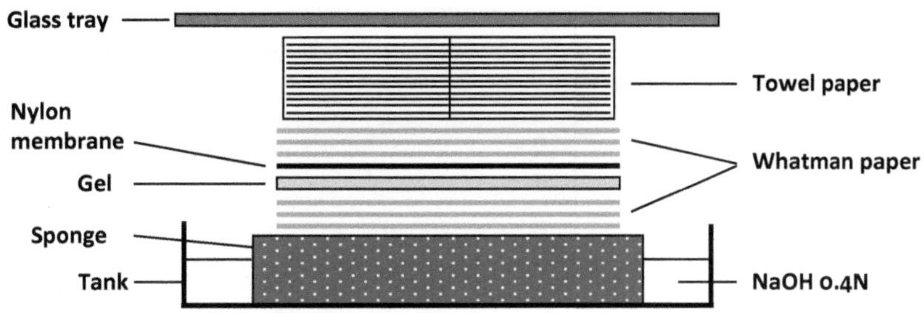

Fig. 4 Setup of the southern blot stacking

Whatman paper, ~10 cm thickness of paper towels, and top with a heavy glass tray (Fig. 4).

6. Let the DNA transfer overnight.
7. The next day, rinse the nylon membrane with (2×) SSC in a tank for 5 min.

3.4.7 Hybridization

1. Roll up the nylon membrane with the face containing DNA inside. Pour ~50 mL of (2×) SSC in a hybridization oven bottle and insert the nylon membrane. Close the bottle and carefully place the membrane in contact with the glass surface by rolling the bottle horizontally. Drain the (2×) SSC from the bottle and replace with 20 mL of prehybridization buffer. Place the bottle in the hybridization oven at 65 °C (*see* **Note 19**). Prehybridize for 6 h minimum up to overnight.
2. In a 1.5 mL eppendorf tube with screw cap, mix 50–100 ng of the probe you want to use (*see* **Note 20**), 5 μL of the primers furnished in the random-prime labelling kit and distilled water to a final volume of 35 μL. Denature by heating 5 min at 95 °C and chill on ice immediately. Add 10 μL of the labelling buffer, 2 μL of Klenow enzyme and 3 μL of [α^{32}P]-dCTP (specific activity 3,000 Ci/mmol) (*see* **Note 21**). Mix by gently pipetting, spin briefly to collect the contents at the bottom of the tube and incubate at 37 °C for 45 min. Stop the reaction by adding 450 μL of TE.
3. Denature by heating 5 min at 95 °C (*see* **Note 22**). Meanwhile, take out the hybridization bottle from the oven and drain the prehybridization solution. Prepare a disposable plastic tube with 10 mL of hybridization solution. Add the denatured probe and pour into the hybridization bottle. Replace the bottle in the oven at 65 °C and incubate overnight.
4. The next day, drain the hybridization solution from the bottle (*see* **Note 21**). Pour 80 mL of the S1 washing solution into the bottle and replace in the oven for 15 min (*see* **Note 23**). Drain this solution and repeat this step with 50 mL of the S1 solution for 30 min. Repeat again with 50 mL of the S2 solution for

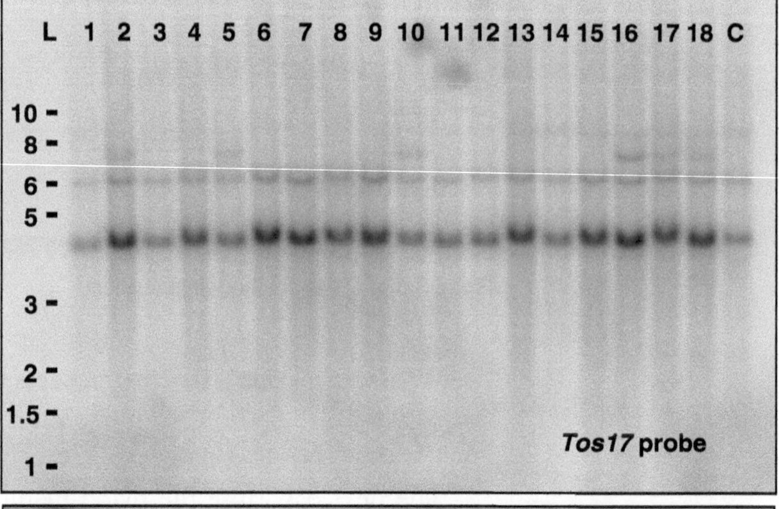

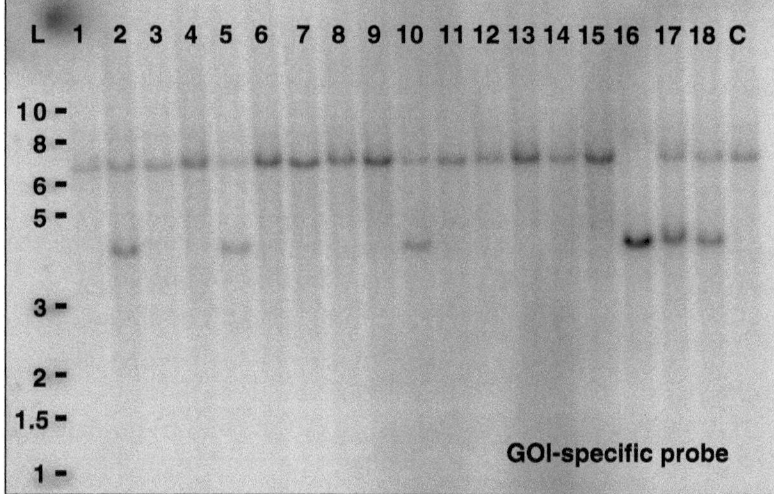

Fig. 5 Result of Southern blotting of *Xba*I digests of 18 progeny plants of a *Tos17* insertion line hybridized to a *Tos17* reverse transcriptase probe (*upper panel*) and to a probe specific to the gene of interest (GOI, *lower panel*): *Upper panel*: The insertion line under study contains three insertions of *Tos17*, the two endogenous ones plus a newly inserted copy that segregates in the progeny. *Lower panel*: Plants 2, 5,10, 17, and 18 are heterozygous for the *Tos17* insertion disrupting the gene of interest. Plant 16 is homozygous for the insertion. *L*, DNA ladder with size; 1–18, genomic DNA samples; *C*, genomic DNA of the control plant of the same background cultivar without rearranged *Tos17*. Note that here the control plant is Nipponbare which contains two endogenous *Tos17* copies

15 min. Finally, drain the last washing solution, remove the nylon membrane from the bottle and dry it between towel papers. Cover the nylon membrane with a plastic wrap.

5. Reveal your blot by using a phosphorimager screen or an X-ray system (*see* Fig. 5).

3.4.8 De-hybridization of Nylon Membrane

1. Pour the boiling 0.1 % SDS solution in a tank containing the nylon membrane (*see* **Note 24**). Agitate gently until the solution cools down to room temperature (~30 min).
2. Discard this solution and rinse the nylon membrane with (2×) SSC for 5 min.
3. Dry the membrane between paper towels. The membrane can be stored for months at room temperature.
4. To hybridize another probe to this membrane, repeat the protocol beginning with **Subheading** 3.4.7.

4 Notes

1. Be careful to always mix concentrated acid and water by adding the acid to a large volume of water. Mixing them the wrong way (water into acid) could cause a violent reaction and is hazardous.
2. This solution must be prepared fresh. Keep at 65 °C until use, but no more than one night.
3. The *Tos17* and GOI probes are amplified by PCR:
   (*Tos17* forward primer: GCGTGCACTTTATGGTCTCA and *Tos17* reverse primer: AGCAATATCCGGCCTAGTGA, the size of the amplified DNA fragment is 490 bp.)
4. Unlike the Nottingham Arabidopsis Stock Centre, which serves as a single repository for Arabidopsis, there is no unique repository for rice insertion line seeds.
5. The MTA must be signed by the principal investigator and authorized representative from your institution.
6. As the movement of rice seeds across countries is prone to phytosanitary restrictions, an import permit should be secured by the importing laboratory from its national quarantine authorities. Seeds are generally treated before shipment with hot water (57–60 °C for 15 min) to eliminate nematodes; fungicides and pesticides may affect germination ability.
7. Whereas NIAS *Tos17* lines are non-GMO, OTL, and RMD lines also bear T-DNA insert(s) and therefore have to be handled as GMOs in containment facilities.
8. Plant genotyping can be done at several stages of growth. At the plantlet stage, after 2 weeks of growth in soil, Protocol 1 (PCR) is preferred because minimal leaf material is needed for DNA extraction. Whereas, for Protocol 2 (DNA Southern blot), let the plants grow for at least 1 month to get enough material for genomic DNA extraction.
9. Ethidium bromide is mutagenic and must be handled with care. Wear gloves and dispose of waste properly.

10. If the size of the fragment exceeds 15 kb, choose another restriction enzyme.
11. For PCR amplification of both probes, use your wild type control plant.
12. Alternatively, samples can be frozen in liquid nitrogen and powdered with a pestle in a mortar. In between each sample, mortar must be clean with 70 % ethanol.
13. From this step of the protocol, all manipulations have to be done in a fume hood.
14. Include genomic DNA of a control plant of the same variety.
15. This buffer is provided with the *Xba*I enzyme by the manufacturer.
16. If the number of samples is large, prepare the digestion mix (distilled water, 10× restriction enzymes buffer, *Xba*I and BSA) in a 13 or 50 mL tube depending on the number of samples, then aliquot 192 μL of this mix into 1.5 mL eppendorf tubes and add 8 μL of DNA to each tube.
17. The sponge must be slightly larger than the gel.
18. To get even contact between all layers, be careful to always avoid bubbles. If bubbles are formed, eliminate them by rolling the top with a 50 mL Falcon tube or a 10 mL pipette.
19. Place the bottle in the right orientation to keep the nylon membrane adhered to the bottle glass surface.
20. Hybridize first the gene-specific probe and second, the *Tos17* probe.
21. Manipulate radioactivity behind a plexiglass screen and dispose of waste properly.
22. Do not tighten the cap completely or the tube could explode.
23. The duration of washes and the stringency of the washing solutions can be adapted according to the intensity of the radioactive signal.
24. Do not pour the solution directly on the membrane but to the side.

Acknowledgments

These protocols have been designed with the support of the French genomics initiative Génoplante and of the Agence Nationale de la Recherche (ANR) (Grant ANR08- GENM-021). Parts of these protocols have been developed on the RicE FUnctional GEnomics international hosting platform funded by Agropolis Fondation.

The technical assistance of Florence Artus is greatly acknowledged. The authors also thank Dr. Pietro Piffanelli, PTP, Lodi, Italy and Dr Emmanuelle Bourgeois, AFSSA, Paris, France for their past inputs in the *Tos17* program.

References

1. Hirochika H et al (1996) Retrotransposons of rice involved in mutations induced by tissue culture. Proc Natl Acad Sci USA 93: 7783–7788
2. Petit J et al (2009) Diversity of the Ty-1 copia retrotransposon *Tos17* in rice (*Oryza sativa* L.) and the AA genome of the *Oryza* genus. Mol Genet Genomics 282:633–652
3. Cheng C, Daigen M, Hirochika H (2006) Epigenetic regulation of the rice retrotransposon *Tos17*. Mol Genet Genomics 276: 378–390
4. La H et al (2011) A 5-methylcytosine DNA glycosylase/lyase demethylates the retrotransposon *Tos17* and promotes its transposition in rice. Proc Natl Acad Sci 108:15498–15503
5. Hirochika H (2001) Contribution of the *Tos17* retrotransposon to rice functional genomics. Curr Opin Plant Biol 4:118–122
6. Hsing YI et al (2007) A rice gene activation/knockout mutant resource for high throughput functional genomics. Plant Mol Biol 63: 351–364
7. Zhang J et al (2006) RMD: a rice mutant database for functional analysis of the rice genome. Nucleic Acids Res 34:D745–D748
8. Piffanelli P et al (2007) Large-scale characterization of *Tos17* insertion sites in a rice T-DNA mutant library. Plant Mol Biol 65:587–601
9. Miyao A et al (2003) Target site specificity of the *Tos17* retrotransposon shows a preference for insertion within genes and against insertion in retrotransposon-rich regions of the genome. Plant Cell 15:1771–1780
10. Lorieux M et al (2012) In-depth molecular and phenotypic characterization in a rice insertion line library facilitates gene identification through reverse and forward genetics approaches. Plant Biotechnol J 10:555–568
11. Droc G et al (2006) OryGenesDB: a database for rice reverse genetics. Nucleic Acids Res 34:736–740
12. Droc G et al (2009) OryGenesDB 2008 update: database interoperability for functional genomics of rice. Nucleic Acids Res 37: D992–D995

Chapter 16

Identification and Applications of the Petunia Class II *Act1/dTph1* Transposable Element System

Tom Gerats, Jan Zethof, and Michiel Vandenbussche

Abstract

Transposable genetic elements are considered to be ubiquitous. Despite this, their mutagenic capacity has been exploited in only a few species. The main plant species are maize, Antirrhinum, and Petunia. Representatives of all three major groups of class II elements, viz., hAT-, CACTA- and Mutator-like elements, have been identified in Petunia. Here we focus on the research "history" of the Petunia two-element *Act1–dTph1* system and the development of its application in forward- and reverse-genetics studies.

Key words hAT elements, Insertion, Excision, Petunia, Transposon tagging

1 Introduction

Transposable elements (TEs) are DNA sequences capable of moving from one position in the genome to another. Since the development of this concept by McClintock [1], based upon observations in Zea mays, TEs have been demonstrated in a wide range of organisms [2] and they are presumably ubiquitous [3]. In plants, extensive research has been devoted to the so-called Class II elements, which transpose directly from DNA to DNA [4–6], probably via a conservative excision/insertion process [7]. Because of their mobility and intrinsic ability to cause mutations these elements have been employed in applications such as Transposon Tagging [8–10], Site-selected Transposon Mutagenesis [11, 12], and Cell Lineage analysis [13–15].

In so-called one element systems, the element itself produces the protein(s) necessary for transposition. Two-element systems consist of a nonautonomous element which is dependent on the presence of an active autonomous element to enable its transposition. Nonautonomous elements lack the capacity to deliver the needed proteins but can still respond to them by transposing.

Thomas Peterson (ed.), *Plant Transposable Elements: Methods and Protocols*, Methods in Molecular Biology, vol. 1057, DOI 10.1007/978-1-62703-568-2_16,

Insertion of a transposable element (TE) into a gene may have no effect (e.g., in case the TE has landed in an intron and is spliced out as part of the intron) or it can induce changes in expression pattern, in level (from enhanced to knockdown or knockout), timing, or place. A basic feature of the insertion process is the formation of a target-site duplication (TSD); in *Ac/Ds*-like systems, the TSD is 8 bp (see below). Excision of a Class II element typically leaves behind a footprint which in literature usually is stated to equal the TSD. Thus, if excision of an *Ac/Ds*-like element from a position in a reading frame would leave the exact 8 bp TSD behind, the resulting sectors most often would retain the mutant phenotype due to the induced frameshift. Petunia line W138 has been and still is the main source of our insertion mutants. Roughly two out of three mutants, selected from self-progenies of this line are stable, non-sectored loss-of-function mutants. These are normally considered to represent cases where, upon excision, the resulting revertant sector retains its mutant phenotype. This creates one of the biggest problems in forward mutant screenings: the uncertainty about the presence/absence of the causative element in stable mutant phenotypes (see below).

Petunia varieties, as sold commercially and as used in research, have originally been derived mostly from interspecific crosses between the white-flowering *P. axillaris* and the purple flowering *P. integrifolia*. The first specific descriptions of variegated flower-color phenotypes in Petunia, indicative of transposable element-induced mutations, appeared in reports from Malinowski and Sachs [16] and Dale [17], although one can find occasional remarks on unstable phenotypes throughout the older literature. There is some evidence that transposons were activated as the result of interspecific hybridization [18]; if so, the transposable elements of Petunia can be cited as an example of Barbara McClintock's "genomic shock" hypothesis [19] as the trigger for transposon activation. This in turn suggests a natural role for transposons as a kind of ultimate barrier against unwanted interspecific hybridizations: transposon activity can be expected to produce mainly negative mutations and therefore should drive selection against hybrids.

Among the classic mutable lines in Petunia are an unstable dwarf line, in which reversions in the different tunica layers of the flower lead to changes in flower size, most notably in petals: somatic revertants in L1 had large petals, while L2 revertants had intermediate sizes (compare a hand in a too small glove) [20]. A set of lines bearing radiation-induced genic instabilities, for the flower color genes *Anthocyanin-2* (*An2*) and *Rhamnosyltransferase* (*Rt*), was maintained at the INRA, Dijon [21]. In the mutant collection at the Genetics Institute of the University of Amsterdam unstable alleles were also maintained for the flower color genes *Anthocyanin-3* (*An3*), *Anthocyanin-6* (*An6*), and *pH-4* (*Ph4*).

In the late 1960s, Hess and coworkers claimed that white-flowering Petunia plants could be transformed to give rise to sectored red-flowering plants by administering DNA from red-flowering plants to their white-flowering counterparts [22]. When Bianchi tried to repeat these experiments, he obtained sectored red-flowering plants from the acceptor line even in the absence of any exogenous DNA. This discovery intrigued him so much that he started to analyze a comparable case, in which the red-colored line "Roter Vogel" ("Red Bird") upon selfing, spontaneously gave rise to some progeny that were white-flowering with occasional red spots and sectors at a low frequency (less than one per flower). The subsequent basic description of the reversion behavior of *an1*-unstable alleles in derivative lines like W17 and W28 can be regarded as the true start of Petunia "transposonology" [23], the report appearing just ahead of the Big Bang of plant molecular biology.

2 Methods

2.1 Genetic Definition of the Act1–dTph1 System

In this pioneering period, classical genetic experiments were performed to elucidate the basic features of the *an1* unstable system. Many sublines that exhibited genetic changes in reversion frequencies could be selected, as proven by the faithful transmission of the novel reversion patterns to subsequent generations. Indeed, lines with more than 10,000 revertant spots per flower were recovered [23, 24]. Mulder et al. [25] showed that reversion frequency may be modulated in a dosage-dependent manner by one or more host factors, as suggested by the observation that a primary trisomic plant exhibited twice the number of spots compared to its diploid counterparts. In contrast, the reversion frequencies of homozygous versus heterozygous W138 (with the second allele being a stable white) are not significantly different. When the influence of environmental conditions on reversion frequency was analyzed, it was found that when plants were shifted from 18 to 25 °C the original unstable *an1* allele reverted less frequently. Surprisingly, a newly selected unstable allele derived from a revertant of the original unstable allele, exhibited the opposite response: a clear increase in reversion frequency under the higher temperature. In both cases, sporogenic reversion frequencies remained unchanged, at a level roughly 100-fold higher than the somatic frequency [26]. This last result contrasted with the otherwise surprisingly parallel Antirrhinum work as described, for example, by Harrison and Fincham [27].

Based upon the analysis of the positions and sizes of revertant sectors, ranging from a single cell to complete flowers/branches, it was concluded that reversion can occur at any time during the lifecycle of the plant and that it is linked to cell division, more specifically to DNA replication; temperature-shift experiments showed that in the final 5–6 days before maturity, reversion patterns could not be modulated, indicating that no more cell divisions take place during this period in flower development. Consequently, flower-bud opening is primarily, if not exclusively, a matter of differential cell elongation. Moreover, based upon exactly coinciding abaxial and adaxial revertant sectors at the corolla's distal end, it was concluded that two loosely defined rings of meristematic cells independently give rise to the lower and upper epidermis of the corolla and that these rings frequently flip and exchange positions [28].

While it was remarked that the occurrence of somatic and sporogenic reversions were reminiscent of McClintock's transposable elements, it was also stated that "...the postulation of transposition of the controlling element is unnecessary," with the addition that "...[this] does not permit the conclusion that in Petunia no transpositions take place" [23]. At that time, genomes were believed to be extremely stable, and transposition was generally considered a deviant process, no more than an exceptional peculiarity, occurring in just a few systems.

The stability of fully colored homozygous revertants was tested by crossing a number of them to a stable white *an1/an1* mutant. Among the 40,000 *An1/an1* heterozygous red-flowering progeny plants analyzed, nine white-flowering mutant plants were recovered, two of which exhibited a new reversion pattern. One of the latter has been maintained as W138, a line upon which most subsequent work has been based [23]. The second unstable allele (W109) gave rise to very complex progenies as could be predicted from its flower color: a pink background, with spots ranging in color from white to full red, classified in 27 classes (see cover photo). Even more striking, approximately 90 % of alleles transmitted from W109 had undergone a heritable change [29].

In the 1980s, two major points were established: (1) in 15–20 % of the progeny derived from self-fertilization of W138 plants new mutant phenotypes were encountered [30, 31], and (2) the unstable system behaved like a two-element system, similar to McClintock's *Ac/Ds* and McClintock's/Petersons' *En/Spm* systems [32, 33]. Some of the earliest transposon-induced mutants recovered were *yg3* (*yellow green3*), *px* (*phoenix*, giving rise to new flowers that emerge from senescing flowers), and *alf* (*aberrant leaf and flower*, the Petunia counterpart of *leafy/floricaula* [34]). The high incidence of new mutations finally led to the recognition that transposons were on the move.

Crosses between line W138 and a divergent set of varieties and species revealed that the ability of *an1* unstable alleles to revert (leading mostly to red spots on a white background) depended on the presence of a single, unlinked Mendelian factor, named *Activator* (*Act1*), which, surprisingly, appeared to occupy a fixed position in the genome. Mapping experiments invariably located the *Act1* element on chromosome I [18]. Although an active *Act1* element was never found to be present in any of the accessions of pure species tested, it is, with one exception, ubiquitously present in all cultivars tested. This finding argues for the presence of the system in an inactive state in at least one of the species that gave rise to the wide array of current cultivars. Indeed it was eventually shown that a collection of *P. axillaris* accessions from Uruguay harbor a variable number of *dTph1*-related sequences (Stuurman, personal communication, and unpublished results). Hence by 1985 it was clear that endogenous transposable element(s), like in other species, were capable of inducing mutations at a high frequency and in a broad range of genes in Petunia, at least in *Petunia hybrida*.

2.2 Isolation of the dTph1 Element and Its Basic Characteristics

Subsequently, it took some years to obtain a molecular entry point into the system. The molecular identification of *dTph1* (*defective Transposable element petunia hybrida 1*), which was isolated from the *Dihydroflavonol Reductase C* (*DFR-C*) gene, was published in 1990. It was argued that *DFR-C* should logically coincide with *An1*, as the presence/absence of the element was reported to correlate with the phenotypes underlying the different *An1* variants [35].

This claim could not be substantiated and has caused confusion, as *An1* later was identified as a transcription factor of the basic helix-loop-helix (bHLH) type [36]. Whatever the reason for the confusion, the *dTph1* element appeared unmistakably responsible for most of the phenotypes that have been defined and analyzed subsequently.

The various elements that have been described for Petunia include representatives of five families of elements, including *copia*, *Mutator*, *Ps*1, *Ac/Ds*, and *En/Spm* [37]. Without question, it is the *Ac/Ds*-like family of elements that has been exploited most successfully for both forward- and reverse-genetics investigations in Petunia. For that reason the *dTph* family and its applications are the focus of the remainder of this chapter.

The first isolated *dTph1* element was 284 bp in length. Based on sequence composition, specifically the terminal inverted repeats, internal conserved motifs, and the typical 8-bp TSD, the *Act1–dTph1* system is assigned to the *hAT*, family of elements (Fig. 1). Till today, the *Act1* "controlling" element remains to be identified at the molecular level. While in cultivar crosses *Act1* appears to be a unique activator, Stuurman and Kuhlemeier [38] described the action of a second activator-like element, identified from a *P. inflata* introgression in W138. There is a proposed PCR approach which can be used to identify autonomous *hAT* elements in a range of species to help clarify the origin and diversity of *Act* elements [39].

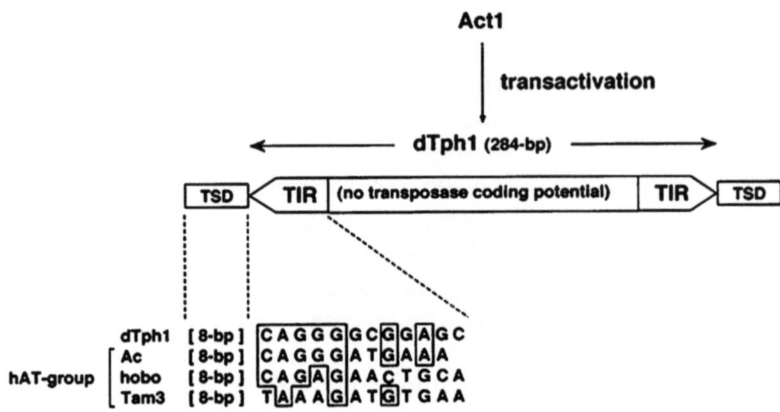

Fig. 1 The Petunia hybrida *Act1/dTph1* transposable element system. *Act1* Activator-1 encodes the transposase, *dTph1:* defective Transposon petunia hybrida-1: transposes in the presence of *Act1*, *TIR* terminal inverted repeat; note the conserved A2G5 motif (*see* ref. 52), *TSD* target-site duplication

In all examples identified to date, *dTph* integration sites contain the 8-bp TSD that is typical of *hAT* element insertions. No sequence preference for integration has been reported, though an exhaustive study has not yet been performed. A broad variation in TSD-derived footprints left behind upon excision, however, has been reported in most studies that looked into this aspect. As in maize [40] there thus is at least some variation in the DNA footprint. Not only is there variation within the TSD excision product, excision can remove the whole TSD or add some extra nucleotides, and even part of the genomic environment can be excised, or left-over bits of an element may be retained. This can lead to the selection of large allelic series of mutants, especially in the case of an easy-to-score visible phenotype like that associated with the *An1* gene [36].

3 Materials

3.1 Transpositional Behavior of dTph1 as Revealed by Transposon Display

Most cultivated lines and the species from which *P. hybrida* is thought to be derived, contain between 5 and 25 hybridizing fragments that are highly homologous to the *dTph1* element [18]. In contrast, the genomes of line R27 and of derivative lines like W138 contain between 100 and 200 *dTph1*-hybridizing fragments [18, 35]. Probably as a consequence of this high copy number, progenies of selfed plants of the line W138 display mutant

Transposon Display reveals the apical lineage of somatic insertion events

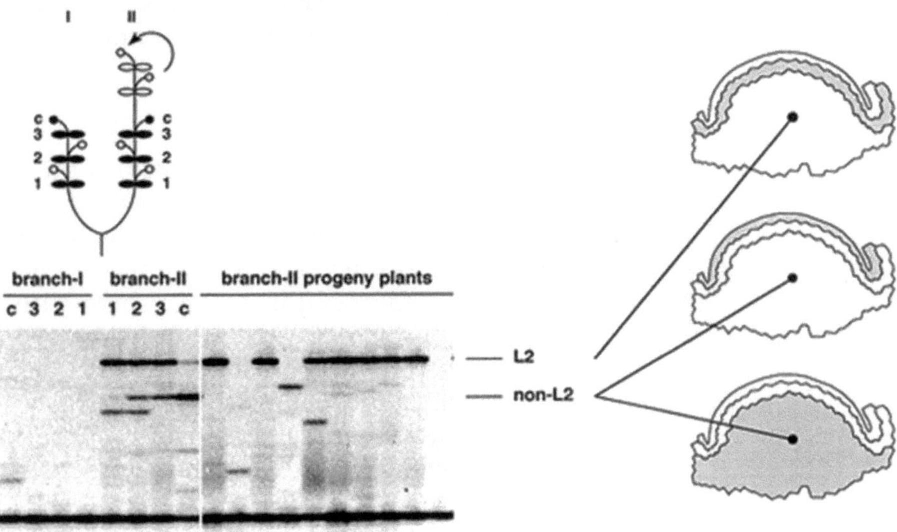

Fig. 2 Transposon display reveals the apical lineage of somatic insertion events. *Upper left*: sampling diagram. *Lower left*: TD on branch-specific samples (*left*) and on individual progeny plants from a branch II selfing (*right*). On the *far right* the different tunica layers are depicted in *gray*

phenotypes at a high frequency [23, 30, 31]. For this reason, the line W138 has become the standard line for Petunia insertion mutagenesis and for the analysis of the transpositional behavior of the *dTph1* element. Because of the high copy number, individual elements cannot be distinguished satisfactorily by conventional Southern blot analysis. To solve this problem, a strategy called *Transposon Display* was developed that allows the simultaneous detection of individual *dTph1* elements in high copy number lines [41]. This approach is the method of choice for the forward cloning of *dTph1* tagged mutants and is described in detail in an accompanying chapter (*see* Chapter 17). It also forms an essential component in the protocol for the massive indexed parallel amplification and sequencing of transposon flanking sequences at the population level (*see* Chapter 18). Here we present some of the results obtained by transposon display to analyze the behavior of the *dTph1* element.

To study somatic events, we sampled leaves and the distal parts of petals and compared the obtained patterns with those in progenies. In Fig. 2 a serendipitous finding is presented: the ratio of L2 tissues over non-L2 is >1 in leaves, but <<1 in distal parts of the petal. Thus we observed that somatic events that were transmitted (and thus derived from an L2 event) exhibited a leaf/corolla signal intensity ratio >1, while non-transmitted fragments have a signal intensity ratio of <1 (Fig. 2). Since somatic sectors can range from a single cell up to the whole sample (actually one of three tunica layers), we never tried to quantify the non-L2 events, simply because it is easy to start but very difficult to stop.

In an in-depth study of the behavior of the *dTph1* family as a whole, the frequency and timing of insertion events throughout the lifecycle and several generations of plants were followed [42]. Petunia plants flower as a cyme, which means that the next flower develops from a side meristem of the previous flower, leading to a zig-zagging stem. Such a flowering stem typically reveals 2–4 new insertions per five anthoclades (Fig. 3). The very detailed analysis enabled by TD is demonstrated in Fig. 4, where a single plants analysis shows that starting out with a plant holding 73 positions carrying 117 copies of *dTph1*, a progeny plant can be defined holding 92 positions (with 24 new ones) and 141 copies. The only hard indication for excision to occur can be derived from the 4 out of 44 homozygous parental fragments that became heterozygous. Of the 29 heterozygous parental fragments (Fig. 4, lower gray panel on the right), 5 have become homozygous, 16 remain heterozygous, and 6 disappear, either by Mendelian segregation or by excision. An astonishing 24 new insertions took place between zygote formation and maturity. A puzzling outcome was the finding that the number of *dTph1* elements increased regularly over five unselected generations; simple extrapolation argues that such an increase cannot be maintained. Thus transposition frequencies almost certainly cycle over time. Alternatively, selection for maintenance of overall plant vigor, as has been practiced over time simply to keep line W138 alive, may have been a factor in controlling the numbers of insertions and thus minimizing the mutational damage caused by enhanced numbers of transpositional events.

One of the outcomes of De Keukeleire's work [39, 42] was confirmation that reversion can take place throughout the lifetime of a plant. The work also revealed that, in line W138, insertions occur with a frequency of 10–20 sporogenic events per generation. About 60 TD fragments did not change position in material tested over five generations. An analysis of plants of the ancestral line, Roter Vogel, by then separated from W138 for approximately 50 generations, revealed the presence of most of these fragments in Roter Vogel, indicating that they represent stably positioned elements (defined as fragments capable of amplification by the two-stage TD procedure). These fragments might represent degenerate elements or elements that are trapped by some other means in their sequence environment. Overall it can be concluded that individuals of line W138 harbor a minimum of 100–150 active *dTph1* elements, of which 10–20 occupy new genomic positions in the next generation. This indicates that a library of 1,000 plants can be expected to carry 10,000–20,000 first-generation insertion events. Unpublished results indicate these to be minimum numbers [43].

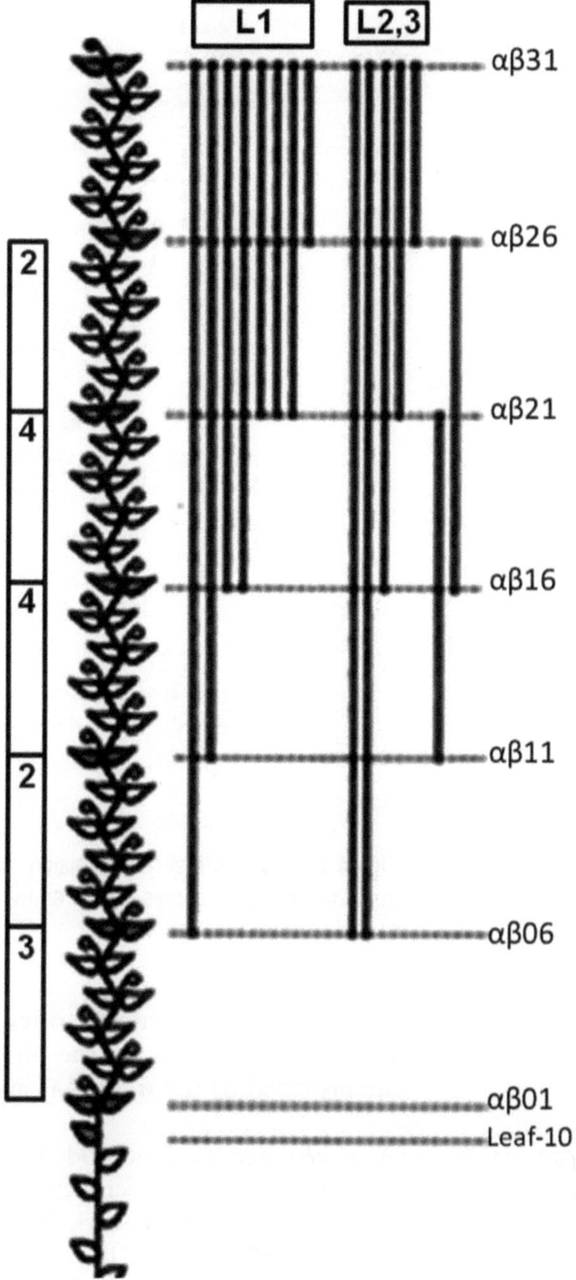

Fig. 3 Diagram of an average somatic insertion pattern in 31 anthoclades; around two insertions per five anthoclades, equally divided over L1 and L2 + L3

So far about 200 insertions, mostly *dTph1* or closely related elements, have been characterized in approximately 100 genes [37]. In most cases, the system was used in a reverse way, whereby sequence information was available and mutants were recovered by selection as a means of providing an entry to their functional analysis. Although a literature search argues that forward genetic

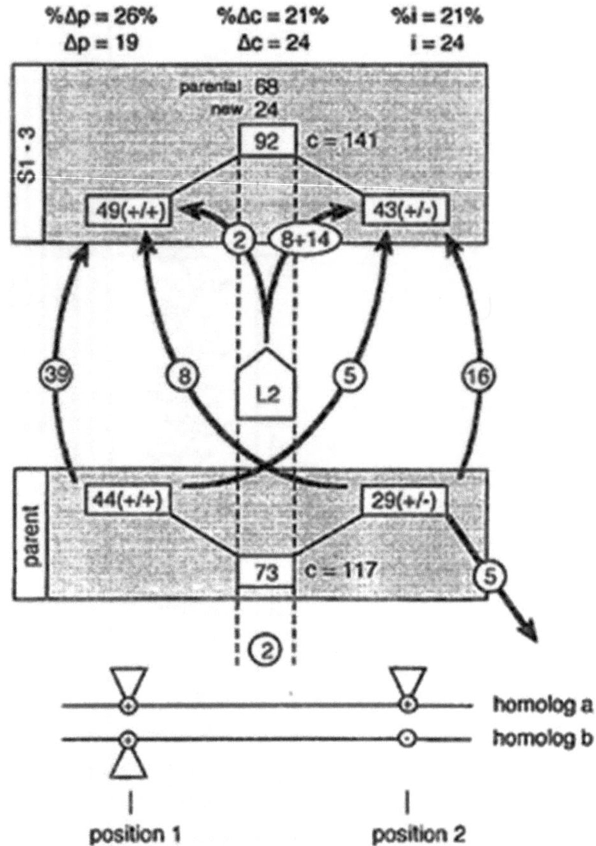

Fig. 4 Example of insertion/excision events from parent to offspring. Element positions and total copies (c) are shown. +/+: homozygous; +/−: heterozygous; for detailed explanation see text

screens are employed less often than reverse screens, they have been successful, despite the high *dTph1* copy number. Examples are provided by the *nam* [44] and *blind* [45] mutants.

3.2 Reverse Insertional Mutagenesis

A reverse-genetics method for efficient detection of *dTph1* and related elements in specific genes was published by Koes et al. [12], based on analogous work in *Drosophila* [46, 47]. Initially the method employed PCR reactions, combining gene-specific and *dTph1*-specific primers with sets of DNA samples. The material for these had been pooled in a 3-D matrix, such that each plant was represented in three pools, one for each dimension. These experiments for identifying insertions in genes of interest typically covered libraries of 1,000 individuals in an array of 10 block, 10 row and 10 column samples, each harvested from 100 plants. Originally, the method involved separation of PCR products on agarose gels, which were then blotted and the filters hybridized with gene-specific probes (Fig. 5).

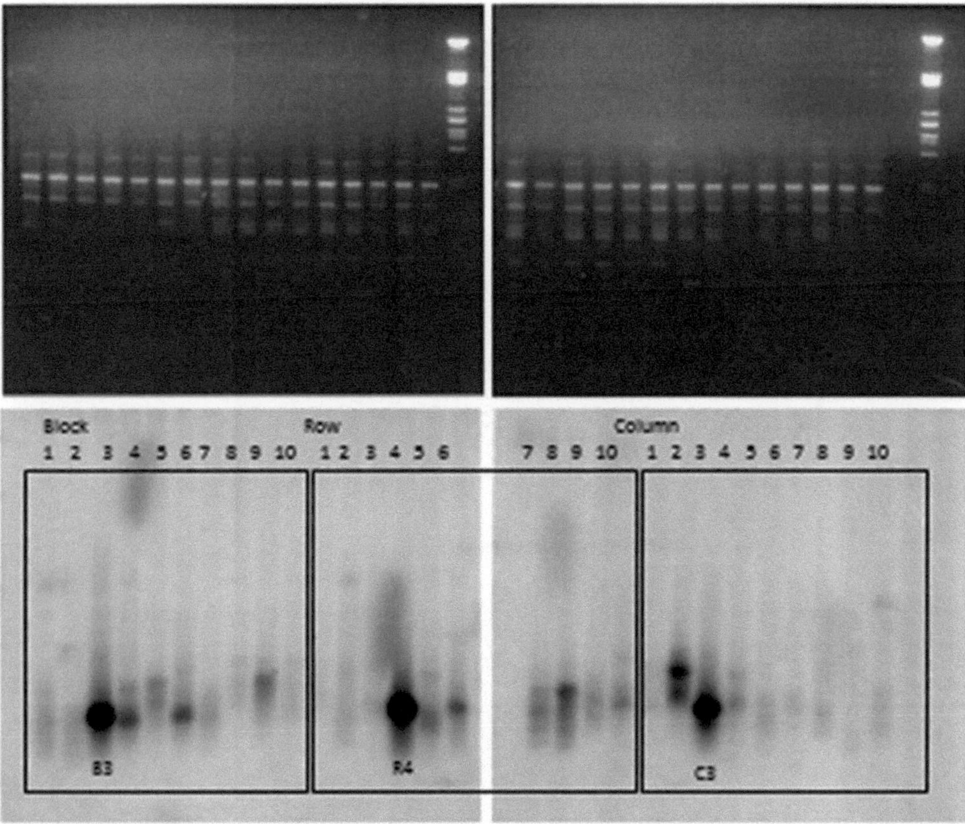

Fig. 5 The first reverse screening setup (*see* ref. 12). Sampling in 3D (block, row, column) uniquely identifies 1000 plants. *Upper panels*: PCR samples with a gene-specific and a *dTph1*-specific primer, run on agarose gel; *lower panels*: Southern blotting and hybridization with a gene-specific probe

If signals were detected, a second gel was run, part of which was Southern blotted to identify the location of the desired product; the fragment was subsequently isolated from the non-blotted part of the gel, re-amplified and subcloned for further analysis [12]. This laborious method was applied, for example, to identify and characterize insertions in *Apetala2* genes [48]. A comparable method involving inverse-PCR and differential screening was proposed by Souer and coworkers [49]; like its predecessor, it was abandoned following the development of more sensitive methods.

In a bid for higher efficiency, several technical improvements were proposed, tested, and eventually adopted. These included the use of labeled gene-specific primers to circumvent tedious blotting and hybridization procedures. A second critical improvement was the adoption of polyacrylamide gel electrophoresis (PAGE). PAGE not only provided much higher resolution of fragment sizes but also allowed the isolation of fragments of interest directly from the

Table 1
Comparison of theoretical and experimental gene tagging efficiencies

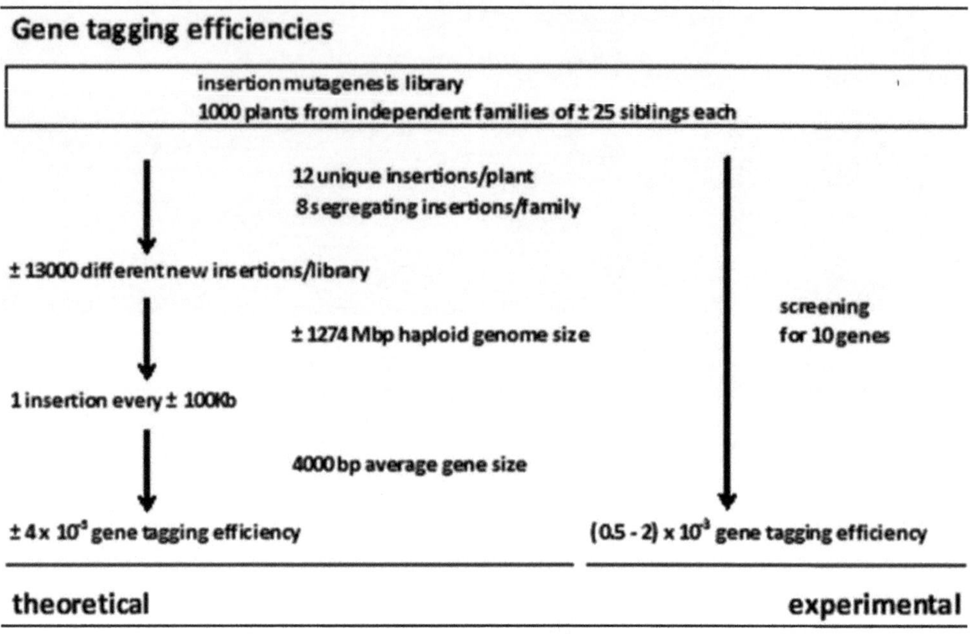

gel. A third significant step forward was the replacement of gene-specific with gene family-specific primers. Such primers are based on a specific sequence motif which should be highly conserved throughout the gene family of interest. This approach allows one to catch several independent insertions simultaneously and, moreover, permits identification of hitherto unknown family members [50].

While insertion screens were thus already rather efficient, we searched for possibilities to enhance the system further. This was realized when "454 pyrosequencing" technology hit the market [51]. Core to this method was the current capability of sequencing 400,000 DNA fragments of about 100 bp in a single experiment. We aimed to adapt this technology, replacing "DNA fragments" with "insertion-site sequence tags obtained by Transposon Display." As developed, the procedure applies the TD approach to pooled DNA samples [43]. This approach (detailed in Chapter 17) allows researchers to amplify and sequence transposon flanking sequences from an entire population in a highly efficient and cost-effective manner. Finally, in our experience, insertions are not randomly distributed throughout the genome, but exhibit a 100- to 200-fold preference for insertion into genic regions (Table 1).

4 Notes

As demonstrated in this chapter, methods developed for exploiting transposable elements of Petunia have proven highly useful for gene identification and isolation. Further improvements to this technology will make it feasible to construct an essentially saturated insertion library. This will be a valuable addition to the already wealthy set of materials and procedures presently available for functional gene analysis.

Acknowledgments

It is with great pleasure and thankfulness that I acknowledge the contribution of so many to the development of the Petunia transposon system described here; in chronological order: Marcel Beld, Eli Vrijlandt, Eric Souer, Wim Veerman, Henk Huits, Ronald Koes, Jan Zethof, Tamara Maes, Dirk van den Broeck, Peter de Keukeleire, and Michiel Vandenbussche, to mention those most heavily involved. Funding for parts of this research was provided by the Netherlands Organization for Scientific Research (H.H. and M.V.), the Flemish IWT (T.M., D.vd B. P.K.), the EC Biotech Program Bio4-CT97-2217 (M.V.), and a HORIZON grant (050-71-036) (M.V.) from the Netherlands Genomics Initiative; the development of the "454-TD" approach was performed in cooperation with Keygene (Wageningen).

References

1. McClintock B (1948) Mutable loci in maize. Carnegie Inst Washington Year Book 47:155–169
2. Berg DE, Howe MM (eds) (1989) Mobile DNA. American Society for Microbiology, Washington, DC
3. Finnegan DJ (1989) Eukaryotic transposable elements and genome evolution. Trends Genet 5(4):103–107
4. Coen ES, Carpenter R (1986) Transposable elements in *Antirrhinum majus*: generators of genetic diversity. Trends Genet 2:292–296
5. Gierl A et al (1989) Maize transposable elements. Annu Rev Genet 23:71–85
6. Bennetzen JL et al (1993) Specificity and regulation of the mutator transposable element system in maize. Crit Rev Plant Sci 12(1–2):57–95
7. Gierl A, Frey M (1991) Eukaryotic transposable elements with short terminal inverted repeats. Curr Opin Genet Dev 1(4):494–497
8. Fedoroff NV et al (1984) Cloning of the bronze locus in maize by a simple and generalizable procedure using the transposable controlling element activator (AC). Proc Natl Acad Sci USA 81(12):3825–3829
9. Gierl A, Saedler H (1992) Plant-transposable elements and gene tagging. Plant Mol Biol 19(1):39–49
10. Osborne BI, Baker B (1995) Movers and shakers—maize transposons as tools for analyzing other plant genomes. Curr Opin Cell Biol 7(3):406–413
11. Das L, Martienssen R (1995) Site-selected transposon mutagenesis at the hcf106 locus in maize. Plant Cell 7(3):287–294
12. Koes R et al (1995) Targeted gene inactivation in petunia by PCR-based selection of transposon insertion mutants. Proc Natl Acad Sci USA 92(18):8149–8153
13. Finnegan EJ et al (1989) Transposable elements can be used to study cell lineages in transgenic plants. Plant Cell 1(8):757–764
14. Dellaporta S et al (1991) Cell lineage analysis of the gynoecium of maize using the transposable element Ac. Dev Suppl 1: 141–147

15. Scheres B et al (1995) Mutations affecting the radial organization of the Arabidopsis root display specific defects throughout the embryonic axis. Development 121(1):53–62
16. Malinowski E, Sachs M (1916) Die vererbung einiger blumenfarben und blumengestalten bei petunia. Comptes Rendus Se. Soc. Sciet, Varsovie
17. Dale EE (1941) A reversible variegation in petunia. J Hered 32:123–126
18. Huits HSM et al (1995) Genetic characterization of *Act1,* the activator of a non-autonomous transposable element from *Petunia hybrida.* Theor Appl Genet 91:110–117x
19. McClintock B (1983) The significance of responses of the genome to challenge. Science 226:792–801
20. Bianchi F, De Boer R, Pompe AJ (1974) Investigation into spontaneous reversions in a dwarf mutant of Petunia-*hybrida* in connection with interpretation of results of transformation experiments. Acta Bot Neerl 23: 691–700
21. Cornu A (1977) Induced unstable systems in petunia. Mutat Res 42:235–248
22. Hess D (1973) Versuche zur transformation an hoheren pflanzen: Untersuchungen zur realization des exosomen-modells der transformation bei *Petunia hybrida.* Z Pflanzenphysiol 68:432–440
23. Bianchi F et al (1978) Regulation of gene action in *Petunia hybrida*: unstable alleles of a gene for flower colour. Theor Appl Genet 53:157–167
24. Doodeman M et al (1984) Genetic analysis of instability in *Petunia hybrida.* 1. A highly unstable mutation induced by a transposable element inserted at the *An1* locus for flower colour. Theor Appl Genet 67:345–355
25. Mulder RJP et al (1981) Dosage effect of a gene with a regulating effect on anthocyanin synthesis in a trisomic *Petunia hybrida.* Genetica 55(2):111–115
26. Doodeman M et al (1984) Genetic analysis of instability in *Petunia hybrida.* 4. The effect of environmental factors on the reversion rate of unstable alleles. Theor Appl Genet 69:489–495
27. Harrison BJ, Fincham JRS (1964) Instability at the *Pal* locus in *Antirrhinum majus.* 1. Effects of environment on frequencies of somatic and germinal mutation. Heredity 19: 237–258
28. Martin C, Gerats T (1993) Control of pigment biosynthesis genes during petal development. Plant Cell 5:1253–1264
29. Gerats AGM, Farcy E, Wallroth M, Groot SPC, Schram A (1984) Control of anthocyanin synthesis in Petunia hybrida by multiple allelic series of the genes An1 and An2. Genetics 106(3):501–508
30. Doodeman M et al (1984) Genetic analysis of instability in *Petunia hybrida.* 2. Unstable mutations at different loci as the result of transpositions of the genetic element inserted at the *An1* locus. Theor Appl Genet 67:357–366
31. Gerats A et al (1989) Gene tagging in *Petunia-hybrida* using homologous and heterologous transposable elements. Dev Genet 10:561–568
32. Gerats AGM et al (1985) A two-element system controls instability at the *An3* locus in *Petunia hybrida.* Theor Appl Genet 70:245–247
33. Wijsman HJW (1986) Evidence for transposition in Petunia. Theor Appl Genet 71:791
34. Souer E et al (1998) Genetic control of branching pattern and floral identity during Petunia inflorescence development. Development 125:733–742
35. Gerats AG et al (1990) Molecular characterization of a nonautonomous transposable element (*dTph1*) of petunia. Plant Cell 2:1121–1128
36. Spelt C et al (2002) ANTHOCYANIN1 of Petunia controls pigment synthesis, vacuolar pH, and seed coat development by genetically distinct mechanisms. Plant Cell 14:2121–2135
37. Gerats T (2009) Identification and exploitation of Petunia transposable elements: a brief history. In: Strommer J, Gerats T (eds) Petunia. Springer, New York, pp 365–379
38. Stuurman J, Kuhlemeier C (2005) Stable two-element control of *dTph1* transposition in mutator strains of Petunia by an inactive *ACT1* introgression from a wild species. Plant J 41:945–955
39. De Keukeleire P et al (2004) A PCR-based assay to detect HAT-like transposon sequences in plants. Chromosome Res 12:117–123
40. Wessler S (1988) Phenotypic diversity mediated by the maize transposable elements *Ac* and *Spm.* Science 242:399–405
41. Van den Broeck D et al (1998) Transposon display identifies individual transposable elements in high copy number lines. Plant J 13:121–129
42. De Keukeleire P et al (2001) Analysis by transposon display of the behaviour of the *dTph1* element family during ontogeny and inbreeding of *Petunia hybrida.* Mol Genet Genom 265:72–81
43. Vandenbussche M, Janssen A, Zethof J et al (2008) Generation of a 3D indexed petunia insertion database for reverse genetics. Plant J 54:1104–1114
44. Souer E, van Houwelingen A, Kloos D, Mol J, Koes R (1996) The *no apical meristem* gene of Petunia is required for pattern formation in embryos and flowers and is expressed at meristem and primordial boundaries. Cell 85: 159–170

45. Cartolano M et al (2007) A conserved microRNA module exerts homeotic control over *Petunia hybrida* and *Antirrhinum majus* floral organ identity. Nat Genet 39:901–905
46. Ballinger DG, Benzer S (1989) Targeted gene mutations in *Drosophila*. Proc Natl Acad Sci USA 86:9402–9406
47. Kaiser K, Goodwin SF (1990) "Site-selected" transposon mutagenesis of *Drosophila*. Proc Natl Acad Sci USA 87:1686–1690
48. Maes T et al (2001) Petunia *Ap2*-like genes and their role in flower and seed development. Plant Cell 13:229–244
49. Souer E et al (1995) A general method to isolate genes tagged by a high copy number transposable element. Plant J 7:677–685
50. Vandenbussche M et al (2003) Toward the analysis of the petunia MADS box gene family by reverse and forward transposon insertion mutagenesis approaches: B, C, and D floral identity functions require SEPAllATA-like MADS box genes in petunia. Plant Cell 15:2680–2693
51. Margulies M et al (2005) Genome sequencing in micro-fabricated high-density picolitre reactors. Nature 437:376–380
52. Warren WD, Atkinson PW, O'Brochta DA (1995) The Australian bush fly *Musca vetustissima* contains a sequence related to transposons of the *hobo*, *Ac* and *Tam3* family. Gene 154(1):133–134

Chapter 17

Transposon Display: A Versatile Method for Transposon Tagging

Michiel Vandenbussche, Jan Zethof, and Tom Gerats

Abstract

Transposon tagging has been used successfully in a range of organisms for the cloning of mutants of interest. In species containing high copy numbers of transposable elements combined with a high transposition rate, forward cloning can be quite challenging and requires specific high-resolution methods. Here we detail an updated version of the Transposon Display technique, which allows visualization of large numbers of transposon-flanking sequences simultaneously in a highly robust and reproducible manner. This strategy was developed for the analysis of the transpositional behavior of the *dTph1* transposon and for the forward cloning of mutants, particularly in the Petunia W138 background, individuals of which can contain >200 copies of the endogenous *dTph1* element. The method is derived from the AFLP technique and can in principle easily be adapted to any system.

Key words Forward genetics, Transposon Display, Transposon tagging, High copy number lines, dTph1, Petunia

1 Introduction

Progenies of selfed plants of the line W138 display mutant phenotypes at a high frequency [1–3]. Therefore, the line W138 has become the standard line for Petunia insertion mutagenesis. However, because of its high copy number, forward cloning becomes quite challenging and requires specific high-resolution methods. Therefore a strategy called *Transposon Display* was developed that allows the simultaneous detection of individual *dTph1* elements in high copy number lines [4]. Transposon Display (TD) (*see* Fig. 1) is a variation on the AFLP technique, a highly prolific source of molecular markers [5] typically used for cloning and mapping. The TD approach is based on the use of two restriction endonucleases: (a) a "six-cutter" (*Mun*I) that cleaves within the *dTph1* element and (b) a "four-cutter" that cuts in the neighboring genomic DNA but not in the element (*Bfa*I or *Mse*I; note that *dTph1* has an *Mse*I site, precluding TD in that direction).

Thomas Peterson (ed.), *Plant Transposable Elements: Methods and Protocols*, Methods in Molecular Biology, vol. 1057, DOI 10.1007/978-1-62703-568-2_17, © Springer Science+Business Media New York 2013

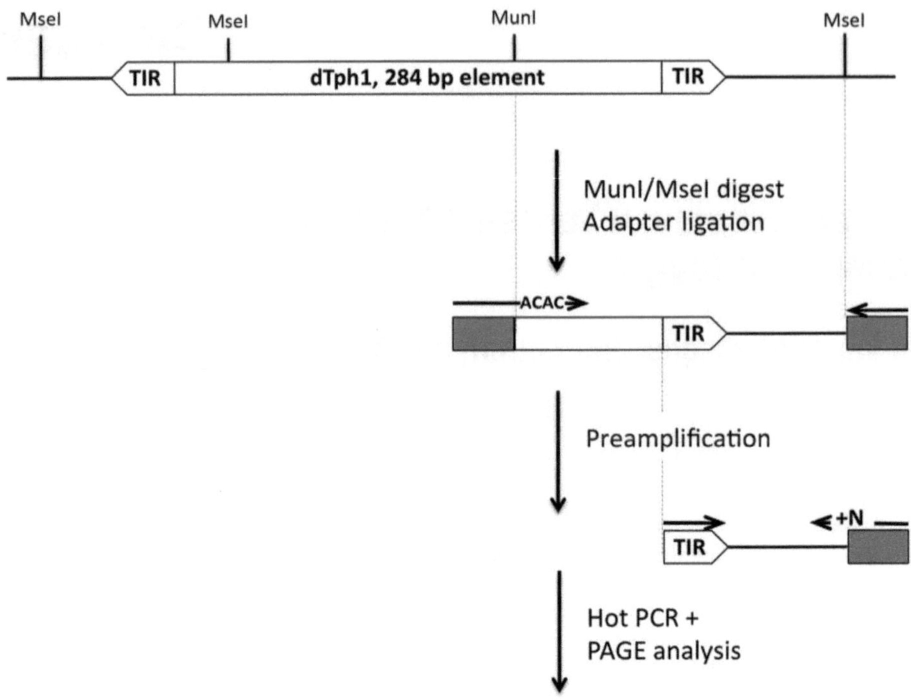

Fig. 1 Strategy for recovery of *dTph1* insertion-flanking sequences by means of Transposon Display (TD). Genomic fragments adjacent either 5′ or 3′ to the site of insertion can be recovered. Here the procedure only shows recovery 3′ of *dTph1*, using *Mse*I as a tetracutter. When *Bfa*I is used instead, both sides can be amplified

Adaptors are ligated as in standard AFLP protocols; in contrast with the standard AFLP procedure; however, the selective nucleotides of the six-cutter adaptor are not randomly chosen but defined by the four *dTph1* nucleotides next to the restriction site: ACAC to amplify in one direction, AACC if one needs (or prefers) to amplify in the other direction (only possible if the four-cutter used does not cut in the remainder of the element, e.g., *Bfa*I). The adaptor at the four-cutter end carries no selective nucleotides at this stage. In this way PCR products are enriched for sequences neighboring *dTph1*-insertion sites. In a subsequent nested PCR, a labeled primer based on the terminal inverted repeat of *dTph1* is used, in combination with the four-cutter-based primer. The latter is extended by an extra nucleotide, such that if all four possible extensions are used, the amplified fragments are partitioned into four subsets, which allows a better interpretation of the results.

The nested PCR approach basically ensures that all amplified fragments are derived from true *dTph1* insertion sites. On the other hand, elements that lack either one of the necessary conditions (presence of the internal six-cutter site, sequence conservation at sites critical for PCR), will not be recognized or, if an outside restriction site is too close by or too far away, will not be amplified sufficiently for recovery (*see* Subheading 3). The technique thus is restricted to those elements that meet these conditions. Based on comparisons with Southern blot analyses, we guesstimate that more than 90 % of *dTph1* elements are amenable to this approach. While deviant *dTph1* elements might be missed, distinct but related elements responsive to the TD procedure can be recovered.

Here we provide a detailed protocol as to how we currently perform transposon display experiments in Petunia. Compared to the original protocol [4], we have modified two steps. First of all, we have omitted the beads extraction step. We found that this greatly simplifies the procedure without losing specificity in amplifying *dTph1* flanking fragments. Further, in two independent transposon display experiments, we initially did not find a candidate fragment, perfectly co-segregating with the mutant phenotype, when using only four different hot PCRs as originally described. The reason for that in both cases was the presence of another transposon-flanking fragment with exactly the same size (thus co-migrating) and masking the segregation pattern of the transposon insertion co-segregating with the mutation of interest. We did find, however, the correct flanking fragment after adding another selective nucleotide on the transposon-flanking primer. We now routinely perform 16 different hot PCRs on the preamplification products instead of the original 4, using also one extra selective nucleotide on the transposon primer during the hot-PCR (*see* Fig. 2). This further decreases the probability for comigration of different fragments of the same length.

Finally, a correct choice of the plant material to be used for transposon display is a critical factor in successfully identifying the one insertion that is causing the mutation of interest amongst the sometimes >200 other copies. This high copy number, often combined with a low recombination frequency in certain chromosomal regions, may easily lead to the identification of multiple transposon-flanking fragments that all seem to be tightly linked with the phenotype of interest. However, by choosing the right genetic material, these problems can be largely avoided. In the materials section, we discuss how to obtain an informative selection of plant material to be used for cloning interesting mutants by Transposon Display.

2 Materials

2.1 Choice of Plant Material

In a typical transposon display experiment, 24 individuals are analyzed for the identification of a single recessive mutation. With 16 different primer combinations, this allows the researcher to run the complete transposon display set on four polyacrylamide gels, using 96-well combs.

A W138 plant on average acquires approximately 20 new insertions per generation. If distributed randomly, every chromosome

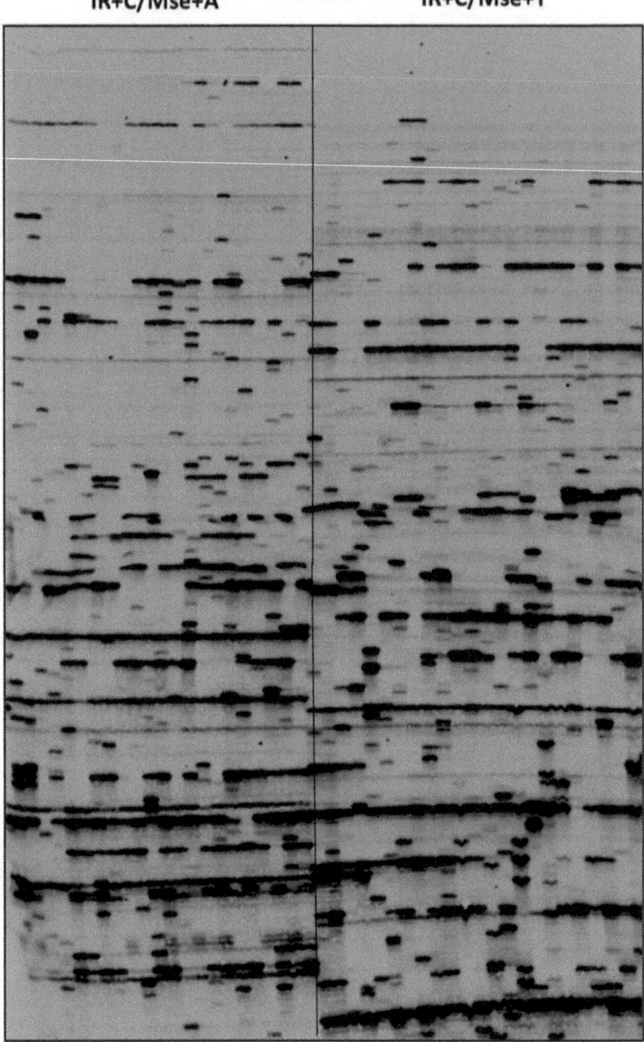

Fig. 2 Transposon Display example using a selective nucleotide on both amplification primers during the Hot PCR. Two out of the 16 possible primer combinations are shown

gets around three of these, thus (close) linkage between insertions is to be expected. As mentioned in the introduction, often the difficulty thus is not to find (a) perfectly co-segregating fragment(s), but to reduce this to only one possible candidate. This is partly a consequence of the limited number of individuals on which a TD is performed. To be able to exclude closely linked insertions, the presence of sufficient homozygous WT individuals in the TD experiment is crucial. Therefore, for each TD profile we normally select 14 homozygous WT plants and 10 homozygous mutants. Ideally, the homozygous WT plants should be genetically as closely related as possible to the homozygous mutants. Within this constraint, the origin of homozygous mutants and WT plants should be as diverse as possible. The following setup takes all these criteria into account and has led to several successful TD experiments [6, 7] and unpublished results.

We start by screening families of 20–25 plants for a mutant of interest, which ideally segregates in a classic Mendelian recessive fashion. The usually 3–6 available homozygous mutants are all selected for DNA isolation and TD. The remaining plants of the family are phenotypically WT; they are all self-pollinated for progeny analysis. Heterozygous parents will yield ¼ of homozygous mutants in their offspring, a selection of which (e.g., 1–2 per heterozygous parent) will be used to complete the set of ten homozygous mutants. Homozygous WT parents will yield 100 % WT offspring and a selection of these plants (1–2 individuals from each homozygous WT line) will be used to compose the homozygous WT set in the TD experiment. In case the pedigree of the original family segregating for the mutation of interest is well documented, siblings from the parent of this family can also be used as WT samples. These plants are very closely related and thus share most of the segregating transposons, except for new insertions, which arose for the first time as a heterozygous insertion in the parent of the mutant family. These plants are thus ideal to exclude the majority of closely linked insertions. When available we take 3–4 plants of this type to complete the WT set.

More rarely (*see* **Note 1**), revertant branches arise on (some of) the homozygous mutants. If seeds can be produced that inherit the reversion (for which the reversion has to occur in the L2 layer), then the progeny of these revertants provides the perfect genetic material to generate homozygous WT material for TD comparison. This is achieved in the same way as described above, by analyzing the offspring of the ¾ phenotypically WT plants derived from selfing of revertant flowers. The selected homozygous WT plants will have most transposons in common with the mutant, except for at least the transposon that caused the mutation.

2.2 Transposon Display

1. Digestion/Ligation:
   (a) *Mse*I (10 U/μl) *Mun*I (or *Mfe*I, isoschizomer), NEB4 buffer (all from New England Biolabs, www.NEB.com).
   (b) 100× BSA stock solution, ATP 10 mM stock solution, T4 DNA ligase (5 U/μl).
2. Adapters:
   (a) *Mun*I (*Mfe*I) adapter top strand (100 μM stock): 5′-CTCGTAGACTGCGTACG-3′.
   (b) *Mun*I (*Mfe*I) adapter bottom strand (100 μM stock): 5′-AATTCGTACGCAGTC3′.

(c) *Mse*I/*Bfa*I (*see* **Note 2**) adapter top strand (100 μM stock): 5′-GACGATGAGTCCTGAG-3′.

(d) *Mse*I/*Bfa*I (*see* **Note 2**) adapter bottom strand (100 μM stock): 5′-TACTCAGGACTCAT-3′.

3. Oligos for Preamplification PCR:
   *Mse*I digestion (amplification only possible of 3′ flanking sequence).
   (a) *Mun*I + ACAC: 5′-AGACTGTGTACGAATTGACAC-3′.
   (b) *Mse* + 0: 5′-GACGATGAGTCCTGAGTAA-3′.
4. *Bfa*I digestion:
   (a) *Mun*I + ACAC: 5′-AGACTGTGTACGAATTGACAC-3′ (for 3′ flanking sequence).
   (b) *Mun*I + AACC: 5′-AGACTGCGTACGAATTGAACC-3′ (for 5′ flanking sequence).
   (c) *Bfa* + 0: 5′-GACGATGAGTCCTGAGTAG-3′.
5. HOT PCR:
   (a) Primer labeling: γ-^{33}ATP (370 MBq/μl), T4 Polynucleotide Kinase, T4 buffer.
6. Oligos for Hot PCR:
   (a) IR + A: 5′-GGAATTCGCTCCGCCCCTGA-3.
   (b) IR + T: 5′-GGAATTCGCTCCGCCCCTGT-3.
   (c) IR + C: 5′-GGAATTCGCTCCGCCCCTGC-3.
   (d) IR + G: 5′-GGAATTCGCTCCGCCCCTGG-3.
   (e) *Mse*I + A: 5′-GACGATGAGTCCTGAGTAAA-3′.
   (f) *Mse*I + T: 5′-GACGATGAGTCCTGAGTAAT-3′.
   (g) *Mse*I + C: 5′-GACGATGAGTCCTGAGTAAC-3′.
   (h) *Mse*I + G: 5′-GACGATGAGTCCTGAGTAAG-3′.
   (i) *Bfa*I + A: 5′-GACGATGAGTCCTGAGTAGA-3′.
   (j) *Bfa*I + T: 5′-GACGATGAGTCCTGAGTAGT-3′.
   (k) *Bfa*I + C: 5′-GACGATGAGTCCTGAGTAGC-3′.
   (l) *Bfa*I + G: 5′-GACGATGAGTCCTGAGTAGG-3′.

2.3 PAGE Analysis Components

1. Formamide loading dye: 98 % Formamide, Bromophenol Blue, and Xylene cyanol, 10 mM EDTA.
2. 4.5 % stock gel solution:
   In a 2 l beaker, combine:
   (a) 450 g of ureum.
   (b) 112.5 ml acrylamide stock solution (40 %, 19:1 acrylamide:bisacrylamide).

(c) 100 ml TBE (10×).

(d) Nanopure H_2O to (a little bit less than) 1,000 ml.
Put on a magnetic stirrer until dissolved. Filter the solution (0.45 μm filter), add nanopure H_2O to exactly 1,000 ml, mix, and store in the dark at 4 °C. The solution can be kept for at least 6 months.

3. 10 % APS stock solution (ammonium persulfate dissolved in nanopure H_2O), to be stored at 4 °C, and not for longer than 2 months.

4. TEMED (*N*,*N*,*N'*,*N'*-tetramethyl-ethylenediamine) (Sigma), store at 4 °C.

(a) Sequencing gel system.

(b) Whatmann paper (3MM).

(c) Saran wrap.

3 Methods

The protocol provided here targets the amplification of 3′ flanking fragments using *Mse*I as a tetracutter. The recognition sequence for *Mse*I is TTAA, which is a very frequently occurring sequence motif. Therefore, for the large majority of the insertion sites, a TTAA site will be in close enough vicinity to allow amplification by the transposon display procedure. In rare cases, such a site might be absent. Alternatively, the *Mse*I site might be very close to the insertion site, yielding a flanking sequence of only a few nucleotides. In both cases, it might be worthwhile to try a TD using *Bfa*I, which allows amplification of both sides of the transposon, starting from separate preamplifications. The procedure is otherwise exactly the same as for *Mse*I, except that other *Mse/Bfa* adapter primers are used, which are provided also in Subheading 2.

3.1 Adapter Preparation

1. For the preparation of 100 μl *Mun*I-adapter (5 μM), combine

(a) 5 μl Mun top strand (100 μM stock)

(b) 5 μl Mun bottom strand (100 μM stock)

(c) 90 μl H_2O.

2. For the preparation of 100 μl *Mse*I-adapter (50 μM), combine

(a) 50 μl Mse top strand (100 μM stock)

(b) 50 μl Mse bottom strand (100 μM stock)
Incubate the mixtures in a PCR machine for 5 min at 80 °C, gradually cool down to RT over the period of 1 h, put on ice.

3.2 Digestion/Ligation

1. Start with 50–200 ng genomic (high quality) DNA, dissolved in 10 μl H_2O.
2. Add 20 μl mix, containing:
   (a) 5 U *Mun*I (20 U/μl)
   (b) 5 U *Mse*I (10 U/μl)
   (c) 3 μl NEB 4 (10× stock)
   (d) 0.3 μl BSA (100× stock)
   (e) H_2O to 20 μl
   Incubate: 1 h at 37 °C (*see* **Note 3**).
3. To the digestions, add 10 μl mix, containing:
   (a) 1 μl *Mun*-Adapter (5 μM stock)
   (b) 1 μl *Mse*-Adapter (50 μM stock)
   (c) 1 μl NEB 4 (10× stock)
   (d) 0.1 μl BSA (100× stock)
   (e) 1 μl ATP (10 mM stock)
   (f) 1 Weiss U T4 DNA ligase (5 U/μl)
   (g) H_2O to 10 μl
   Incubate: 3 h at 37 °C.
4. Add 160 μl H_2O to the ligation mixture and continue with the selective preamplification. Alternatively, store at −20 °C (not at 4 °C).

3.3 Selective Preamplification

1. Take 5 μl template DNA (diluted digestion/ligation mixture) and add:
   (a) 0.7 μl *Mun*+ACAC (10 μM)
   (b) 0.7 μl *Mse*+0 (10 μM)
   (c) 0.5 μl dNTP (10 mM)
   (d) 2.5 μl 10× PCR buffer (with 20 mM $MgCl_2$)
   (e) 0.1 μl Taq DNA polymerase (Dream taq, 5 U/μl) H_2O to 20 μl
2. 20 μl PCR mix, containing:
3. Incubate the samples in a PCR machine according to the following PCR profile:
   (a) 13 cycles (94 °C for 15 s, 65 °C for 30 s −0.7 °C/cycle, 72 °C for 60 s)
   (b) 22 cycles (94 °C for 15 s, 56 °C for 30 s, 72 °C for 60 s) Check 5 μl on 1.5 % agarose gel (*see* **Note 4**), should give a low molecular weight smear (50–700 bp) (Fig. 3).
4. Dilute the remainder 10–20× with H_2O (or TE) and store at −20 °C.

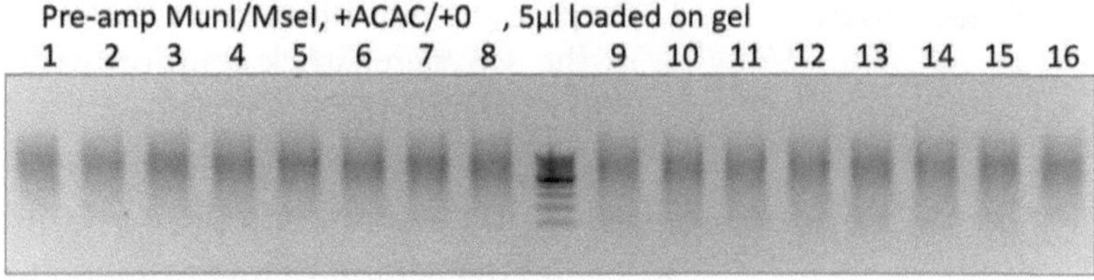

Fig. 3 Transposon Display preamplification products run on a 1.5 % agarose gel, typically showing a smear (50–700 bp)

3.4 Primer Labeling

1. For 25 PCR reactions, mix together in a 1.5 ml microcentrifuge tube placed on ice (*see* **Note 5**):
   (a) IR+N primer (10 pmol/µl): 2.5 µl
   (b) T4 PNK 10× buffer : 2.5 µl
   (c) H_2O:16.25 µl
   (d) T4 PNK: 1.25 µl
   (e) (γ-^{33}P)-ATP: 2.5 µl
   (f) Total volume: 25 µl
2. Incubate the mixture for 30 min at 37 °C.
3. Inactivate the enzyme by incubating the tube for 10 min at 80 °C.
4. Centrifuge the mixture briefly to collect any condensate that has formed on the top of the tube. The labeled primers can be used immediately or stored for maximum 1 week at −20 °C.

3.5 Hot PCR

1. Take 5 µl diluted pre-amp and add 15 µl PCR mix, containing:
   (a) 1 µl labeled IR+N primer
   (b) 0.5 µl *Mse*+N (10 µM)
   (c) 0.4 µl dNTP (10 mM)
   (d) 2 µl 10× PCR buffer (with 20 mM $MgCl_2$)
   (e) 0.08 µl Taq DNA polymerase (Dream taq, 5 U/µl)
   (f) H_2O to 15 µl
2. Incubate the samples in a PCR machine according to the following PCR profile:
   (a) 13 cycles (94 °C for 15 s, 65 °C for 30 s −0.7 °C/cycle, 72 °C for 60 s)
   (b) 22 cycles (94 °C for 15 s, 56 °C for 30 s, 72 °C for 60 s)
3. Add 20 µl formamide loading dye to the samples (*see* **Note 5**), close with an aluminum cover, mix well or freeze ON.
4. Denature (5 min at 94 °C in a PCR machine) and put directly on ice prior to loading (once denatured, the DNA stays denatured when kept cool).

3.6 PAGE Analysis

1. Prepare the gel solution just before casting, by mixing 100 ml of the 4.5 % gel stock solution with 500 μl of 10 % APS, and 100 μl of TEMED.
2. Using a sequencing gel system, cast the gel minimally 2 h before use to allow complete polymerization.
3. Prerun the gel for 30 min using preheated 1× TBE (microwave 2–4 min, max power) as a running buffer at constant power (40–50 V/cm).
4. Load 2 μl of the reaction products after cleaning the slots (with a syringe using the 1× TBE in buffer tank).
5. Run the gel at constant power (40–50 V/cm) until the bromophenol blue reaches the bottom of the gel.
6. Open the gel system and lift the gel from the glass plate with Whatmann (3MM) paper.
7. Cover the gel on the Whatmann paper with Saran wrap and dry the gel on a standard gel slab dryer.
8. Remove the Saran wrap, let the gel further air-dry for 2–3 min and expose the completely dried gel to a phosphor imager screen or expose to film overnight.

3.7 Reamplification of a Candidate Transposon-Flanking Fragment

1. Visually inspect the digital gel images obtained by phospor image screen exposure for the presence of a co-segregating transposon-flanking fragment (present in all homozygous mutants (*see* **Note 6**) and absent in all homozygous WT samples).
2. Expose the dried gel overnight to a film (*see* **Note 7**).
3. Excise the candidate fragment from the gel and put the piece of gel (including the Whatmann paper) in 200 μl dH_2O (*see* **Note 8**).
4. Allow the DNA to elute for about 45–60 min at room temperature and vortex occasionally.
5. Centrifuge the tube(s) for 1 min at maximum speed to spin down the paper and gel particles just before proceeding to the next step. The water phase contains the template DNA.
6. Use 5 μl of this template in a standard PCR reaction using the same pair of selective primers, and using the following PCR profile:
   (a) 13 cycles (94 °C for 15 s, 65 °C for 30 s −0.7 °C/cycle, 72 °C for 60 s)
   (b) 30 cycles (94 °C for 15 s, 56 °C for 30 s, 72 °C for 60 s)
7. Load 5–10 μl of the PCR products on an agarose gel to check the purity of the bands and proceed to cloning and sequencing.

4 Notes

1. Upon excision of the *dTph1* element, usually an 8 bp footprint is left behind, while in frame footprints or perfect excisions occur much more rarely. As a consequence, *dTph1* excisions from insertions into coding regions in the majority of cases do not result in restoration of gene function and phenotypic reversion. Note that the numerous red revertant spots and sectors in the famous W138 flower arise from excision of a *dTph1* transposon inserted at the Intron6/Exon7 junction of the *An1* gene [8].
2. The same adapter can be used for either *Mse*I or *Bfa*I digestions.
3. Once the digestion reactions have been started the protocol must be continued through adapter ligations, because the restriction enzyme activity is needed to prevent fragment–fragment back ligation.
4. Use a loading dye which does not contain bromophenol blue (because that masks the PCR result).
5. Use filtertips to prevent pipetman contamination.
6. Although rare, it is possible that in one or few of the homozygous mutant samples the transposon has fully excised, while the mutant phenotype is maintained by the presence of an out-of-frame footprint. A fragment that is completely absent in all WT samples and is present in the majority (but not all) of the mutant samples should therefore not be directly discarded as a possible candidate.
7. Put the film on the gel in such a way that you can later reposition the film on the gel exactly in the same way. Punch film and gel together with staples in several places.
8. The gel can be reexposed to check whether bands have been properly cut out. Try to cut out the fragments (dried gel + paper) as small and as accurate as possible.

References

1. Bianchi F et al (1978) Regulation of gene action in *Petunia hybrida:* unstable alleles of a gene for flower colour. Theor Appl Genet 53:157–167
2. Doodeman M et al (1984) Genetic analysis of instability in *Petunia hybrida.* 2. Unstable mutations at different loci as the result of transpositions of the genetic element inserted at the *An1* locus. Theor Appl Genet 67:357–366
3. Gerats A et al (1989) Gene tagging in *Petunia-hybrida* using homologous and heterologous transposable elements. Dev Genet 10:561–568
4. Van den Broeck D et al (1998) Transposon Display identifies individual transposable elements in high copy number lines. Plant J 13:121–129
5. Vos P et al (1995) AFLP: a new technique for fingerprinting. Nucl Acids Res 23:4407–4414

6. Cartolano M et al (2007) A conserved microRNA module exerts homeotic control over *Petunia hybrida* and *Antirrhinum majus* floral organ identity. Nat Genet 39:901–905
7. Vandenbussche M et al (2009) Differential recruitment of WOX transcription factors for lateral development and organ fusion in Petunia and Arabidopsis. Plant Cell 21:2269–2283
8. Spelt C et al (2000) anthocyanin1 of petunia encodes a basic helix-loop-helix protein that directly activates transcription of structural anthocyanin genes. Plant Cell 12:1619–1632

Chapter 18

Massive Indexed Parallel Identification of Transposon Flanking Sequences

Michiel Vandenbussche, Jan Zethof, and Tom Gerats

Abstract

The large scale sequencing of insertion element flanking sequences has revolutionized reverse genetics in plant research: Insertion mutants can now simply be identified *in silico* by BLAST searching the resulting flanking sequence databases. The development of next-generation sequencing technologies has further facilitated the creation of flanking sequence collections derived from entire mutant populations. Here we describe a highly efficient and widely applicable method that we developed to amplify, sequence, and identify *dTph1* transposon flanking sequences from a library of 1000 Petunia W138 individuals simultaneously.

Key words Reverse genetics, Insertion mutagenesis, Next-generation sequencing, Transposon, dTph1, Transposon display, Petunia

1 Introduction

The methodology we detail here allows amplification and sequencing of insertion element-flanking fragments from an entire population simultaneously. Importantly, with this setup, each flanking sequence can be traced back to its individual plant of origin, or to specific families, in case the insertion is segregating within a family. We have used this strategy (*see* Fig. 1) to sequence transposon flanking sequences from a Petunia *dTph1* transposon insertion library of 1,000 individuals [1]. From this population, we have harvested DNAs according to the classical three-dimensional (3D) pooling principle as applied before in PCR-based Petunia insertion mutagenesis screens [2, 3], yielding 30 DNA samples each prepared from 100 pooled leaf samples. Next, we have mass-amplified transposon flanking sequences from each of the 30 samples separately using the TD methodology (*see* Chapter 17). For the nested PCR (corresponds to the "Hot PCR" step in the TD protocol), we have used a different amplification primer for every PCR reaction,

Thomas Peterson (ed.), *Plant Transposable Elements: Methods and Protocols*, Methods in Molecular Biology, vol. 1057, DOI 10.1007/978-1-62703-568-2_18, © Springer Science+Business Media New York 2013

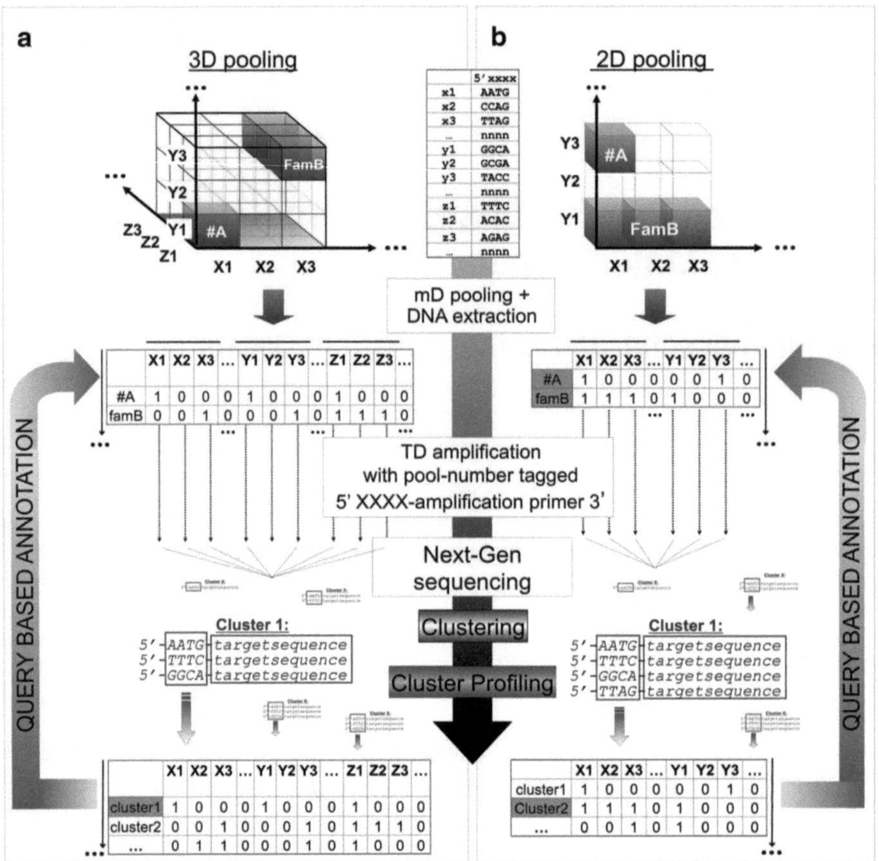

	X1	X2	X3	...	Y1	Y2	Y3	...	Z1	Z2	Z3	...
#A	1	0	0	0	1	0	0	0	1	0	0	0
famB	0	0	1	0	0	0	1	0	1	1	1	0

	X1	X2	X3	...	Y1	Y2	Y3	...
#A	1	0	0	0	0	0	1	0
famB	1	1	1	0	1	0	0	0

	X1	X2	X3	...	Y1	Y2	Y3	...	Z1	Z2	Z3	...
cluster1	1	0	0	0	1	0	0	0	1	0	0	0
cluster2	0	0	1	0	0	0	1	0	1	1	1	0
...	0	1	1	0	0	0	1	0	1	0	0	0

	X1	X2	X3	...	Y1	Y2	Y3	...
cluster1	1	0	0	0	0	0	1	0
Cluster2	1	1	1	0	1	0	0	0
...	0	0	1	0	1	0	0	0

Fig. 1 Massive indexed parallel identification of transposon flanking sequences flow chart. DNA templates are obtained by organizing the source material in a virtual 3D cubic setup (**a**), or a 2D setup (**b**), and pooling material from each dimension before DNA extraction. For example in the 3D setup, all individuals in the X1 dimension are pooled together, and this is similarly performed for other planes in all dimensions. Each individual DNA source in such a cube is therefore represented in three pools, once in every dimension, where the pool numbers reflect the spatial positioning of the DNA source in the 3D grid, e.g., DNA from plant #A will be present in pools X1, Y1, and Z1. Such 3D (or 2D) coordinate sets can be converted to binary code as a series of 1 s and 0 s in a database query line, showing the presence or absence of individual DNA templates in every pool of the library. Small groups of related individuals (e.g., famB), representing a family, can be profiled in a similar fashion. Templates are subsequently PCR-amplified from every DNA pool separately, using the TD protocol with amplification primers containing pool number information encrypted as a "barcode" at its 5′ end. These amplification products are finally pooled together in one sample and sequenced by next-generation sequencing. The resulting sequences are then clustered according to homology in the region between the amplification primers. Next, these clusters are converted to single database lines containing a consensus sequence representing all sequences within each cluster, and the 5′ barcodes are converted to a series of numbers indicating the pool numbers encountered within each cluster. These database lines are cross-searched against the originally defined profiles of individuals and known families within the population, leading to automatic annotation of the resulting clusters

carrying a unique four-nucleotide sample identification tag at the 5′ end. The 30 resulting PCR products are subsequently pooled and digested again with *Mun*I and *Mse*I (*see* **Note 1**), restriction sites of which have been incorporated in the amplification primers. This allows ligation of the GS20/454 adaptors A and B, required for 454 sequencing: We modified these adaptors to have sticky *Mse*I and *Mun*I ends instead of blunt ends, allowing unidirectional ligation of the sequencing annealing site in adaptor A, immediately adjacent to the four-nucleotide tag in the transposon-specific primer. As a consequence, all sequences originating from the same template have the same orientation, facilitating subsequent computational clustering. The adaptor-ligated mixture is then used as a template for a final PCR amplification step using amplification primers A and B. An aliquot of this mixture is then used for unidirectional sequencing using a modified GS20 protocol. Finally, the resulting sequences are clustered according to sequence identity and reduced to a single database line by "reading" the various sequence identification tags encountered within each cluster. Querying the resulting database with the predefined profiles of individuals and families allows automatic assignment of sequences to their source. Using this approach, which demanded only 30 DNA isolations, 70 PCR reactions, and two GS20 454 sequencing runs, we were able to assign approximately 10,000 transposon flanking sequences to specific plants in the library.

Since 2005, next-generation sequencing technologies have evolved rapidly and alternative technologies have become available (such as Illumina/Solexa sequencing) which by now generate sufficiently long sequence reads as required for this purpose. Especially their highly increased sequencing capacity make these new technologies much better suited to completely resolve the high complexity of a sample derived from thousands of different individuals. Our protocol can be adapted easily to these newer sequencing approaches by substituting the correct adaptor sequences in the last step of the workflow.

The protocol described in this chapter includes a normalization step. This normalization is particularly useful in the case of Petunia *dTph1* mutant populations, since a large proportion of the *dTph1* elements are immobile and thus shared between every member of the population. By normalizing the sample before sequencing, we aim to enrich for unique insertions. This normalization step might not be needed for other systems. We have therefore included also an alternative protocol that skips the normalization step.

2 Materials

2.1 DNA Extraction from Pooled Leaf Samples

1. Extraction buffer: (0.1 M Tris–Cl pH 8, 0.5 M NaCl, 0.01 M β-mercapto-ethanol).
2. $T_{01}E$ buffer (10 mM Tris–Cl pH 8, 0.1 mM EDTA).
3. 5/3 Kac: (60 ml 5 M Kac, 11.5 ml glacial acetic acid, 28.5 ml H_2O).
4. Iso-Propanol.
5. 8 M LiCl.
6. Ethanol (96 and 70 %).
7. H_2O (Dnase free).
8. 10 mg/ml Rnase.
9. 0.05 M EDTA.
10. 10 % SDS.

2.2 "Transposon Display" Amplification of DNA Pools

1. Digestion/ligation:
   (a) *Mse*I (10 U/μl); *Mun*I (or *Mfe*I, isoschisomer), NEB4 buffer (New England Biolabs, www.NEB.com).
   (b) 100× BSA stock solution, ATP 10 mM stock solution, T4 DNA ligase (5 U/μl).
2. Adapters:
   (a) *Mun*I-bio-adapter top strand (100 μM stock): 5′-**BIOTIN**-CTCGTAGACTGCGTACG-3′.
   (b) *Mun*I-adapter bottom strand (100 μM stock): 5′-AATTC GTACGCAGTC3′.
   (c) *Mse*I-adapter top strand (100 μM stock): 5′-GACGA TGAGTCCTGAG-3′.
   (d) *Mse*I-adapter bottom strand (100 μM stock): 5′-TACTCA GGACTCAT-3′.
3. Beads extraction:
   (a) Dynabeads® *MyOne Streptavidin C1* (Invitrogen).
   (b) Magnet (e.g., DynaMag™-2 magnet, Invitrogen).
   (c) STEX (10 mM Tris–Cl pH 8, 1 M NaCl, 1 mM EDTA, and 0.1 % Triton X-100).
   (d) Binding Buffer (10 mM Tris–Cl pH 8, 2 M NaCl, 1 mM EDTA, and 0.1 % Triton X-100).
   (e) $T_{01}E$ buffer (10 mM Tris–Cl pH 8, 0.1 mM EDTA).
4. Oligos for Preamplification PCR:
   (a) *Mun*I + ACAC: 5′-AGACTGTGTACGAATTGACAC-3′.
   (b) *Mse*I + 0: 5′-GACGATGAGTCCTGAGTAA-3′.

5. Oligos for specific amplification of *dTph1* flanking sequences using barcoded primers:
   (a) NNNN-IR primer: 5′-CATATATTAA**NNNN**GTAGCTC CGCCCCTG-3′ (*see* **Note 2**).
   (b) For each pool sample order a differently barcoded NNNN-IR primer and substitute NNNN with 1 of 30 following barcodes: (ACAC, ACAG, ACGA, ACGT, ACTC, ACTG, AGAC, AGAG, AGCA, AGCT, AGTC, AGTG, ATCG, ATGC, CACA, CACT, CAGA, CAGT, CATC, CATG, CGAT, CGTA, CTAC, CTAG, CTCA, CTCT, CTGA, CTGT, GACA, GACT).
   (c) *Mun*I/*Mse*I+0 primer: 5′-CATATA**CAATTG**GACGAT GAGTCCTGAGTAA-3′ (if normalization step is omitted).
   (d) *Mse*I+0: 5′-GACGATGAGTCCTGAGTAA-3′ (if normalization is included).

2.3 Normalization (Optional)

1. Formamide.
2. Buffer A (0.1 M Tris–Cl pH 8.0, 1.2 M NaCl, 50 mM EDTA).
3. Tween 20.
4. 1 M Na-Pi stock solution: mix 46.3 ml Na_2HPO_4 with 53.7 ml NaH_2PO_4 (pH 6.8).
5. HAP: DNA Grade Bio-Gel HTP, Hydroxyapatite, Bio-rad (130-0520).
6. A 2 ml syringe contained in a holder kept at 60 °C by connection with a hot water circuit (connect the syringe holder with the pump of a hot water bath). Attach a rubber tube to the end of syringe that can be opened (for elution) or closed with a clamp. Put glass wool at the bottom of the syringe to allow HAP loading.
   (a) Column preparation:
      - Clean the syringe and fill with 40 mM Pi buffer (pre-warmed at 60 °C).
      - In a falcon tube, equilibrate a spoonful of HAP one time in 0.1 M Pi buffer.
      - Wash the HAP five times with 40 mM Pi buffer (pour off supernatants, every time after 2 min). The fifth time, leave ca. 1 ml of liquid.
      - Drain the syringe by opening the clamp until ca. 0.5 ml of the Pi buffer is left.
      - Pipet equilibrated HAP into the syringe (in several steps) while draining the syringe until a HAP layer is formed of around 150 μl.
      - Wash the HAP column six times with pre-warmed 40 mM Pi buffer. Nucleobond AX20 columns (Macherey Nagel; http://www.mn-net.com).

(b) Oligo for primer extension after normalization:
- *Mun*I/*Mse*I + 0 primer: 5′-CATATACAATTGGACGA TGAGTCCTGAGTAA-3′.

2.4 Template Preparation for Pyro-Sequencing

1. GS20 adapter oligos if the normalization step is omitted:
   (a) GS20 *Mun*I Biotin-TEG adapter B:
   - GS20-*Mun*I-top: 5′-**BIOTIN-TEG-CCTATCCCC TGTGTGCCTTGCCTATCCCCTG TTGCGTGTCTCAG**-3′.
   - GS20-*Mun*I-bottom: 5′-AATTCTGAGACACGCAA CAGGGGATAGGCAAGGCACACAGGGGA-3′.

   (b) GS20 *Mse*I-adapter A:
   - GS20-*Mse*I-top 5′-CCATCTCATCCCTGCGTGTCC CATCTGTTCCCTCCCTGTCTCAG-3′.
   - GS20-*Mse*I-bottom 5′-TACTGAGACAGGGAGGGA ACAGATGGGACACGCAGGGATGAG-3′.
2. GS20 adapter oligos if the normalization step is performed:
   (a) GS20 *Mun*I-adapter B:
   - GS20-*Mun*I-top: 5′-CCTATCCCCTGTGTGCCTTG CCTATCCCCTGTTGCGTGTCTCAG-3′.
   - GS20-*Mun*I-bottom: 5′-AATTCTGAGACACGCAA CAGGGGATAGGCAAGGCACACAGGGGA-3′.

   (b) GS20 *Mse*I-adapter A:
   - GS20-*Mse*I-top 5′-CATCTCATCCCTGCGTGTCCC ATCTGTTCCCTCCCTGTCTCAG-3′.
   - GS20-*Mse*I-bottom 5′-TACTGAGACAGGGAGGGA ACAGATGGGACACGCAGGGATGAG-3′.

   (c) Amplification adapter primers A and B:
   - BioTEG-adapter primer B: 5′-**BIOTIN-TEG**-CCTAT CCCCTGTGTGCCTTG-3′.
   - Adapter primer A: 5′-CCATCTCATCCCTGC GTGTC-3′.

3 Methods

3.1 Population Setup and DNA Extraction

Grow a mutant population and harvest leaf samples (each time one young non-expanded leaf/per plant) according to a three or two-dimensional matrix (*see* Fig. 1) and freeze pooled leaf samples in liquid nitrogen for storage at −80 °C. Extract high quality DNA from each sample using the following protocol:

1. Grind the frozen leaves in liquid nitrogen using a cooled mortar and pestle (or grinding ball/vortex).

2. Transfer 1–2 g of the powdered tissue to a 10 ml screw-cap tube.
3. Add 6 ml Extraction buffer and 0.8 ml 10 % SDS.
4. Extract at 65 °C for 60 min (mix the solution periodically).
5. Add 2 ml of an ice cold 5/3 Kac solution and mix by inversion.
6. Incubate on ice for 30 min.
7. Centrifuge for 15 min at 3,000×*g* (at 4 °C).
8. Transfer the supernatant (carefully) to a 50 ml Falcon tube.
9. Add 6 ml (=equal volume) of Iso-Propanol and mix by inversion.
10. Centrifuge for 10 min at 3,000×*g*.
11. Discard the supernatant.
12. Wash the pellet with 1 ml 70 % ethanol.
13. Recentrifuge for 5 min at 3,000×*g*.
14. Discard the supernatant.
15. Leave the pellet as dry as possible (without using the speedvac!).
16. Dissolve the pellet in 600 μl $T_{01}E$.
17. Add 400 μl 8 M LiCl.
18. Mix by inverting the tube and incubate o/n at 4 °C.
19. Centrifuge for 20 min at 3,000×*g* at 4 °C.
20. Transfer the supernatant to a 2 ml Eppendorf tube.
21. Add 1 ml iso-propanol, mix by inverting the tube.
22. Spin for 10 min (Eppendorf centrifuge, max. speed).
23. Wash the pellet with 70 % ethanol.
24. Re-spin and remove the ethanol (leave the pellet as dry as possible, without using the speedvac!).
25. Dissolve the DNA pellet in a 400 μl $T_{01}E$.
26. Check 1 μl on Agarose gel (=quality and quantity control).
27. Resuspend the DNA at a concentration of ~100 ng/μl and store at −20 °C.

3.2 "Transposon Display" Amplification of DNA Pools

1. Digestion/ligation:
   (a) Prepare the *Mun*I-bio-adapter and *Mse*I-adapter. Note that here a biotinylated *Mun*I-adapter is used to allow a beads extraction purification step after the ligation step.
   - For the preparation of 100 μl *Mun*I-adapter (5 μM), combine
     - 5 μl *Mun*I top strand (100 μM stock).
     - 5 μl *Mun*I bottom strand (100 μM stock).
     - 90 μl H_2O.

- For the preparation of 100 µl *Mse*I-adapter (50 µM), combine
  - 50 µl *Mun*I top strand (100 µM stock).
  - 50 µl *Mun*I bottom strand (100 µM stock).

  Incubate the mixtures in a PCR machine for 5 min at 80 °C, gradually cool down to RT over the period of 1 h, put on ice.

(b) Digest 5 µg DNA (dissolved in 50 µl) of each pool sample with *Mun*I and *Mse*I by adding 20 µl of the following restriction mix:
- 2 µl *Mun*I (10 U/µl).
- 2 µl *Mse*I (10 U/µl).
- 7 µl NEB 4 (10×).
- H_2O to 20 µl.

(c) Incubate for 1.5 h at 37 °C.

(d) To the digestions, add 30 µl mix, containing:
- 8 µl *Mun*I-bio-adapter (5 pmol/µl).
- 8 µl *Mse*I-adapter (50 pmol/µl).
- 3 µl NEB 4 (10×).
- 0.3 µl BSA (100×).
- 3 µl ATP (10 mM).
- 3 µl T4 DNA ligase (5 Weiss U/µl).
- H_2O to 30 µl.

(e) Incubate for 5 h at 37 °C.

2. Selection of biotinylated DNA fragments:

(a) Remove first the excess of biotinylated adapters using the QIAquick PCR purification kit (or equivalent) and elute the DNA in a total volume of 55 µl (elution buffer supplied by the kit).

(b) Per sample, wash 25 µl streptavidin beads (ca. 0.1 mg MyOne beads, streptavidin C1) once in 200 µl STEX, remove the STEX using a magnet and resuspend in 100 µl binding buffer.

(c) Add the 100 µl of Streptavidin beads (in binding buffer) into the 50 µl restriction/ligation mixture and incubate for 60 min on a rotator, at RT.

(d) Collect the beads with a magnet and remove the supernatant. Resuspend the beads in 200 µl STEX buffer, and transfer the solution to a new tube.

(e) Perform three additional washes with 200 μl STEX buffer using the magnet (remove the STEX well after the last wash).

(f) Resuspend the beads in 50 μl $T_{01}E$ and transfer to another tube.

3. Selective preamplification:

(a) Pipet 2 μl template beads DNA in a PCR tray (*see* **Note 3**) and add 18 μl mix containing:

- 0.6 μl *Mun*I + ACAC primer (10 μM).
- 0.6 μl *Mse*I + 0 primer (10 μM).
- 0.8 μl dNTP (5 mM).
- 2 μl 10× PCR buffer.
- 2 μl $MgCl_2$ (25 mM).
- 0.12 μl standard Taq DNA polymerase (5 U/μl).
- H_2O to 18 μl.

(b) Incubate the samples in a PCR machine according to the following PCR profile:

- 13 cycles (94 °C for 15 s, 65 °C for 30 s −0.7 °C/cycle, 72 °C for 60 s).
- 22 cycles (94 °C for 15 s, 56 °C for 30 s, 72 °C for 60 s).

(c) Check 5 μl of the resulting PCR products by electrophoresis on a 1.5 % agarose gel: a weak low molecular-weight smear should be visible.

(d) Dilute the remaining 15 μl ten times with H_2O.

4. Specific amplification of *dTph1* flanking sequences using bar-coded primers:

(a) Pipet 5 μl of the ten-fold diluted pre-amplified material into a PCR tray and add 45 μl of the following mix to each sample:

- 1.5 μl NNNN-IR primer (10 μM, a differently bar-coded primer for each reaction).
- 1.5 μl *Mun*I/*Mse*I + 0 primer(10 μM), **OR** *Mse*I + 0 primer (10 μM) (*see* **Note 4**).
- 2 μl dNTP (5 mM).
- 5 μl 10× PCR buffer.
- 5 μl $MgCl_2$ (25 mM).
- 0.2 μl Red Hot standard Taq DNA polymerase (5 U/μl).
- H_2O to 45 μl.

(b) Incubate the samples in a PCR machine according to the following PCR profile:
   - 13 cycles (94 °C for 15 s, 65 °C for 30 s –0.7 °C/cycle, 72 °C for 60 s).
   - 22 cycles (94 °C for 15 s, 56 °C for 30 s, 72 °C for 60 s).

(c) Pool the resulting PCR products from the ten samples from each dimension together to create three "super pool" samples: an X, Y and Z sample.

(d) Purify the "super pool" samples using the QIAquick PCR purification kit (or equivalent) and elute the DNA in a total volume of 200 μl (elution buffer supplied by the kit) (*see* **Note 5**).

   If normalization is not needed, proceed immediately with the next step below. Otherwise, skip the next step, and continue with Subheading 3.3.

5. Digestion and directional 454 adapter ligation.

(a) Precipitate the DNA with Ethanol/NaAc and resuspend in 70 μl H_2O.

(b) Digest 70 μl of each DNA superpool sample obtained above and add 20 μl of the following mix:
   - 2 μl *Mfe*I (10 U/μl stock).
   - 2 μl *Mse*I (10 U/μl stock).
   - 9 μl NEB Restriction Enzyme Buffer 4 (10× stock).
   - 0.9 μl BSA (100× stock).
   - H_2O to 20 μl.

(c) Add 20 μl of the following mix to the digestions and incubate for another 3 h at 37 °C:

(d) Incubate for 2 h at 37 °C (*see* **Note 6**).
   - 4 μl BioTEG *Mun*I-adapter primer B (50 pmol/μl stock).
   - 4 μl *Mse*I-adapter A (50 pmol/μl stock).
   - 2 μl NEB 4 (10× stock).
   - 0.2 μl BSA (100× stock).
   - 3 μl ATP (10 mM stock).
   - 3 μl T4 DNA ligase (5 Weiss U/μl stock).
   - H_2O to 20 μl.

(e) Purify all the samples using a PCR purification kit (e.g., Qiagen) to remove non-ligated adapters. Elute with 55 μl of elution buffer.

(f) Combine all samples together and use approximately 4 μg for GS20 sequencing. These samples are ready for the "library immobilization step" of the GS20 sequencing protocol, so all preceding steps in the official GS20 library preparation protocol (shearing, end-polishing, phosphorylation or ligation of the A and B adapters, Bst DNA polymerase fill-in step) should be omitted.

3.3 Normalization

To optimize the amount of unique fragments against a background of fragments shared by many to all individuals, we separately normalize the three pooled samples by hydroxyapatite chromatography [4]. The approach takes advantage of the differential re-annealing kinetics between highly abundant and rare DNA templates, by which the single-stranded fraction becomes enriched for low copy number templates.

1. Precipitate the pooled and purified PCR products (*x*, *y*, and *z* samples) with Ethanol/NaAc and resuspend in 30 μl formamide in 1.5 ml Eppendorf tubes (*see* **Note 7**).
2. Add to each tube 9 μl TE and 6 μl H_2O, and cover with a small layer of mineral oil.
3. Heat the tubes at 80 °C for 3 min.
4. Add 6 μl buffer A (kept on ice before addition) and 9 μl H_2O to each tube.
5. Incubate for 16 h at 30 °C.
6. Dilute the samples with 500 μl pre-warmed 40 mM Pi buffer (60 °C), incubate 1–2 min at 60 °C, and apply to the prepared column (*see* Subheading 2.3).
7. Wash four times with 500 μl pre-warmed 40 mM Pi buffer.
8. Elute with eight times 100 μl pre-warmed 0.12 M Pi buffer. Keep fractions 2–8, which contain the single-stranded normalized DNA fraction (*see* **Note 8**).
9. Remove the Pi buffer from the eluted DNA by purification over a Machery Nagel AX20 column:
   (a) Add first 0.02 % Tween 20 to the buffers N2, N3, and N5.
   (b) Equilibrate the column with 1 ml N2 and load the samples directly after.
   (c) Wash with 3 ml 1:1 diluted N2 buffer.
   (d) Elute with four times 200 μl N5 buffer (wait for about 5 min after second load).
10. Precipitate the DNA with Iso-Propanol and dissolve the sample in 50 μl H_2O. The sample can be stored at −20 °C.
    1. Template preparation for pyro-sequencing:
       (a) Primer extension for the conversion of ssDNA to dsDNA:

- Mix 50 μl of single-stranded DNA with 50 pmol extension primer in a 1× PCR buffer (20 mM Tris–Cl pH 8.4, 50 mM KCl, 1.5 mM $MgCl_2$) with 1 U Platinum Taq DNA polymerase (Invitrogen), 0.2 mM dNTPs, and H_2O to a final volume of 75 μl.
- Incubate the samples in a PCR machine according to the following PCR profile:
  - Cycle (94 °C for 2 min, 56 °C for 1 min, 72 °C for 10 min).

2. Digestion and directional 454 adapter ligation:
   (a) Use 70 μl of the dsDNA obtained above and add 20 μl of the following mix:
   - 2 μl *Mun*I (10 U/μl stock).
   - 2 μl MseI (10 U/μl stock).
   - 9 μl NEB 4 (10× stock).
   - 0.9 μl BSA (100× stock).
   - H2O to 20 μl.
3. Incubate for 2 h at 37 °C (*see* **Note 6**).
4. Add 20 μl of the following mix to the digestions, and incubate for another 3 h at 37 °C:
   (a) 4 μl GS20 MunI-adapter B (50 pmol/μl stock).
   (b) 4 μl GS20 MseI-adapter A (50 pmol/μl stock).
   (c) 2 μl NEB Restriction Enzyme Buffer 4 (10× stock).
   (d) 0.2 μl BSA (100× stock).
   (e) 3 μl ATP (10 mM stock).
   (f) 3 μl T4 DNA ligase (5 Weiss U/μl stock).
   (g) H2O to 20 μl.
5. Reamplification:
   After normalization, we re-amplify the ligation mixture to obtain sufficient biotinylated DNA for GS20 sequencing.
   (a) Use 5 μl of ligation mixture and add 45 μl of the following mixture:
   - 1.5 μl adapter primer A (10 μM).
   - 1.5 μl BioTEG-adapter primer B (10 μM).
   - 2 μl dNTP (5 mM).
   - 5 μl 10× PCR buffer.
   - 5 μl MgCl2 (25 mM).

- 0.2 μl standard Taq DNA polymerase (5 U/μl).
- H2O to 45 μl.

(b) Incubate the samples in a PCR machine according to the following PCR profile:

- 13 cycles (94 °C for 15 s, 65 °C for 30 s –0.7 °C/cycle, 72 °C for 60 s).
- 22 cycles (94 °C for 15 s, 56 °C for 30 s, 72 °C for 60 s).

(c) Combine all samples together and use approximately 4 μg for GS20 sequencing. These samples are ready for the "library immobilization step" of the GS20 sequencing protocol, so all preceding steps in the official GS20 library preparation protocol (shearing, end-polishing, phosphorylation or ligation of the A and B adapters, Bst DNA polymerase fill-in step) should be omitted.

4 Notes

1. *Mse*I and *Mun*I were initially used for genomic DNA digestion, guaranteeing that no *Mse*I and *Mun*I sites will be present within the amplified transposon flanking sequences during digestion.
2. The NNNN-IR primer contains 5′ a *Mse*I restriction site (TTAA), preceded by six spacer nucleotides (CATATA). The latter have been added to increase digestion efficiency by *Mse*I. Here, we have proposed 30 different 4 nucleotide barcode variants, which have been tested and work well. If the population sampling requires more barcodes, carefully check that these do not create new restriction sites for either *Mse*I or *Mun*I.
3. Mix the beads solution well just before pipetting (the DNA fragments are still connected to the beads).
4. Use the *Mun*I/*Mse*I + 0 primer if the normalization step is omitted; use the *Mse*I + 0 primer if a normalization will be performed.
5. We divide every super-pooled sample into two batches (corresponding to 250 μl PCR product each) and purify these separately on two columns to avoid column saturation. After this, we again combine the two eluted batches.
6. Once the restriction digestion reactions have been started you need to continue with the adapter ligation, because the restriction enzyme activity is required to prevent fragment–fragment back ligation.

7. Do not let the pellets dry completely before dissolving in formamide.

8. The ds DNA fraction can be eluted with eight times 100 µl pre-warmed 0.4 M Pi buffer (keep fraction 2–8) and be further purified as for the single-stranded fraction. The effectiveness of the normalization can be verified by a PCR competition experiment between a high copy number flanking sequence and a low copy sequence amplified from the single- and double-stranded DNA fractions or by a transposon display experiment on normalized and non-normalized samples.

References

1. Vandenbussche M et al (2008) Generation of a 3D indexed Petunia insertion database for reverse genetics. Plant J 54:1105–1114
2. Koes R et al (1995) Targeted gene inactivation in petunia by PCR-based selection of transposon insertion mutants. Proc Natl Acad Sci USA 92:8149–8153
3. Vandenbussche M et al (2003) Toward the analysis of the Petunia MADS box gene family by reverse and forward transposon insertion mutagenesis approaches: B, C, and D floral organ identity functions require SEPALLATA-like MADS box genes in Petunia. Plant Cell 15:2680–2693
4. Bonaldo M, Lennon G, Bento Soares M (1996) Normalization and subtraction: two approaches to facilitate gene discovery. Genome Res 6:791–806

Chapter 19

Use of Next Generation Sequencing (NGS) Technologies for the Genome-Wide Detection of Transposition

Moaine Elbaidouri, Cristian Chaparro, and Olivier Panaud

Abstract

Plant transposable elements are ubiquitous in eukaryotes. Their propensity to densely populate the genomes of many plants and animal species has put them in the focus of both structural and functional genomics. Although a number of bioinformatic software have been recently developed for the annotation of TEs in sequenced genomes, there are very few computational tools strictly dedicated to the identification of active TEs using genome-wide approaches. In this paper, we describe SearchTESV, a pipeline that we have developed to detect Transposable Elements-associated structural variants (TEASVs) using Next Generation Sequencing (NGS) technologies.

Key words Transposable elements, NGS, Genomics

1 Introduction

The availability of the full genome sequence of many model organisms has revealed the abundance of repeated elements [1]. Transposable elements contribute to a large portion of the nuclear DNA of most eukaryotic species. Around 44.7 % of the human genome corresponds to sequences derived from transposable elements, while genes constitute only 1.5 % [2]. A fraction of TEs can even reach more than 90 % of the genome sequence of some species like hexaploid wheat (*Triticum aestivum*) [3]. The emergence of Next Generation Sequencing strategies over the past few years has opened new perspectives to study the dynamics of transposable elements at the full genome scale. Several reports established that structural variation (SV) is unexpectedly high within species [4]. SVs include insertions, deletions, inversions or translocations, some of which are associated with TE activity. However, the exact

Moaine Elbaidouri and Cristian Chaparro contributed equally to this work.

Thomas Peterson (ed.), *Plant Transposable Elements: Methods and Protocols*, Methods in Molecular Biology, vol. 1057, DOI 10.1007/978-1-62703-568-2_19, © Springer Science+Business Media New York 2013

extent of the impact of transposition in genome dynamics remains to be elucidated. Although TEs are ubiquitous and densely populate most eukaryotic genomes, for a vast majority, they are under epigenetic silencing via TGS and/or PTGS pathways [5]. Mutants affected for such pathways show reactivation of TE transposition causing new insertions in the genome [6].

The detection of active mobile DNA is not always straightforward. Many of the examples reported to date were obtained through the cloning of genes that were mutated by TE insertions which caused specific and detectable alterations in phenotype [7]. In addition, PCR-based methods have been developed to screen for insertions of known TEs at particular loci, but these methods usually did not allow a complete characterization of the transpositional landscape at a whole genome scale. A number of computational tools and software are available to detect genomic structural variations (SVDetect, Break-dancer …), but the majority of them are not designed to specifically detect TE-related insertions [8].

In this chapter we will describe the use of TESV detect, a pipeline that we developed for the detection of TE insertion using NGS. This pipeline was recently used to identify active TEs in rice [9]. Our strategy is based on the detection of homology breakpoints between paired end reads in a given genome when compared to a reference genome. The presence of SV events can be deduced from paired end reads from a mutant genome which do not map as expected on the reference genome. Because the identification of TE-related breakpoints is complicated by the repetitive nature of most TE sequences, only those TEs inserted in a non-repeated region can be reliably detected. Thus, the TE neo-insertion signature corresponds to uniquely mapped pair-end reads in which the second pair maps to a known TE at a distant location (500 bp away from the unique read). Although this strategy is very efficient to detect neo-insertions of mobile DNA, it requires a high-quality reference sequence as well as good TE annotation for a given species.

2 Materials

2.1 Reference Genome Sequence

Since the putative breakpoints are detected through the identification of abnormally mapped paired-ends (*see* Subheading 3), the procedure requires a high quality reference sequence. In this regard, high quality refers to the final assembly of the pseudomolecules of the completed genomic sequence. So far in plants, Paired-End Mapping (PEM) for the detection of transposition was successfully applied on the two species for which a high quality assembly based on a physical map is available, i.e., rice [9] and *Arabidopsis thaliana* [6].

Table 1
Plant TE databases available

Name	Description	URL and reference
Repbase	Database of repetitive DNA from different eukaryotic species	http://www.girinst.org/ Jurka et al. [11]
Retroryza	Rice LTR Retrotransposons database	http://retroryza.fr/ Chaparro et al. [12]
Maize database	Maize transposable elements database	http://maizetedb.org/~maize/
SoyTEdb	Soybean genome transposable elements	http://www.soybase.org/soytedb/ Du et al. [13]
TREP	Triticeae Repeat Sequence Database	http://wheat.pw.usda.gov/ITMI/Repeats/
TIGR	Plant Repeat Database	http://blast.jcvi.org/euk-blast/index.cgi?project=plant.repeats Ouyang and Buell [14]
IS Finder	IS elements database from eubacteria and archaea	https://www-is.biotoul.fr// Siguier et al. [15]

2.2 Sequencing Data

Various NGS technologies have been developed during the past few years, allowing for very fast and low cost DNA sequencing. The Illumina (Solexa) technology provides short paired-end reads. These paired-end sequences are produced by sequencing both ends of small linear DNA fragments generated from randomly sheared double-stranded DNA. The Illumina sequencing yields two FASTQ files corresponding to the paired-end reads (i.e., one forward and one reverse). In addition to the nucleotide sequences, the FASTQ files also provide information about the quality score of the sequencing data.

2.3 Mapping Tools

The first step in the procedure is to map the sequencing reads onto a reference genome (*see* Subheading 3). This is achieved using mapping software especially designed for NGS data (large dataset and short length reads). The pipeline described below uses Bowtie (an ultrafast memory-efficient short read aligner) that can align a large set of paired-end reads very quickly on a large genome [10].

2.4 TE Database

The growth of genome sequencing projects for both plant and animal genomes has made available large amounts of data related to transposable elements. Table 1 lists some of these TE databases.

3 Methods

3.1 In Silico Analysis

The fundamental principle of TE associated structural variants (TEASV) discovery by Paired-End Mapping (PEM) is schematized in Fig. 1a. The two reads of a paired-end generated by Illumina sequencing technology are physically separated by a given distance (the insert size of the DNA library used for sequencing, *see* Fig. 1b). This distance, compared with the average size of the DNA library, can be used to infer the presence of potential structural variations caused by TE insertions. As shown in Fig. 1 the sequencing and mapping of paired-end reads of a mutant genome (containing a novel TE insertion) onto the reference sequence (which lacks the TE insertion) can lead to several different results:

(a) Both reads of a pair are "normally" mapped (i.e., the paired-end reads are separated by a distance which is compatible with the average insert size of DNA library) at a unique site onto the reference sequence (pairs 1 and 5).

(b) Both reads correspond to TE sequences (pairs 3).

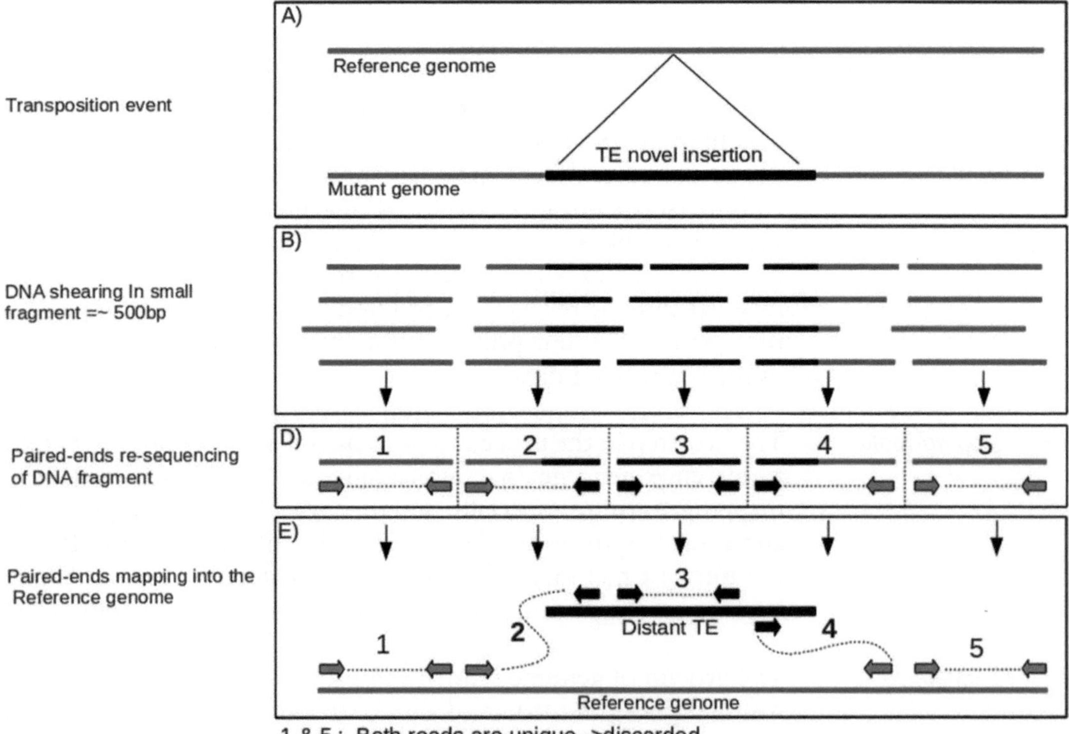

Fig. 1 Principle of TEASV detection using Paired-end mapping

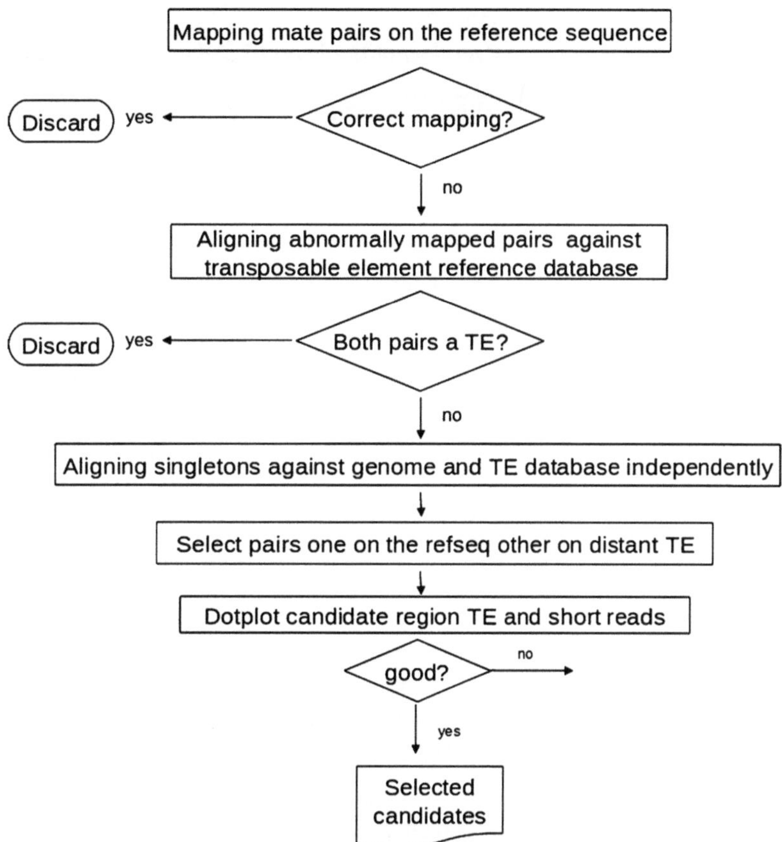

Fig. 2 Workflow of TEASV detection using Paired-end mapping

(c) "Abnormally" mapped reads, i.e., one read maps onto a unique sequence in the reference genome, while the mate read maps onto a distant TE (pairs 2 and 4).

Therefore the information about the relative distances between paired-end reads can be used in an automated procedure to detect potential new insertions of transposable elements into non-repeated regions.

3.1.1 Description of the Pipeline Procedure

Figure 2 represents the workflow of the automatic procedure.

(a) The paired-end reads are mapped against the reference genome using Bowtie software (*see* Subheading 3). This step allows the reads to be sorted into two classes: "normally" mapped pairs and "abnormally" mapped pairs. Reads in the first class (normal reads) are discarded because they are separated by the approximate insert size of the library (expected distance), i.e., there is no indication of structural variation between the mutant genome and the reference sequence at this locus. Reads in the second class (abnormal reads) correspond to pairs that mapped

Table 2
Overview of the SearchTESV options sorted by categories

Input options	
-in	Base name of reads without _1 or _2 <>
-g	Reference genome multiple fasta filename <path>
-t	Transposable element database <path>
-p	Number of processors <1>
Filter options	
-I	Min distance between reads <100>
-X	Max distance between reads <450>
-C	Min number of reads for TESV detection
? Options	
-keep	keep intermediate results <boolean>
--	Forcibly stop option processing
-help	Print this message
-?	Print this message

at an unexpected distance from each other compared with the insertion size of the library. This may indicate that a TE present at one location in the reference sequence inserted into a new site in the mutant genome (putative transposition).

(b) The "abnormally" mapped read pairs are aligned against a transposable elements database and rejected if both of them are TE (*see* **Note 1**).

(c) The remaining abnormally mapped read pairs are aligned against the reference genome sequence and the TE database to identify pairs that have one read corresponding to an unambiguous unique site in the reference genome while its mate pair maps to a distant TE (*see* **Note 2**).

(d) In order to eliminate some false positive candidates a manual verification step is necessary with a dot plot (dotter software) generated from the paired-end, TE inserted region and the transposable element sequence (*see* **Note 3**).

3.1.2 Pipeline Usage

The pipeline can be downloaded from: http://gamay.univ-perp.fr/~SearchTESV/. Table 2 provides the list of options of the software, sorted by categories.

Input options

(a) -in
The two Illumina FASTQ files should end in _1.fq and _2.fq for paired-end reads. The program will use the base name of the FASTQ files to name all other files it produces. For example, if the two FASTQ files corresponding to the paired-ends reads (forward and reverse) are named Read_Illumina_1.fq

and Read_Illumina_2.fq, then you should enter "-in Read_Illumina" on the command line.

(b) -g

Specify the path to the multi fasta file name of the reference genome where you will map the paired-end reads.

(c) -t

Specify the path to the multi fasta file name of the transposable elements database corresponding to the re-sequenced species genome.

(d) -p

Specify the number of processors used by the pipeline. This number will depend on the number of cores present on the computer used to do the analyses.

Filter options

(a) -I

Specifies the minimum insert size for valid paired-ends alignments.

(b) -X

Specifies the maximum distance for a mapped read to be considered as valid; this value is based on the insertion size of the Illumina library. For example, if the reported average insert size is 400 bp a -X of 500 or 600 could detect all correctly mapping pairs.

(c) -C

Specifies the minimum number of reads required to identify a break-point candidate. For a 20× depth coverage we recommend setting a -C value of 10 to detect heterozygous neo-insertions (*see* **Note 4**).

3.2 Manual Validation

3.2.1 In Silico Validation

Figure 3 represents a dot plot of a "good" candidate locus for TE insertion (note that for simplicity we use a 4× sequence coverage in this figure). When a candidate region of a potential new TE insertion is detected by the pipeline, a dot plot is generated automatically (for manual validation) using the sequence of the TE family (TE) together with the sequence of the locus where the insertion has occurred (RefSeq), and the paired-end reads covering this locus (A1/A2, B1/B2, C1/C2, D1/D2). The "abnormally" mapped pairs A1/A2, B1/B2 and C1/C2 correspond to case No. 2 in Fig. 1 while the D1/D2 reads represent case No. 4 in the same figure (*see* **Note 4**).

3.2.2 Wet Lab Validation

A final step is necessary to confirm the TE insertion candidates previously identified by the in silico procedure. Figure 4 shows the PCR-based validation of a transposition event in a mutant genome

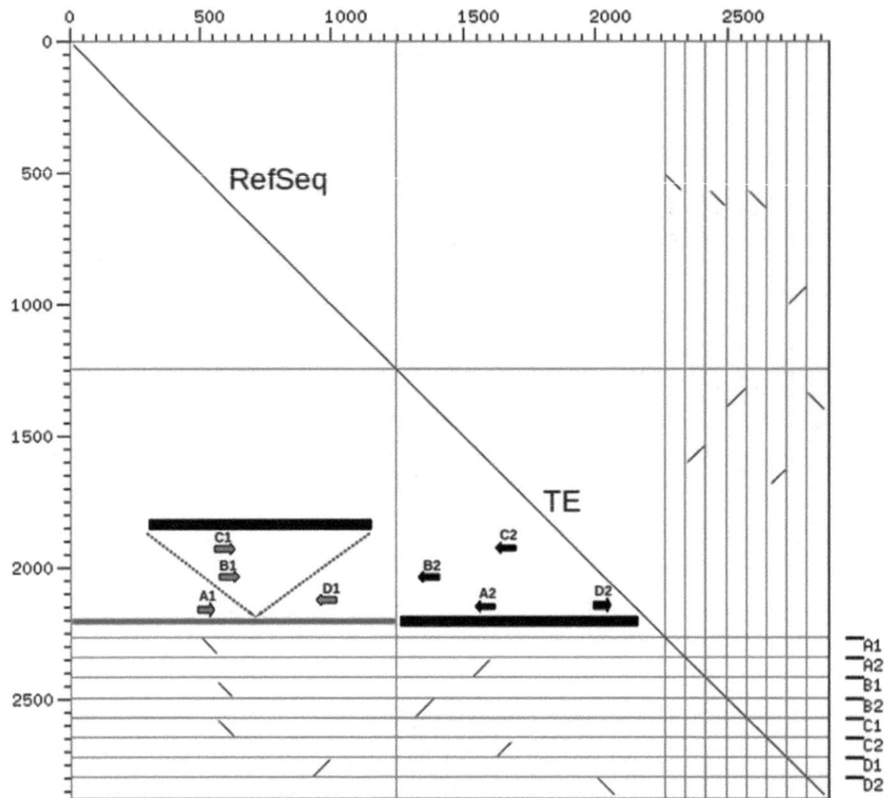

Fig. 3 Reference genome and TE dotplots for manual validation of the breakpoints. Valid breakpoints show a diagonal line in the matrix plot of the reads corresponding to flanking sequence of the TE insertion (*A1*, *B1*, *C1* in the *left* flanking region and *D1* on the *right*) while the paired reads (*A2*, *B2*, *C2*, *D2*) are absent from the target sequence (no homology with the RefSeq) and only present in the TE sequence. *RefSeq* Reference sequence, *A1*/*A2*, *B1*/*B2*, *C1*/*C2*, *D1*/*D2* paired-end reads from the mutant genome

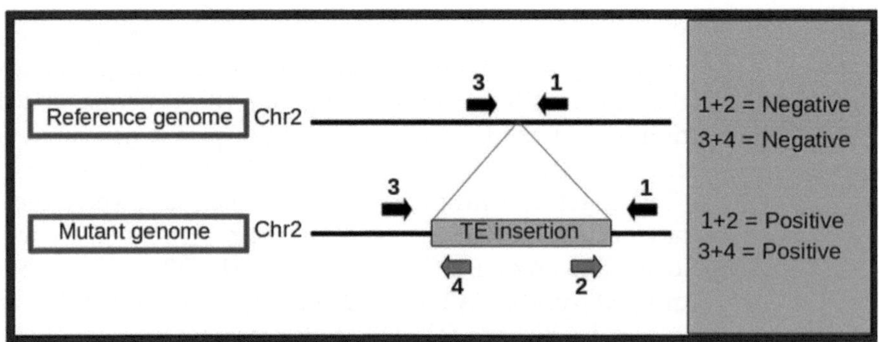

Fig. 4 Strategy for PCR-based wet-lab validation of TEASVs. The *black arrows* (*1* and *3*) correspond to PCR primers of the *right* and *left* flanking region respectively, while the *gray* ones (*2* and *4*) correspond to the newly inserted TE. The 1 + 2 and 3 + 4 sets of primers yield a PCR product only in the mutant genome. *Chr* Chromosome, *TE* Transposable Element

using two sets of primers designed for each putative TE insertion: primer 1 and 3 match the unique right and left flanking regions, respectively, based on the reference genome sequence and primer 2 and 4 match the TE sequence of the candidate (*see* **Note 5**). A PCR assay using primers 1+2, and 3+4 on the mutant genome yields an amplification product, while the same primers give a negative result on the reference genome. The PCR product is sequenced directly, or after cloning into a pGEM vector; if the resulting sequence shows the TE insertion in the target locus the new insertion is confirmed.

4 Notes

1. This second step facilitates the detection of TE neo-insertions. When a TE is inserted into a repeated region, its detection is complicated because of the potential for ambiguous mapping. Therefore, in order to retain only the "good" candidate loci for new TE insertions, only those pairs for which one read unambiguously maps onto a single location are kept.
2. Some chromosomal rearrangements related to transposable elements may occur after their initial transposition. This can potentially lead to detection of false positives. When a large TE, for example a LTR-retrotransposon, is inserted in the genome, the reads A2, B2, C2 and D2 must correspond to the distal regions of the element in order for it to be considered as a true neo-insertion and not some other form of structural variation associated with the TE.
3. A low-quality assembly of the reference genome sequence can lead to false positive candidate detection. In such cases, repeated TE families are often at the origin of mis-assembly. Moreover, false positive cases may be due to insertion polymorphisms found within different individuals of the accession used to generate the reference sequence. In order to circumvent this problem, re-sequencing of the mutant and wild-type genomes will enable the identification and exclusion of insertions common to both genomes.
4. A new TE insertion in a mutant genome can be in a heterozygous state. Re-sequencing of a heterozygous TE insertion with a coverage depth of 20×, should give on average ten paired-end reads corresponding to the TE insertion ("abnormally" mapped pairs) and ten others that mapped "normally" at the same locus.
5. Because of the high copy number and sequence heterogeneity of many TE families, the particular copy responsible for a neo-insertion may not be immediately apparent. It is therefore recommended that a very conserved region of the TE be used to

design primer 2 (wet lab validation). Conserved regions can be identified by performing a multiple alignment of different genomic copies of the same TE family.

Acknowledgments

This work is supported by CNRS and the University of Perpignan Via Domitia. Moaine Elbaidouri is funded by a joint CNRS/ Région Languedoc Roussillon PhD grant.

References

1. Bennetzen JL (2007) Patterns in grass genome evolution. Curr Opin Plant Biol 10:176–181
2. International Human Genome Sequencing Consortium (2004) Finishing the euchromatic sequence of the human genome. Nature 431: 931–945
3. Philippe R et al (2012) Whole Genome Profiling provides a robust framework for physical mapping and sequencing in the highly complex and repetitive wheat genome. BMC Genomics 13:47
4. Cao J et al (2011) Whole-genome sequencing of multiple Arabidopsis thaliana populations. Nat Genet 43:956–963
5. Slotkin RK, Martienssen RA (2007) Transposable elements and the epigenetic regulation of the genome. Nat Rev Genet 8:272–285
6. Mirouze M et al (2009) Selective epigenetic control of retrotransposition in Arabidopsis. Nature 461:427–430
7. Tsugane K et al (2006) An active DNA transposon nDart causing leaf variegation and mutable dwarfism and its related elements in rice. Plant J Jan 45:46–57
8. ElBaidouri M, Panaud O (2012) Genome-wide analysis of transposition using Next Generation Sequencing technologies. Top Curr Genet 24:59–70
9. Sabot F et al (2011) Transpositional landscape of rice genome revealed by Paired-End Mapping of high-throughput resequencing data. Plant J 66:241–246
10. Langmead B et al (2009) Ultrafast and memory-efficient alignment of short DNA sequences to the human genome. Genome Biol 10:R25
11. Jurka J et al (2005) Repbase Update, a database of eukaryotic repetitive elements. Cytogenet Genome Res 110:462–467
12. Chaparro C et al (2007) RetrOryza: a database of the rice LTR-retrotransposons. Nucleic Acids Res 35:D66–D70
13. Du J et al (2010) SoyTEdb: a comprehensive database of transposable elements in the soybean genome. BMC Genomics 11:113
14. Ouyang S, Buell CR (2004) The TIGR plant repeat databases: a collective resource for the identification of repetitive sequences in plants. Nucleic Acids Res 32:360–363
15. Siguier P et al (2006) ISfinder: the reference centre for bacterial insertion sequences. Nucleic Acids Res 34:D32–D36

Chapter 20

Overview of Repeat Annotation and De Novo Repeat Identification

Ning Jiang

Abstract

The availability of a large amount of genomic sequences has provided unique opportunities for understanding the composition and dynamics of transposable elements (TEs) in plants. As the cost of sequencing declines, the genomic sequences of most crop plants will be available within the next few years. Thus, the annotation of genomic sequences, rather than sequence availability, will become the "bottleneck" for genome study. Since TEs are the largest component of most plant genomes, the automation of TE identification and classification is essential for future genome annotation as well as characterization of TEs. In this chapter, the functions and mechanisms of different repeat finding tools are reviewed, with a focus on de novo repeat identification programs. In addition, this chapter covers the further processing of results from de novo identification programs and the construction of repeat libraries for downstream genome analyses.

Key words Transposons, Retrotransposons, Automation, Repeat classification, Repeat library

1 Introduction

Most plant genomes are rich in repetitive sequences. These include gene families, satellite repeats, centromeric repeats, telemetric repeats, and transposable elements (TEs), with TEs comprising the largest portion of the genome. TEs are DNA fragments that are capable of moving from one genomic locus to another, thereby increasing their copy numbers. Based on their transposition mechanisms, TEs fall into two classes. Class I elements, or retrotransposons, use the element–encoded mRNA as the transposition intermediate and consist of two subclasses, the long terminal repeat (LTR) retrotransposons, and the non-LTR retrotransposons, which include long interspersed nuclear elements (LINEs) and short interspersed nuclear elements (SINEs) [1]. Non-LTR retrotransposons are often associated with a poly-A tail at the 3′end of the element as a consequence of transcription. Class II elements, or DNA transposons, are characterized by transposition through a DNA intermediate. With the exception of *Helitrons* [2],

Thomas Peterson (ed.), *Plant Transposable Elements: Methods and Protocols*, Methods in Molecular Biology, vol. 1057, DOI 10.1007/978-1-62703-568-2_20, © Springer Science+Business Media New York 2013

plant DNA elements contain terminal inverted repeats (TIRs), which are often the binding sites for the corresponding transposase. When TEs insert into a specific locus, they often duplicate a small piece of their flanking genomic sequences. This duplication is called target site duplication (TSD), and their length and sequence vary with different TE families [3], so they can be used to classify the relevant TEs. Again, *Helitron* is an exception in that it does not generate a TSD upon insertion.

Despite differences in transposition mechanisms, both classes of TEs can be divided into autonomous and nonautonomous elements. Autonomous elements are those encoding functional transposases (or reverse transcriptases), while nonautonomous elements rely on the transposases from autonomous elements for their transposition [2]. Although nonautonomous elements do not encode any functional transposase, a subset of them appears to have coding capacity, either containing known protein domains or "unknown proteins." In most cases, the coding capacity of nonautonomous elements results from their ability to capture and amplify genomic sequences including genes or gene fragments. A recent study in maize indicated that almost all types of TEs carry gene or gene fragments in their internal regions, with Pack-MULEs (*Mutator-like* elements containing gene fragments) and *Helitrons* the most frequent types [4]. In maize (*Zea mays*), a total of 1,194 *Helitron* elements harbor gene fragments [5]. In rice (*Oryza sativa*, a monocot) and *Lotus japonicus* (a dicot), thousands of Pack-MULEs are present in the genome, suggesting that the mechanism of gene duplication by *Mutator*-like elements in higher plants is ancient and widespread [6, 7].

Because of their abundance in the genome, identification of repeats is not only essential for those who study repeats, but also important for genome annotation. This is because the presence of repeats complicates gene prediction; hence they must be identified and "masked" prior to gene annotation. If repeats are left unmasked, they seed millions of spurious sequence alignments, producing false evidence for gene annotations. Moreover, the open reading frames (ORFs) of autonomous elements appear similar to *bona fide* genes to most gene-prediction programs, causing portions of these repeats to be listed as additional exons within gene predictions, thereby corrupting the final gene annotations. Finally, the large numbers of TEs that carry genes or gene fragments should be distinguished in gene annotation, despite the fact that some of them might represent *bona fide* genes [8].

Plant genomes differ dramatically in their collection of TEs. In general, LTR retrotransposons make up the largest fraction of the repetitive portion in most characterized plant genomes [1]. In maize, 3/4 of the genome is composed of LTR retrotransposons, while the remaining TEs account for less than 10 % of the DNA fraction [4]. Despite the relatively small contribution to genome size, DNA elements are very important to genome evolution as well as gene annotation since many DNA elements are very abundant numerically and frequently insert in genic regions or directly reside within genes [2, 9]. Due to the unique composition of TEs in each genome, and the fact that transposon sequences evolve faster than normal host genes, the majority of TE sequences are species-specific at the nucleotide level. For example, a rice TE library only masks 25 % of the maize genome (Jiang, unpublished), suggesting that the majority of maize TEs would be missed if TE sequences from related species were used as a repeat library. As a result, a species-specific repeat library is required for accuracy in annotation of TEs and genes in distinct plant genomes.

2 Methods

2.1 Overview of Approaches in Repeat Annotation

All programs for TE identification fall into one of two categories: homology-based and structure-based. Homology-based programs can be further divided into two subgroups: those that rely on homology between a query sequence and known TEs in a database (e.g., RepeatMasker, http://www.repeatmasker.org/), and those that exploit the homology among the query sequences themselves. The latter are also termed de novo repeat identification programs, because they do not require any additional information about the query sequences. As a result, these tools may identify novel repeat sequences that have not been previously described.

If a TE library or repeat database is available, the identification of similar genomic sequences is relatively straightforward. This can be achieved using any sequence alignment tool, such as BLAST (basic local alignment search tool) [10], which will report all sequences with homology to the TEs in databases. Prior to gene annotation, genomic sequence is usually masked using a repeat library. The most popular program for this task is RepeatMasker, which uses BLAST or cross_match as a search engine to identify sequences similar to known repeats and convert them into "NNN…" In this way, the putative repeats will be ignored by gene prediction programs. In addition to masking repeat sequences, RepeatMasker also generates an output file (the .out file) containing additional information about repeats that are masked. This includes the position of the repeats in the genome sequences, the length of the masked sequence, the similarity between the masked sequences and known repeats, and the coordinates in known repeats matching the corresponding genomic sequences. As a result, the output of RepeatMasker is very helpful in deducing the relative relationship of local repeats (e.g., nested insertions) as well as estimation of copy numbers of repeats in a set of given sequences. In fact, some TE identification programs, such as LTR_MINER and REannotate [11, 12], use the output of RepeatMasker to further elaborate the identification and annotation of individual TEs.

In contrast to homology-based strategies, structure-based approaches search for the structural features of known TEs in query sequences. These include LTR for LTR elements, TIR for most DNA elements, unique TSDs for distinct transposon families, and the most terminal nucleotides of elements [3]. For example, the LTR_STRUC program identifies LTR-retrotransposons using the presence of LTR, the terminal sequence of LTR (5′ TG...CA 3′), and the 5-bp TSD [13]. Unlike homology-based search programs, structure-based searches do not require that a TE be repetitive in order to be discovered. As a result, it favors identification of TEs with low copy numbers. The drawback of structure-based approaches is that they do not recover TEs with novel structures and do not recognize ancient TEs that lack canonical structural features. A comprehensive list of various types of TE-related programs can be found in Lerat [14] and in a Web site constructed by Bergman's lab (http://bergmanlab.smith.man.ac.uk/?page_id=295).

2.2 De Novo Identification of Repeats

De novo identification approaches utilize the repetitive features of TEs and other repetitive sequences to identify novel repeats. Since most TEs are species-specific at the nucleotide level and are more or less repetitive, de novo identification of repeats is the most effective way to collect repeats by a single program. There are many de novo identification tools (*see* Table 1) using different methodologies with different forms of output. Some of them define repeat families while others generate consensus sequences for individual repeat families. Nevertheless, most of these programs employ two fundamental mechanisms for detecting repetitive sequences. One is based on the alignment among query sequences to identify homologous sequences (self-comparison), while the other involves searching for the repeated occurrence of short motifs (word counting) that can be extended to larger sequences (seed/string extension). The technical aspects of these programs have been evaluated previously [15], so this section will be devoted to the practical aspects, and will focus on the most popular programs.

2.3 Self-Comparison of Query Sequences

This approach defines repeat families if they share a certain level of sequence similarity over a certain length. This can be achieved, for example, by aligning the entire genomic sequence with itself (all vs. all searches) and in this case multiple matches will be found for repetitive sequences. Some programs use existing alignment tools, mostly BLAST or BLAST-like software, as the initial search engines. Other programs use custom tools for initial alignment. The results from initial alignment are further clustered to form repeat families.

RECON is one of the most widely used de novo search programs, and can be used for both raw reads and assembled sequences [16]. RECON uses WU-BLAST as the initial alignment tool. An important issue that was resolved by RECON is the

Table 1
De novo repeat identification programs

Detection mechanism	Program	URL and Reference	Features
Self-comparison	Repeat Pattern Toolkit	Agarwal and States [28]	Scoring system based on BLAST search. Families defined by graph representation
	RECON	http://selab.janelia.org/recon.html, http://www.repeatmasker.org/ Bao and Eddy [16]	Clustering sequences based on multiple alignment. Families defined by graphic representation
	PILER	http://www.drive5.com/piler/ Edgar and Myers [19]	Uses PALS for local alignment. Families defined by clustering that distinguishes different types of repeats
	Blaster suite	http://urgi.versailles.inra.fr/Tools/Blaster Quesneville (unpublished)	Composed of three programs (BLASTER, MATCHER and GROUPER). Families defined by clustering
K-mer	REPuter	http://bibiserv.techfak.uni-bielefeld.de/reputer/ Kurtz et al. [29]	Families defined by string extension. Interactive visualization of repeats provided
	FORRepeats	Lefebvre et al. [30]	Detection of exact repeats followed by approximate repeats
	RAP	Campagna et al. [31]	Identify repeats by word counting
	ReAS	ftp://ftp.genomics.org.cn/pub/ReAS/software/ Li et al. [17]	For unassembled sequences. Identifying repeat family by string extension. Generate a repeat library.
	RepeatScout	http://bix.ucsd.edu/repeatscout/ Price et al. [21]	For assembled sequences. Identifying repeat family by string extension. Generate a repeat library.
	P-Clouds	http://www.evolutionarygenomics.com/PClouds.html Gu et al. [32]	Identifying repetitive sequences through oligonucleotide excess probability. Computationally efficient
	Anonymous algorithm	Singh et al. [22]	Reconstruct relatively ancient repetitive sequences by progressively reducing the size of k-mer

differentiation between segmental duplication and typical interspersed repeat such as TEs. RECON resolves this issue through the different distribution graphics of the two types of repeats upon multiple sequence alignment. With its algorithms to deal with segmental duplications, RECON excels in reporting clusters that correspond well with what experts consider to be biologically meaningful families. RECON recovers much more novel repeat families than another de novo repeat identification program [15], ReAS [17]. The first active miniature inverted repeat transposable element (MITE), *mPing*, was identified from the rice genomic sequence through RECON [18]. Given that de novo search approaches are intended for discovering novel repeats, RECON fulfills its task very well. The drawback of RECON is its computational requirement; it requires 60 h to analyze 18 Mb rice genomic sequences in a single compilation [15]. Moreover, it cannot operate in parallel. The computational requirement can be addressed by an incremental approach. For example, one should use only a few Mb sequences in the initial compilation. For genomes that are highly repetitive (such as maize), 3–5 Mb sequences are suggested. For genomes with less repetitive sequences (such as *Arabidopsis*), 10–15 Mb sequences are appropriate. After the initial compilation, the repetitive sequences should be masked from the target genomic sequences for next round of repeat collection (e.g., 20–50 Mb), and the masked sequences are used for compilation. In this way, the time required for each iteration can be controlled within a few hours and the entire maize genome, which is over 2 GB in size, can be processed.

In contrast to RECON, the PILER package uses PALS (Pairwise Alignment of Long Sequences) as the initial search engine for self-comparison [19]. PILER identifies repeats using characteristic patterns of local alignments that are typical for certain classes of repeats. As a result, PILER is a tool with both homology-based and structure-based components. The PILER package retrieves different types of repeats using distinct methods. PILER-TA is for tandem repeats such as satellite DNAs; PILER-DF is for the dispersed repeats, which are probably TEs; PILER-PS is for pseudosatellites, which are repeats clustered in certain locations in the genome but not exactly in tandem arrays; PILER-TR searches for terminal repeats, including both direct (e.g., LTR elements) and inverted repeats (e.g., DNA elements). A consensus sequence is generated for each repeat family that can be used with RepeatMasker, and provides convenience for end users. Compared with other self-comparison search tools, PILER is very efficient computationally, for it can search 27 Mb rice sequences in less than 4 min [15]. However, its relatively low sensitivity with default parameters [15] implies that substantial amount of repeats will be missed by this program.

2.4 K-Mer and Spaced Seed Approaches

Since the self-comparison tools require other alignment programs for the initial comparison, they often have the same limitations as these alignment tools do. These include low sensitivity, lack of accuracy in the definition of boundaries, and intermediate documents with large sizes. An alternative and increasingly used approach is the k-mer method, which enumerates the repeated short motif or word (a substring w of length k, k-mer) in a certain sequence. The short motif is then used as a seed to extend to flanking sequences to form a larger sequence that represents a repeat family.

ReAS is a k-mer based program that was designed for unassembled sequence reads from whole genome shotgun sequencing so as to avoid errors introduced by incorrect sequence assembly [17]. The ReAS algorithm randomly selects a high frequency k-mer as a "seed" to retrieve sequence reads containing that k-mer. The resulting sequence reads for a given k-mer are aligned by ClustalW [20] to generate an initial consensus sequence with the k-mer in the center. If there is another high copy k-mer adjacent to the initial consensus sequence, it is used to retrieve additional sequences from the input dataset, and a new consensus sequence is built. The new consensus sequence and the initial consensus are then merged if they overlap with similarity of 95 % or higher. The extension process can be repeated up to five times so that consensus sequences of large transposable elements can be recovered. In a comparison with RECON, ReAS conferred faster search and higher sensitivity. Therefore, it was considered to be the best program for unassembled sequences [15]. Nevertheless, it recovers less novel repeat families than RECON, and difficulties in installation and utilization have been reported [14].

RepeatScout is similar to ReAS in terms of detection mechanism but it uses assembled sequences [21]. It first identifies a set of highly repetitive k-mers with fixed length, then extracts each k-mer and its surrounding regions to yield a consensus sequence for the repeat family representing the k-mer. RepeatScout is significantly faster than RECON and slightly more sensitive. It has been found to be the best de novo repeat identification program for assembled sequences.

Most of the de novo identification programs are associated with two major issues: the fragmentation of large TE sequences and the low sensitivity for highly divergent (or ancient) sequences. The fragmentation is mainly caused by nested insertions that interfere with the definition of boundaries. Alternatively, elements often have both highly conserved regions and less conserved regions. The presence of less conserved regions makes a large TE discontinuous to the program, so a large TE is often split into pieces. For this reason, many sequences retrieved by de novo identification programs are TE fragments, not entire TEs. Exceptions are the small TEs, such as MITEs, which are small and homogenous in sequence, so the entire element can be recovered as a single sequence. The low sensitivity for ancient TE families is either derived from the initial alignment tool (for the self-comparison approach) or due to the lack of identical k-mers with certain size as seeds.

To address these issues, a modified k-mer approach was developed [22], to identify repeats through a series of iterative runs. In the initial iteration, most conserved (or highly similar) repetitive sequences are identified from the query sequence using a relative long k-mer (e.g., 14 bp) and their flanking sequences are extended and retrieved. The resulting sequences are used to mask the query sequences to exclude them from further iteration. Since the most similar sequences usually represent the youngest TEs, this process also removes the top insertion in a nested structure so that the structure of the TE that was the target of other elements is recovered in this process. In the next round, the size of the k-mer will be reduced so that more degenerated elements will be recovered. The progressive iteration will be conducted until the k-mers become so short (e.g., 7–8 bp) that their representation is largely due to chance. In this way, the degraded, more ancient sequences will be recovered with better boundary definition. In a comparative study [22], it was shown that the relevant program is dramatically more efficient in recovering old and fragmented repetitive sequences than RepeatScout. The tradeoff is that the specificity of the program is low.

As new programs are published, more de novo search tools will be available. However, verifying whether a new program can confer the benefits that are intended by their authors is important. Based on previous experience, no single program is sufficient to detect all repeats; repeat libraries from different programs are always complementary, at least to some degree. A safe choice for biologists who intend to build a comprehensive repeat library is to use a combination of two (or more if possible) distinct programs that have been evaluated or used by others. Needless to say, these programs should be readily accessible.

3 Methods

3.1 The Classification and Processing of Repeat Sequences from De Novo Identification Programs

As mentioned above, the repeats collected by de novo repeat identification programs consist of all types of repeats, including gene families as well as segmental duplications. As a result, these sequences need to be further processed prior to being used as a repeat library for genome annotation or other purposes.

1. Gene sequences must be excluded from a repeat library so that the genes are not mistakenly masked during genome annotation. For TEs that carry genes, only the true element sequence,

such as the TIR of Pack-MULEs, should be included in the library, and the acquired gene-related sequences should be either eliminated or retained as an independent file. A simple way to exclude gene sequences is to compare the repeat collection with a protein database and exclude all sequences that match non-transposase proteins with a certain cutoff (e.g., $E = 10^{-5}$). However, complications are common. For instance, many transposase or transposon-related proteins are annotated as "unknown proteins" or "hypothetical proteins," which will cause false positives in the exclusion process. To address this issue, the repeat library should be first searched against a reliable transposase database; the matching sequences are likely true TEs and should be retained. A group of representative autonomous elements have been manually curated that could serve for this purpose [23]. Among the remainder of repeat collections, sequences that match known genes are probably true gene sequences and should be excluded. Caution should be taken with sequences that match "unknown proteins" or "hypothetical proteins" because some of them represent *bona fide* genes while others are mis-annotated TEs. This set of sequences should be classified based on additional evidence.

2. Under optimal conditions, a compiled set of manually curated sequences for TEs would be most reliable. However, manual curation is very time-consuming and such an investment is not feasible for every new genome. As a result, automated classification of TEs would minimize the requirement for manual curation. Unfortunately, development in this area has been slow and there are only a few programs (or pipelines) that are tailored to systematically classify various types of TEs and other repeats.
3. TEclass is a tool that was conceived based on the assumption that distinct classes of TEs are associated with distinguishable oligomer frequencies [24]. It employs machine learning support vector machine (SVM) for classification using oligomer (tetramer or pentamers) frequencies in known TEs for training. TEclass divides unknown TEs into four main taxonomic groups: DNA transposons, LTR elements, LINEs, and SINEs. In a test operation with TE sequences from RepBase [25], it achieves 90–97 % accuracy in the classification of novel DNA and LTR repeats, and 75 % for LINEs and SINEs. However, it remains unknown whether such accuracy will apply to the novel repeats from a new genome that lacks any entries in RepBase. Another issue with TEclass is that it does not distinguish TEs from non-TEs, and assumes that all input sequences are TE consensus sequences. As a result, gene-related sequences must be excluded prior to the application of this tool.

4. REPCLASS
   REPCLASS is another tool that uses three modules to classify TEs with both homology-based and structure-based components [26]. One module compares the query sequences against the known repeats database or the custom database. The second module searches for structural features of different types of TEs including LTR, TIR, and poly-A tail for non-LTR retrotransposons. The third module classifies elements based on the presence of distinct TSDs in each TE family. If a TSD is not detected, the sequence will be further examined for the presence of features characteristic of *Helitron* elements, which are not associated with a TSD. This program was used to classify the repetitive sequences in *Drosophila* that were collected using RepeatScout. A total of 57 % of the repeats were classified and 93 % of the classifications were correct. If this program can achieve similar sensitivity and specificity upon a new genome, it will greatly reduce the requirement for manual curation.
5. RepeatModeler is a package for de novo repeat family identification, modeling, and classification (http://www.repeatmasker.org/RepeatModeler.html). The package utilizes two existing de novo repeat finding programs (RECON and RepeatScout) [16, 21] that employ complementary detection methods for identifying repeat element boundaries and family relationships from sequence data. RepeatModeler facilitates automated operation of RECON and RepeatScout and uses the output to build, refine and classify consensus models of putative interspersed repeats. Specifically, RepeatModeler extends the edges and defines the boundaries of repeats to resolve the fragmentation issue associated with de novo repeat finding programs. RECON was improved in this package with an upgraded version. Furthermore, RepeatModeler classifies the repeats through their homology (nucleotide and protein level) with known repeats, as well as structural features such as terminal repeats and TSDs. RepeatModeler also excludes gene fragments from the repeat collection.

3.2 Library Construction and Availability of Repeat Libraries

A high quality repeat library is important for many downstream analyses, including the dissection of gene functions, because many TE insertions are close to or even inside genes. A good repeat library should be comprehensive, which means that it should include representative sequences for all or most types of repeats. Moreover, the library should have low redundancy to minimize the computational requirement so that it can be readily used by biologists. To reduce redundancy, one can build consensus sequences using a group of related sequences. For example, related sequences can be aligned using a multiple alignment tool such as ClustalW or DIALIGN-2 [20, 27], and consensus sequences can be built from multiple sequence alignment using consensus sequence server

Table 2
Plant repeat databases and resources

Name of database	Target genomes	Target repeats	URL
Repbase	Animal and Plant genomes	All TEs	http://www.girinst.org/repbase/
MSU plant repeat database	Plant genomes	All TEs and other repeats	http://plantrepeats.plantbiology.msu.edu/
Maize TE database	Maize	All TEs	http://maizetedb.org/~maize/
TREP—Triticeae Repeat Sequence Database	Triticeae	All TEs	http://wheat.pw.usda.gov/ITMI/Repeats/index.shtml
SoyTEdb	Soybean	All TEs	http://www.soybase.org/soytedb/
RetrOryza	Rice	LTR retro-transposons	http://www.retroryza.org/
MASiVEdb	Plant Genomes	The Sirevirus Retrotransposon	http://databases.bat.ina.certh.gr/masivedb/
Gypsy database	Animal, plants, and microbial	LTR retro-transposon and relatives	http://gydb.org/index.php/Main_Page

(http://coot.embl.de/Alignment//consensus.html). Alternatively, representative sequences (exemplars) can be selected among related sequences in a step-wise way to reduce redundancy [4]. With this method, all element sequences from the same family can be compared using BLASTN. The element with the most matches (e.g., cutoff at 90 % identity in 90 % of the element length) is considered to be the first exemplar. Thereafter, this element and its matches are excluded from the group and a second round BLASTN search is conducted with the remainder of the elements, leading to the generation of the second exemplar. This process is repeated until all elements are excluded.

For some plant genomes, repeat libraries are available albeit their quality (completeness, correctness of classification, definition of boundaries, and redundancy) varies (*see* Table 2). Caution is advised in using the libraries and interpreting the relevant results. On the one hand, mis-annotation always exists, especially for sequences without manual curation. On the other hand, not all sequences matching known TEs should be considered TEs. For example, some TEs contain certain micro-satellite sequences, so these sequences are considered as the relevant TEs by programs such as RepeatMasker. Furthermore, some TEs carry gene fragments. If these fragments are not eliminated from the relevant library, the parental genes would also be considered as TE or TE fragments.

As sequencing technology advances, new genomes are being sequenced at a far faster rate than they are being annotated. Given this trend, automated repeat collection and classification will be the future direction for robust analysis such as genome annotation. As a result, further improvements in speed, sensitivity, and specificity of repeat collection and classification programs will certainly benefit the utilization of genomic sequences. On the other hand, the diversity of TEs and the relatively small community dedicated to TE annotation make it very difficult, if not impossible, to generate an annotation engine that is as mature and robust as that for genes in the near future. In this case, manual curation is still essential for construction of a high quality repeat library if the foci of the study are TEs.

Acknowledgments

I thank Dr. Frank Dennis (Michigan State Univ.) for critical reading of the manuscript. This work was supported by NSF grant IOS-1126998.

References

1. Kumar A, Bennetzen JL (1999) Plant retrotransposons. Annu Rev Genet 33:479–532
2. Feschotte C, Jiang N, Wessler SR (2002) Plant transposable elements: where genetics meets genomics. Nat Rev Genet 3:329–341
3. Wicker T et al (2007) (2007), A unified classification system for eukaryotic transposable elements. Nat Rev Genet 8:973–982
4. Schnable PS et al (2009) The B73 maize genome: complexity, diversity, and dynamics. Science 326:1112–1115
5. Yang L, Bennetzen JL (2009) Distribution, diversity, evolution, and survival of Helitrons in the maize genome. Proc Natl Acad Sci U S A 106:19922–19927
6. Jiang N et al (2004) Pack-MULE transposable elements mediate gene evolution in plants. Nature 431:569–573
7. Holligan D et al (2006) The transposable element landscape of the model legume *Lotus japonicus*. Genetics 174:2215–2228
8. Hanada K et al (2009) The functional role of pack-MULEs in rice inferred from purifying selection and expression profile. Plant Cell 21:25–38
9. Jiang N et al (2009) Genome organization of the tomato sun locus and characterization of the unusual retrotransposon Rider. Plant J 60:181–193
10. Altschul SF et al (1990) Basic local alignment search tool. J Mol Biol 215:403410
11. Pereira V (2004) Insertion bias and purifying selection of retrotransposons in the *Arabidopsis thaliana* genome. Genome Biol 5:R79
12. Pereira V (2008) Automated paleontology of repetitive DNA with REANNOTATE. BMC Genomics 9:614
13. McCarthy EM, McDonald JF (2003) LTR_STRUC: a novel search and identification program for LTR retrotransposons. Bioinformatics 19:62–67
14. Lerat E (2010) Identifying repeats and transposable elements in sequenced genomes: how to find your way through the dense forest of programs. Heredity 104:520–533
15. Saha S et al (2008) Empirical comparison of ab initio repeat finding programs. Nucleic Acids Res 36:2284–2294
16. Bao Z, Eddy SR (2002) Automated de novo identification of repeat sequence families in sequenced genomes. Genome Res 12:1269–1276
17. Li R et al (2005) ReAS: recovery of ancestral sequences for transposable elements from the unassembled reads of a whole genome shotgun. PLoS Comput Biol 1:e43
18. Jiang N et al (2003) An active DNA transposon family in rice. Nature 421:163–167

19. Edgar RC, Myers EW (2005) PILER: identification and classification of genomic repeats. Bioinformatics 21:i152–i158
20. Thompson JD et al (1997) The CLUSTAL_X windows interface: flexible strategies for multiple sequence alignment aided by quality analysis tools. Nucleic Acids Res 25:4876–4882
21. Price AL, Jones NC, Pevzner PA (2005) De novo identification of repeat families in large genomes. Bioinformatics 21:i351–i358
22. Singh A et al (2010) An algorithm for the reconstruction of consensus sequences of ancient segmental duplications and transposon copies in eukaryotic genomes. Int J Bioinform Res Appl 6:147–162
23. Kennedy RC et al (2011) An automated homology-based approach for identifying transposable elements. BMC Bioinformatics 12:130
24. Abrusan G et al (2009) TEclass–a tool for automated classification of unknown eukaryotic transposable elements. Bioinformatics 25:1329–1330
25. Jurka J et al (2005) Repbase Update, a database of eukaryotic repetitive elements. Cytogenet Genome Res 110:462–467
26. Feschotte C et al (2009) Exploring repetitive DNA landscapes using REPCLASS, a tool that automates the classification of transposable elements in eukaryotic genomes. Genome Biol Evol 1:205–220
27. Morgenstern B (2004) DIALIGN: multiple DNA and protein sequence alignment at BiBiServ. Nucleic Acids Res 32:W33–W36
28. Agarwal P, States DJ (1994) The Repeat Pattern Toolkit (RPT): analyzing the structure and evolution of the C. elegans genome. Proc Int Conf Intell Syst Mol Biol 2:1–9
29. Kurtz S et al (2001) REPuter: the manifold applications of repeat analysis on a genomic scale. Nucleic Acids Res 29:4633–4642
30. Lefebvre A et al (2003) FORRepeats: detects repeats on entire chromosomes and between genomes. Bioinformatics 19:319–326
31. Campagna D et al (2005) RAP: a new computer program for de novo identification of repeated sequences in whole genomes. Bioinformatics 21:582–588
32. Gu W et al (2008) Identification of repeat structure in large genomes using repeat probability clouds. Anal Biochem 380:77–83

Chapter 21

Computational Methods for Identification of DNA Transposons

Ning Jiang

Abstract

The initial identification of transposable elements (TEs) was attributed to the activity of DNA transposable elements, which are prevalent in plants. Unlike RNA elements, which accumulate in the gene-poor heterochromatic regions, most DNA elements are located in the gene rich regions and many of them carry genes or gene fragments. As such, DNA elements have a more intimate relationship with genes and may have an immediate impact on gene expression and gene function. DNA elements are structurally distinct from RNA elements and most of them have terminal inverted repeats (TIRs). Such structural features have been used to identify the relevant elements from genomic sequences. Among the DNA elements in plants, the most abundant type is the miniature inverted repeat transposable elements (MITEs). This chapter discusses the methods to identify MITEs, *Helitrons*, and other DNA transposable elements.

Key words MITEs, *Helitron*, Gene-carrying, Transposase, Terminal inverted repeat

1 Introduction

Transposable elements are classified as either RNA retrotransposons or DNA transposons based on their transposition mechanism. LTR retrotransposons represent the largest component of plant genomes [1]. In all plant genomes characterized so far, retrotransposons account for more sequence mass than DNA TEs [2–15]. Therefore, DNA TEs contribute less to genome size expansion than retrotransposons. Nevertheless, this does not necessarily mean that DNA transposons are less successful or less important. Compared to the genome fraction, the copy numbers of elements are more accurate in reflecting the rate of transposition. In Arabidopsis (*Arabidopsis thaliana*) and soybean (*Glycine max*), the copy number of DNA TEs is comparable to that of retrotransposons [2, 11], suggesting that DNA TEs can amplify as rapidly as their RNA counterparts. In rice (*Oryza sativa*), the number of DNA TEs is more than twice that of retrotransposons, due to the exceptional abundance of

Thomas Peterson (ed.), *Plant Transposable Elements: Methods and Protocols*, Methods in Molecular Biology, vol. 1057, DOI 10.1007/978-1-62703-568-2_21, © Springer Science+Business Media New York 2013

the small nonautonomous DNA elements called MITE [3]. The high transposition activity of DNA elements is also reflected by the fact that TEs that were initially identified (such as *Ac/Ds* and *Spm/dSpm*) were all DNA elements. To date, multiple active DNA elements have been identified from maize (*Zea mays*), rice, snapdragon (*Antirrhinum majus*), Arabidopsis, morning glory (*Ipomoea purpurea*), soybean, petunia (*Petunia hybrida*), and tobacco (*Nicotiana tabacum*) [16–29]. In contrast, only a few active retrotransposons have been isolated from plants [30–33]. In addition, most DNA transposons are located in the gene rich regions, where they have immediate influence on their adjacent genes [34].

In addition to their high activity, plant DNA elements are very diverse in terms of structure, size, and sequences. These characteristics can be utilized to identify the relevant elements. Most DNA elements are associated with TIRs and duplicate a small piece of flanking sequence upon insertion. This duplication is called Target Site Duplication (TSD), which varies among distinct element families. Autonomous elements encode transposase (Tpase) which is required for transposition, whereas nonautonomous elements do not encode Tpase and rely on the cognate autonomous elements for transposition. The structural features of major plant DNA transposons will be discussed briefly (*see* Table 1).

2 Materials

2.1 Ac/Ds/hAT or DTA Elements

The founder elements of this family are *Ac/Ds* from maize, *hobo* from fruit fly (*Drosophila melanogaster*), and *Tam3* from snapdragon [16, 35, 36]. The first letter of the three elements contributes to the name "*hAT*". "DTA" is the systematic nomenclature [37]. Autonomous *hAT* elements are a few kb in length and nonautonomous elements can be as short as less than 200 bp. Most *hAT* elements have short TIRs that are less than 30 bp, starting with "CA/TA" and ending with "TG/TA". The TSD of *hAT* elements is usually 8 bp in length.

2.2 CACTA or DTC Elements

The founder elements of this family are *Spm/dspm* in maize [38]. They are named CACTA elements because their terminal sequences are "CACTA/G…C/TAGTG". In general, autonomous CACTA elements are large in size. The *Tam1* element in snapdragon is 17 kb in length [22]. Like *hAT* elements, CACTA elements are associated with short TIRs, and frequently contain many subterminal repeats. CACTA elements generate 3 bp TSD upon insertion.

2.3 MULE or DTM Elements

Mutator/Mu/MuDR elements in maize are the founder elements for this family [18]. Similar elements in other organisms are called *Mutator*-like elements (MULEs). Most MULEs have extended long TIRs (100–800 bp). Yet there are a group of MULEs called

Table 1
Structural features and classification of DNA transposons

Family			Element size			
Traditional name	Systematic classification	TSD	Autonomous	Nonautonomous	TIR size	Terminal sequence (5′…3′)
Ac/Ds/hAT	DTA	8 bp	3–6 kb	110 bp to 3 kb	5–22 bp	C/TA…TA/G
En/Spm/dSpm/CACTA	DTC	3 bp	6–17 kb	200 bp to 6 kb	12–28 bp	CACTA/G…C/TAGTG
MuDR/Mutator/Mu/MULE	DTM	7–11 bp, mostly 9	3.9–16 kb	120 bp to 3 kb	0–800 bp	G/C…G/C
PIF/Pong/Tourist	DTH	TNA	3–7 kb	80 bp to 3 kb	14–60 bp	GGG/CC…GG/CCC (PIF) GG/AGCA…TGCT/CC (Pong)
Tc1/Mariner/Stowaway	DTT	TA	3–7 kb	80 bp–3 kb	11–120 bp	CTCCCTCC…GGAGGGAG
Helitron	DHH	None	5–17 kb	150 bp–20 kb	None	TC…CTRR

"non-TIR MULEs," which have short (<50 bp) TIR that are not highly similar [39]. Some non-TIR MULEs, such as the TAFT element in maize, specifically target "TATATA..." satellite sequences [40]. Autonomous MULEs can be rather large in size. For example, the CUMULE element is 12 kb in length [41]. The terminal sequences of MULEs vary among elements but most of them start or end with G or C. Most MULEs create a 9 bp TSD upon insertion.

2.4 PIF/Pong/Tourist or DTH Elements

For this family, the nonautonomous elements (*Tourist*) were found prior to the discovery of their autonomous elements (*PIF/Pong*) [19, 42, 43]. Like CACTA elements, *PIF/Pong/Tourist* elements are associated with a 3 bp TSD. Nevertheless, the TSD of CACTA elements can be any combination of the four nucleotides, whereas that for *PIF/Pong/Tourist* elements is mainly TAA or TTA. Compared with other autonomous DNA elements, *PIF/Pong* elements are relatively small in size, usually a few kb in length. *Tourist* elements are usually less than 500 bp. The terminal sequences for *PIF* like elements are "GGG/CCC...GGC/GCC" and that for *Pong* elements are "GA/GGCA...TGCC/TC" [44].

2.5 Tc1/Mariner/Stowaway or DTT Elements

Like *PIF/Pong/Tourist* elements, the nonautonomous *Stowaway* elements were reported prior to their autonomous elements (*Tc1/Mariner*) in plants [45, 46]. *Tc1/Mariner* elements specifically insert into "TA" sequence and the "TA" sequence is duplicated as a TSD. *Tc1/Mariner* elements in animals, such as those in *C. elegans*, are very compact, usually less than 2.5 kb in size [47]. In contrast, it is not unusual to find *Tc1/Mariner* elements in plants that are over 5 kb. The terminal sequences for *Tc1/Mariner/Stowaway* elements are "CTCCCTCC... GGAGGGAG". Like *Tourist* elements, *Stowaway* elements are small in size (<500 bp).

When *Tourist* and *Stowaway* elements were initially identified from plants [42, 45], it was not clear what type of autonomous elements were associated with these elements. Hence it was impossible to assign them to a specific family. As a result, they were classified as "miniature inverted repeat transposable elements" (MITEs) based on their structural features. Technically MITEs can refer to any small TEs (<500 bp) with TIRs, but for historical reasons the term is most frequently used for *Tourist* and *Stowaway* elements. Most *Tourist* and *Stowaway* elements have short TIRs which are less than 30 bps. However, some MITEs do have extended stem-loop structures longer than 50 bp [42, 45]. Due to the fact that MITEs represent a group of TEs with divergent family origins, the term "MITE" lacks specificity from a scientific and taxonomic point of view. Nevertheless, the structural implications of this term are useful for element identification. For this reason, "MITEs" will be still used in this chapter.

2.6 *Helitron* or DTT Elements

All of the above autonomous DNA transposons encode transposases with a DDE motif for "cut and paste" transposition [48]. In contrast, *Helitron* replicates through a rolling-circle mechanism and does not generate a TSD upon insertion [49]. Autonomous *Helitrons* encode a Y2-type tyrosine recombinase resembling that in bacterial rolling-circle transposons, with a helicase domain and replication initiator activity. *Helitron* specifically inserts into "AT" sequence, i.e., flanked by "A" on the 5′ end and "T" on the 3′ end. The terminal sequence of *Helitrons* are "TC…CTRR", where R stands for G or A. In addition, a short GC-rich stem-loop (a 16–20 bp stem separated by 10–12 bp loop) is often found near the 3′ end of the element.

A recent finding in plant genomes is that TEs can duplicate genomic sequences including gene fragments in large scale. To date, it is clear that all major TE families are capable of capturing and amplifying gene fragments, albeit the frequency of gene acquisition varies from family to family. In maize, there are over one million retrotransposons and only 425 (0.04 %) of them are verified to contain gene fragments. In contrast, over 1 % (1,656 out of 143,000) of the described DNA transposons contains host genes or gene fragments [9]. Among DNA TEs, *Helitrons* and MULEs (Pack-MULEs) seem to be the most frequent type that carry genes [50–53]. In most cases, gene capture is associated with nonautonomous elements. Nevertheless, some autonomous elements have also acquired host genes. For example, a peptidase C48 domain derived from a cellular small ubiquitin-like protease gene was detected in some autonomous MULEs or CACTA elements [41, 54], suggesting the presence of the captured domain may benefit the transposition process.

Another phenomenon that blurs the line between genes and TEs is the domestication of Tpase genes to serve a host function. The best-characterized example is the *FAR1* and FHY3 genes in *Arabidopsis*, where mutations in these two genes result in defects in the response to far-red light. *FAR1* and FHY3 genes exhibit a high degree of similarity to the Tpase encoded by MULEs such as *MuDR* and *Jittery* in maize [55], and it has been demonstrated that each of these genes acts as a *bona fide* transcription factor [56]. Since the mutations in *FAR1* and FHY3 genes lead to a visible phenotype, and no other TE features (such TIRs and TSDs) are found with the two genes, it is unambiguous that these genes have well-defined cellular functions and no longer represent TEs. Accordingly, if a Tpase domain is expressed yet without other structural features of TEs, it is likely a cellular gene arisen from TE domestication.

3 Methods

3.1 Detection of Autonomous DNA Elements

Autonomous elements encode Tpase and are usually large in size. Moreover, autonomous elements are often less abundant than their nonautonomous counterparts, so they are not always repetitive. For example, the *Ping* element, the autonomous element for the *mPing* element in rice, has only one copy in Nipponbare, the *japonica* cultivar which was used for sequencing [3, 19]. As a result, it is not always possible to search for autonomous elements based on the feature of being repetitive. For this reason, the most powerful approach to detect the presence of autonomous elements is through the presence of Tpase domains.

1. MAK is a tool kit for MITE analysis [57]. There are several functions associated with the tool kit, including retrieving family members with known MITE sequences. However, the most useful and distinctive function of this kit is to search for larger or autonomous elements associated with a given MITE. The tool kit can be operated in UNIX, LINUX and Window systems. The input for the kit is a known MITE sequence, which is used for a BLAST search against a database or a certain genome. If two hits matching the TIR of the MITE are located in linked sequence with the same orientation as that in the MITE, the "Long" function retrieves the large element. If the "Anchor" function is selected, the large element sequence is used to search for presence of Tpase or other transposon related proteins. In this way, it is possible to find the autonomous element starting with the sequence of a nonautonomous element.
2. TARGeT is a pipeline for identifying and characterizing gene families including autonomous TE families [58]. The input for TARGeT is the protein sequences of the gene family or Tpase for the TE in question. Theoretically, given known Tpase domains, their homologs can be retrieved based on the results from existing alignment tools such as TBLASTN. Unfortunately, the results of TBLASTN are not always reliable for recovering homologs due to the lack of explicit treatment of frame-shifts and introns. To solve this problem, the authors of TARGeT developed a program called putative homolog identifier (PHI), which finds reliable homologs based on both the *e*-value (default 0.01) and the minimal match percentage (MMP, defaults to 70 %). After grouping and refinement, PHI identifies homologs in a certain genome and provides the sequences (in fasta format) of those homologs. Since the sequences retrieved by TARGeT only represent the coding regions, further extension and definition of boundaries are required to obtain the complete element sequences. After retrieving the

sequences, TARGeT conducts a further analysis to generate a phylogenetic tree for the putative homologs. TARGeT is a Web-based pipeline, which is convenient for nonspecialists.

3. TEseeker is a recently developed program that builds consensus TE sequences present in a new genome [59], with a curated library containing known representative TEs. A known TE family is first used in BLAST search (mostly at the protein level) against the genome in question. Matches are then combined, extracted from the genome and assembled with CAP3 [60], a DNA sequence assembly tool. Since the initial search retrieves only the most conserved portion of the elements, the sequences are then trimmed and again used in a BLAST search against the genome to retrieve larger coding regions. The results are then used to produce a multiple sequence alignment with ClustalW2 [61], where an initial consensus is generated. This consensus is used to perform a final BLAST search against the genome, with flanking sequences extracted to capture the putative intact elements. The results are processed by CAP3 again to generate the final consensus sequences for intact autonomous DNA elements.

3.2 Detection of Nonautonomous DNA Elements

Nonautonomous DNA elements do not encode Tpase, but they often share sequence similarity with their autonomous elements at the termini of the element. As a result, one strategy is to identify nonautonomous DNA elements based on their sequence similarity with the putative autonomous elements. However, nonautonomous elements usually out-number their autonomous partners, and it is not unusual that the autonomous elements are lost from the genome while the nonautonomous elements still remain. Furthermore, nonautonomous elements may not arise as simple deletion derivatives of the autonomous elements. For example, a clear-cut sequence-based relationship between autonomous and nonautonomous elements cannot be established for most of the *Stowaway* elements in the rice genome [62]. Therefore, a substantial number of nonautonomous elements would be missed by this strategy. An alternative approach is to use structural features to identify nonautonomous elements, which often leads to a more comprehensive collection.

As was previously mentioned, MITEs are numerically the most abundant elements in many plant genomes and several programs have been developed for their detection. The MAK tool kit (*see* Subheading 3.1) is one of these, yet it requires a known MITE sequence to retrieve its family members. Other programs do not require input MITE sequences. With the exception of TRANSPO, all the programs discussed below are structure-based tools that are capable of retrieving novel elements based on known structural features.

1. FINDMITE was the first program specifically designed for mining MITEs and was used to find novel MITEs in mosquito [63]. It is implemented in UNIX or LINUX system. This program identifies MITEs by locating the short inverted repeats present in adjacent regions as well as the accompanying TSDs. The user is prompted to give the length of TIR and the exact TSD (e.g., TA for *Stowaway*). It is computationally very efficient, exemplified by completion of a search of a 28 Mb rice sequence in less than 1 min [64]. However, due to the fact that there are many short TIRs and fortuitous short TSDs present in the genomic sequences, the false positive rate of FINDMITE is rather high and manual curation is required for further analysis.
2. TRANSPO searches for inverted repeat sequences with similarity to the TIR of known TEs [65]. As a result, it is unable to detect novel elements that lack sequence similarity to known TEs. The cutoff for the sequence similarity of inverted sequences is rather low (75 %) so it is possible to identify relatively ancient copies.
3. MUST (MITE Uncovering SysTem) has an underlying search mechanism that is very similar to that of FINDMITE [66]. Both programs search for the presence of TIRs and TSDs in genomic sequences. The authors of MUST developed a very elegant Web-based interface that can be conveniently used by nonspecialists. This program was used to search for candidate MITEs in two bacterial genomes [66]. Like FINDMITEs, the specificity of MUST is an issue and therefore the results require manual curation.
4. MITE-Hunter is a UNIX pipeline composed of searching, filtering and classification of nonautonomous DNA elements [64]. The initial search step of MITE-Hunter is similar to that of FINDMITEs and MUST (i.e., it searches for TIRs and accompanied TSDs). The unique contribution of MITE-Hunter is its subsequent filtering method, which significantly improves the specificity. After the initial search, all MITE candidates are retrieved along with their putative flanking sequences. For a certain family, candidates are only considered if the homology of the recovered sequences is limited to that of the candidate MITE, and not the flanking sequences. This filtering procedure is very effective in excluding MITE-like structure buried in a larger repeat or located in segmental duplication. In a comparative study, the false positive rate of MITE-Hunter is only 4.4 %, while that for FINDMITE and MUST is 85 % or higher. The function of MITE-Hunter goes beyond its name. It not only identifies *Tourist* and *Stowaway* elements, but also other nonautonomous DNA elements with TIRs (CACTA, MULEs, *hAT*), with a size up to 1.7 kb [64]. As a result, MITE-Hunter is the most robust tool for identification of nonautonomous DNA TEs (Table 2).

Table 2
Programs for identification of DNA TEs

Program	Target element	URL and reference	Features
MAK	MITE members and their related autonomous elements	http://labs.csb.utoronto.ca/yang/MAK/ Yang and Hall [57] Janicki et al. [77]	Requires known MITE sequences, works with all operation systems
TARGeT	Autonomous elements or gene families	http://target.iplantcollaborative.org/ Han et al. [58]	Web-based program to collect elements and perform phylogenetic analysis
FINDMITE	MITEs	http://tu08.fralin.vt.edu/software.html Tu [63]	UNIX or LINUX based, fast, with high false positive rate
TRANSPO	Autonomous or nonautonomous elements	http://alggen.lsi.upc.es/recerca/search/transpo/transpo.html Santiago et al. [65]	Search for related elements with a given TIR sequence, capable of retrieving relatively ancient elements
MUST	MITEs	http://csbl1.bmb.uga.edu/ffzhou/MUST/ Chen et al. [66]	Web-based, limited to 20 kb in one search, high false positive rate
MITE-Hunter	MITEs and other nonautonomous DNA elements	http://target.iplantcollaborative.org/mite_hunter.html Han and Wessler [64]	UNIX or LINUX based, low false positive rate
HelitronFinder	Helitron	http://limei.montclair.edu/HF.html Du et al. [67]	Web-based, high specificity, only for maize
HelSearch	Helitron	http://sourceforge.net/projects/helsearch/files/ Yang and Bennetzen [68]	UNIX/LINUX based, for small genomes

3.3 Detection of Helitrons

The programs described above to detect nonautonomous DNA elements search for TSDs and TIRs. Since *Helitrons* do not bear TIR or TSD, none of the above tools is useful in searches for nonautonomous *Helitrons*. As a result, distinct programs were developed for searching *Helitrons*.

1. HelitronFinder was a Web-based program designed specifically for finding the most abundant type of *Helitrons* in maize, the *HelA* type of *Helitron* [67]. This program searches for the presence of GC-rich stem-loop and CTRR at the 3′ end of the element. For 5′ end, the only known feature is that the *Helitron* elements start with "TC", which is not sufficient to find element ends with acceptable false negative rate. As a result, the author of HelitronFinder developed an empirical method. All the known maize *Helitron* elements were aligned and the first 18 base pairs of the elements were found to have reasonable conservation level. As a result, the consensus of the 18 bp sequence was used to search for the 5′ end of the element. For this reason, the program is only useful for search sequences from maize or its relatives. If a piece of sequence bears both 5′ and 3′ end in the right orientation, inserted into "A...T", and it is less than 25 kb, it is considered a *Helitron* element. The use of HelitronFinder program is very straightforward and convenient. When the fasta format genomic sequence is pasted or loaded as a file, it will report the putative 5′ end and 3′ end as well as the position and sequence of the entire element.
2. Helsearch is a UNIX-based program that was developed in Perl [68]. Like HelitronFinder, HelSearch also defines the 3′ end of the element by the presence of stem-loop and "CTRR". Nevertheless, HelSearch does not require a designated 5′ end, and it can be used for any genomic sequence. After the candidate 3′ ends are determined, they are grouped based on their hairpin structure. Subsequently, the flanking sequences of the three ends are retrieved and multiple alignments conducted by CLUSTALW [69]. The results of multiple alignments need to be manually inspected for boundaries of *Helitron* elements.

3.4 De Novo Search Method

All of the de novo search methods discussed can be used to search for TEs including both autonomous and nonautonomous DNA elements. Since these search methods do not include a classification function, it is obvious further identification is required. This can be achieved through the classification pipeline, or through manual curation. Apparently, the identification of autonomous elements from de novo search results can be assisted by search against Tpase database, which is very helpful for determining the identity of the element. If there are no hits to Tpases, manual curation is often conducted by retrieving the relevant repeats and the flanking sequences from the genomic sequences. Thereafter, multiple alignment or pair-wise alignment is required. A boundary can be defined if homology for most repeats stops at a certain position, where the flanking sequences are divergent. In these cases, one can further look for the presence of the typical terminal sequence and TSD to identify the element family. If there is no obvious boundary, this implies the retrieved repeats represent part of a large element and longer flanking sequences should be retrieved and aligned, and the manual procedure reiterated. Note that TEs of relatively recent origin usually have a "sharp" boundary, where homology stops at a designated point. If the homology declines gradually from the center of the alignment to the flanking sequences, it implies either it is an ancient TE or it is a gene family. In this case, no conclusion can be made. Despite the fact that manual curation is a very time consuming process, it is still considered to be the most sensitive and reliable method used to identify nonautonomous DNA elements.

3.5 DNA Elements Containing Gene Fragments

As mentioned above, DNA elements exhibit a much higher frequency of host gene transduction than do RNA elements. Elements with transduced host genes are recognized by have terminal sequences typical of regular DNA elements, but their internal regions are usually homologous to genes or gene fragments. Moreover, members of a single element family can carry many different types of gene sequences; one example is the nonautonomous *Mutator* element family in maize [70]. In this case, the terminal sequences (representing the *bona fide* TE sequences) are much more abundant than their internal regions. The verification or collection of these elements is important for studying transposon biology as well as gene annotation. This procedure usually includes the following steps (*see* Fig. 1).

1. Collection of sequences of DNA element ends. Since these elements carry typical DNA ends, the first step is to determine the element ends that could harbor gene fragments. These could be the TIR of MULEs or the 5′ end and 3′ end of *Helitron* sequences determined by the methods mentioned above. For reliability, DNA elements identified or classified by automatic tools should be further manually verified.
2. Localization of candidate elements. The element ends collected in **step 1** are further used to mask the genomic sequences or to search genomic sequences with alignment tools such as BLAST. For simplicity, element ends from each superfamily should be tested separately. Subsequently, a pair of element ends (with perfect or almost perfect ends) present in the correct orientation and within a certain distance of each other will be considered to delineate a candidate element. "In the correct orientation" means that the ends are oriented in the same relative orientation as that of a normal element; for example the

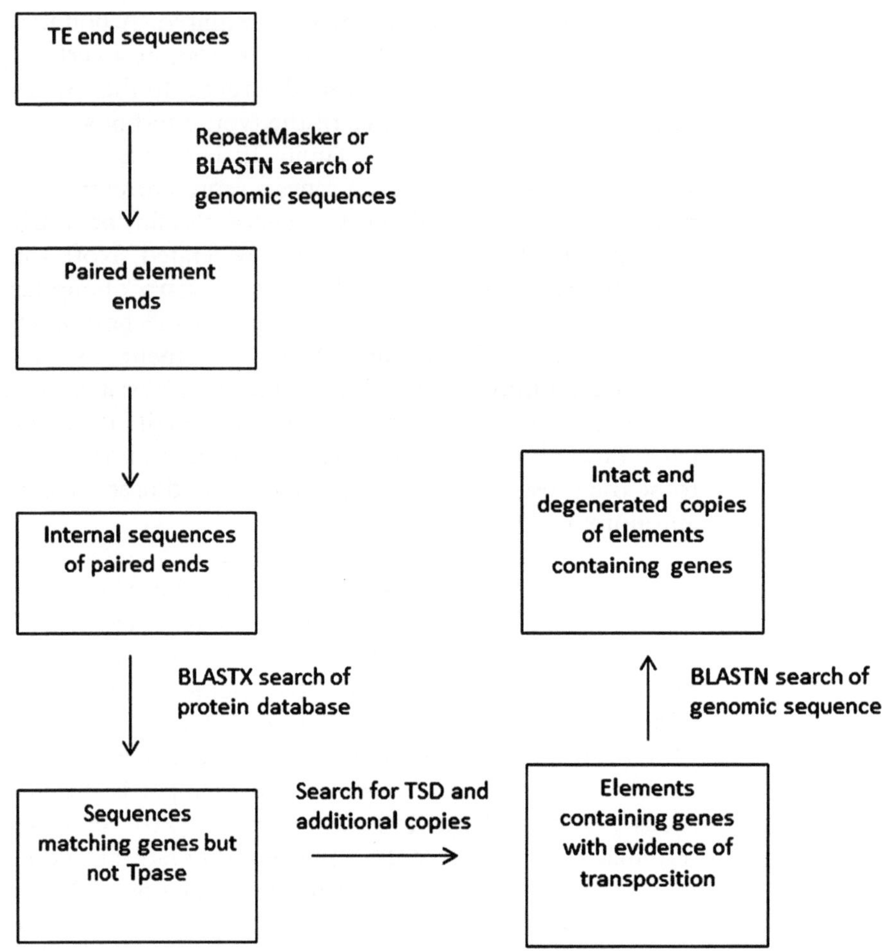

Fig. 1 A generic pipeline for identification of DNA TEs containing gene fragments

very terminal sequences of TIR are outward. For *Helitrons*, the 5′ end is upstream and the 3′ end is downstream in the plus DNA strand, and vice versa for the minus DNA strand. The distance between the ends is critical for specificity and sensitivity, and is usually determined empirically. The distance should be sufficiently large to include most, if not all, possible elements, but not so large as to introduce too many false positives. For most elements, 10–20 kb is the distance limit. Because *Helitrons* are often larger than other DNA elements in size, 20–25 kb is their upper limit.

3. Detection of gene fragments in internal regions. The candidate elements collected in the previous step will include autonomous elements, nonautonomous elements that may or may not contain gene fragments, and false positives. To determine whether the candidate elements contain gene fragments, their internal regions are used to conduct a BLASTX search to test whether they have any homology to known genes. One potential

problem at this step is the presence of transposon-related proteins mistakenly annotated as host genes. To minimize such an artifact, the internal regions should be first used to search against a reliable Tpase database, such as the one associated with RepeatMasker (http://www.repeatmasker.org/), or TEseeker [59]. After any sequences homologous to Tpase are excluded or masked, the remaining internal sequences are used to search against a comprehensive protein library (e.g., the NCBI nonredundant protein database http://www.ncbi.nlm.nih.gov/RefSeq/). Candidate elements should be retained if their internal regions exhibit significant homology with known genes (e.g., $E < 10^{-5}$).

4. Exclusion of false positives. Elements retained in **step 3** will include *bona fide* elements containing genes, as well as false positives. For example, two truncated elements flanking a gene may appear as one element containing the gene. Therefore to confirm the identification of *bona fide* elements, one needs to search for hallmarks of transposition. For elements with long TSDs, such as *hAT* and MULEs, the presence of a TSD flanking the paired ends is reliable indicator of transposition, because the chance for two independent elements to have the same TSD and be located in adjacent regions is very low. For elements with short TSD or without TSD, proof of transposition requires the identification of two or more independent copies which are well aligned (usually >90 % of the element length) and inserted into different flanking sequences. If genomic sequences for multiple cultivars/lines are available, then insertion polymorphism (i.e., empty sites in other lines) can be used as evidence for transposition. In this case, a single copy can be considered as a verified element.
5. Identification of additional gene fragment-containing elements. *Bona fide* gene-containing elements identified in the previous steps can be used as "standard elements" or "seed elements" to search against the genomic sequence to retrieve elements with significant alignments but with slightly truncated ends or lacking TSDs. Such elements likely include degenerate copies of the corresponding gene-carrying elements and should be included in the final collection.

Emerging evidence indicates that some of the gene-containing elements (including both DNA and RNA TEs) may evolve into functional host genes or portions of normal genes [71–74]. TE sequences will degrade over time and thus are only recognizable for a few million years [75, 76]. It follows that the TE ends of many gene-containing elements are partially or completely degenerate and are no longer recognizable. In other words, the total number of gene-containing elements is always underestimated. This is beyond the capacity of computational tools, yet it is one of the driving forces for genome evolution.

Acknowledgments

I thank Dr. Wayne Loescher (Michigan State Univ.) for critical reading of the manuscript. This work was supported by NSF grant MCB-1121650.

References

1. Kumar A, Bennetzen JL (1999) Plant retrotransposons. Annu Rev Genet 33:479–532
2. Arabidopsis-Sequencing-Consortum (2000) Analysis of the genome sequence of the flowering plant *Arabidopsis thaliana*. Nature 408:796–815
3. International-Rice-Sequencing-Project (2005) The map-based sequence of the rice genome. Nature 436:793–800
4. Tuskan GA et al (2006) The genome of black cottonwood, *Populus trichocarpa* (Torr. & Gray). Science 313:1596–1604
5. Jaillon O et al (2007) The grapevine genome sequence suggests ancestral hexaploidization in major angiosperm phyla. Nature 449: 463–467
6. Ming R et al (2008) The draft genome of the transgenic tropical fruit tree papaya (*Carica papaya* Linnaeus). Nature 452:991–996
7. Huang S et al (2009) The genome of the cucumber, *Cucumis sativus* L. Nat Genet 41:1275–1281
8. Paterson AH et al (2009) The Sorghum bicolor genome and the diversification of grasses. Nature 457:551–556
9. Schnable PS et al (2009) The B73 maize genome: complexity, diversity, and dynamics. Science 326:1112–1115
10. The-International-Brachypodium-Initiative (2010) Genome sequencing and analysis of the model grass *Brachypodium distachyon*. Nature 463:763–768
11. Schmutz J et al (2010) Genome sequence of the palaeopolyploid soybean. Nature 463: 178–183
12. Xu X et al (2011) Genome sequence and analysis of the tuber crop potato. Nature 475(7355):189–195
13. Al-Dous EK et al (2011) De novo genome sequencing and comparative genomics of date palm (*Phoenix dactylifera*). Nat Biotechnol 29:521–527
14. Banks JA et al (2011) The *Selaginella* genome identifies genetic changes associated with the evolution of vascular plants. Science 332: 960–963
15. Young ND et al (2011) The *Medicago* genome provides insight into the evolution of rhizobial symbioses. Nature 480:520–524
16. McClintock B (1948) Mutable loci in maize. Carnegie Inst Wash Yr Bk 47:155–169
17. Peterson PA (1960) The pale green mutable system in maize. Genetics 45:115–133
18. Robertson DS (1978) Charaterization of a Mutator system in maize. Mutat Res 51: 21–28
19. Jiang N et al (2003) An active DNA transposon family in rice. Nature 421:163–167
20. Kikuchi K et al (2003) The plant MITE mPing is mobilized in anther culture. Nature 421: 167–170
21. Nakazaki T et al (2003) Mobilization of a transposon in the rice genome. Nature 421: 170–172
22. Bonas U, Sommer H, Saedler H (1984) The 17-kb Tam1 element of *Antirrhinum majus* induces a 3-bp duplication upon integration into the chalcone synthase gene. EMBO J 3: 1015–1019
23. Singer T, Yordan C, Martienssen RA (2001) Robertson's Mutator transposons in *A. thaliana* are regulated by the chromatin-remodeling gene decrease in DNA methylation (DDM1). Genes Dev 15:591–602
24. Hoshino A, Inagaki Y, Iida S (1995) Structural analysis of Tpn1, a transposable element isolated from Japanese morning glory bearing variegated flowers. Mol Gen Genet 247:114–117
25. Zabala G, Vodkin LO (2005) The wp mutation of glycine max carries a gene-fragment-rich transposon of the CACTA superfamily. Plant Cell 17:2619–2632
26. Snowden KC, Napoli CA (1998) Psl: a novel Spm-like transposable element from *Petunia hybrida*. Plant J 14:43–54
27. Grappin P et al (1996) Molecular and functional characterization of slide, an Ac-like autonomous transposable element from tobacco. Mol Gen Genet 252:386–397
28. Tsugane K et al (2006) An active DNA transposon nDart causing leaf variegation and mutable dwarfism and its related elements in rice. Plant J 45:46–57
29. Moon S et al (2006) Identification of active transposon dTok, a member of the hAT family, in rice. Plant Cell Physiol 47:1473–1483
30. Grandbastien MA, Spielmann A, Caboche M (1989) Tnt1, a mobile retroviral-like transposable

element of tobacco isolated by plant cell genetics. Nature 337:376–380

31. Hirochika H (1993) Activation of tobacco retrotransposons during tissue culture. EMBO J 12:2521–2528
32. Hirochika H et al (1996) Retrotransposons of rice involved in mutations induced by tissue culture. Proc Natl Acad Sci U S A 93: 7783–7788
33. Komatsu M, Shimamoto K, Kyozuka J (2003) Two-step regulation and continuous retrotransposition of the rice LINE-type retrotransposon Karma. Plant Cell 15:1934–1944
34. Feschotte C, Jiang N, Wessler SR (2002) Plant transposable elements: where genetics meets genomics. Nat Rev Genet 3:329–341
35. Coen ES, Carpenter R, Martin C (1986) Transposable elements generate novel spatial patterns of gene expression in *Antirrhinum majus*. Cell 47:285–296
36. McGinnis W, Shermoen AW, Beckendorf SK (1983) A transposable element inserted just 5′ to a Drosophila glue protein gene alters gene expression and chromatin structure. Cell 34: 75–84
37. Wicker T et al (2007) A unified classification system for eukaryotic transposable elements. Nat Rev Genet 8:973–982
38. Pereira A et al (1986) Molecular analysis of the En/Spm transposable element system of *Zea mays*. EMBO J 5:835–841
39. Yu Z, Wright SI, Bureau TE (2000) Mutator-like elements in *Arabidopsis thaliana*. Structure, diversity and evolution. Genetics 156: 2019–2031
40. Wang Q, Dooner HK (2006) Remarkable variation in maize genome structure inferred from haplotype diversity at the bz locus. Proc Natl Acad Sci U S A 103:17644–17649
41. van Leeuwen H, Monfort A, Puigdomenech P (2007) Mutator-like elements identified in melon, Arabidopsis and rice contain ULP1 protease domains. Mol Genet Genomics 277: 357–364
42. Bureau TE, Wessler SR (1992) Tourist: a large family of small inverted repeat elements frequently associated with maize genes. Plant Cell 4:1283–1294
43. Zhang X et al (2001) P instability factor: an active maize transposon system associated with the amplification of Tourist-like MITEs and a new superfamily of transposases. Proc Natl Acad Sci U S A 98:12572–12577
44. Zhang X et al (2004) PIF- and Pong-like transposable elements: distribution, evolution and relationship with Tourist-like miniature inverted-repeat transposable elements. Genetics 166:971–986
45. Bureau TE, Wessler SR (1994) Stowaway: a new family of inverted repeat elements associated with the genes of both monocotyledonous and dicotyledonous plants. Plant Cell 6:907–916
46. Feschotte C, Wessler SR (2002) Mariner-like transposases are widespread and diverse in flowering plants. Proc Natl Acad Sci U S A 99:280–285
47. Plasterk RH, Izsvak Z, Ivics Z (1999) Resident aliens: the Tc1/mariner superfamily of transposable elements. Trends Genet 15:326–332
48. Yuan YW, Wessler SR (2011) The catalytic domain of all eukaryotic cut-and-paste transposase superfamilies. Proc Natl Acad Sci U S A 108:7884–7889
49. Kapitonov VV, Jurka J (2001) Rolling-circle transposons in eukaryotes. Proc Natl Acad Sci U S A 98:8714–8719
50. Jiang N et al (2004) Pack-MULE transposable elements mediate gene evolution in plants. Nature 431:569–573
51. Morgante M et al (2005) Gene duplication and exon shuffling by helitron-like transposons generate intraspecies diversity in maize. Nat Genet 37:997–1002
52. Du C et al (2009) The polychromatic Helitron landscape of the maize genome. Proc Natl Acad Sci U S A 106:19916–19921
53. Yang L, Bennetzen JL (2009) Distribution, diversity, evolution, and survival of Helitrons in the maize genome. Proc Natl Acad Sci U S A 106:19922–19927
54. Hoen DR et al (2006) Transposon-mediated expansion and diversification of a family of ULP-like genes. Mol Biol Evol 23:1254–1268
55. Hudson ME, Lisch DR, Quail PH (2003) The FHY3 and FAR1 genes encode transposase-related proteins involved in regulation of gene expression by the phytochrome A-signaling pathway. Plant J 34:453–471
56. Lin R et al (2007) Transposase-derived transcription factors regulate light signaling in Arabidopsis. Science 318:1302–1305
57. Yang G, Hall TC (2003) MAK, a computational tool kit for automated MITE analysis. Nucleic Acids Res 31:3659–3665
58. Han Y, Burnette JM 3rd, Wessler SR (2009) TARGeT: a web-based pipeline for retrieving and characterizing gene and transposable element families from genomic sequences. Nucleic Acids Res 37:e78
59. Kennedy RC et al (2011) An automated homology-based approach for identifying transposable elements. BMC Bioinformatics 12:130
60. Huang X, Madan A (1999) CAP3: a DNA sequence assembly program. Genome Res 9:868–877
61. Larkin MA et al (2007) Clustal W and Clustal X version 2.0. Bioinformatics 23:2947–2948
62. Feschotte C, Swamy L, Wessler SR (2003) Genome-wide analysis of mariner-like transposable

elements in rice reveals complex relationships with stowaway miniature inverted repeat transposable elements (MITEs). Genetics 163:747–758
63. Tu Z (2001) Eight novel families of miniature inverted repeat transposable elements in the African malaria mosquito, *Anopheles gambiae*. Proc Natl Acad Sci U S A 98:1699–1704
64. Han Y, Wessler SR (2010) MITE-Hunter: a program for discovering miniature inverted-repeat transposable elements from genomic sequences. Nucleic Acids Res 38:e199
65. Santiago N et al (2002) Genome-wide analysis of the emigrant family of MITEs of *Arabidopsis thaliana*. Mol Biol Evol 19:2285–2293
66. Chen Y et al (2009) MUST: a system for identification of miniature inverted-repeat transposable elements and applications to *Anabaena variabilis* and *Haloquadratum walsbyi*. Gene 436:1–7
67. Du C et al (2008) Computational prediction and molecular confirmation of Helitron transposons in the maize genome. BMC Genomics 9:51
68. Yang L, Bennetzen JL (2009) Structure-based discovery and description of plant and animal Helitrons. Proc Natl Acad Sci U S A 106: 12832–12837
69. Thompson JD et al (1997) The CLUSTAL_X windows interface: flexible strategies for multiple sequence alignment aided by quality analysis tools. Nucleic Acids Res 25:4876–4882
70. Lisch D (2008) Mutator transposons. Trends Plant Sci 7:498–504
71. Xiao H et al (2008) A retrotransposon-mediated gene duplication underlies morphological variation of tomato fruit. Science 319: 1527–1530
72. Hanada K et al (2009) The functional role of pack-MULEs in rice inferred from purifying selection and expression profile. Plant Cell 21:25–38
73. Jiang N et al (2011) Pack-Mutator-like transposable elements (Pack-MULEs) induce directional modification of genes through biased insertion and DNA acquisition. Proc Natl Acad Sci U S A 108:1537–1542
74. Elrouby N, Bureau TE (2010) Bs1, a new chimeric gene formed by retrotransposon-mediated exon shuffling in maize. Plant Physiol 153:1413–1424
75. SanMiguel P et al (1998) The paleontology of intergene retrotransposons of maize. Nat Genet 20:43–45
76. Ma J, Devos KM, Bennetzen JL (2004) Analyses of LTR-retrotransposon structures reveal recent and rapid genomic DNA loss in rice. Genome Res 14:860–869
77. Janicki M, Rooke R, Yang G (2011) Bioinformatics and genomic analysis of transposable elements in eukaryotic genomes. Chromosome Res 19:787–808

Chapter 22

TEnest 2.0: Computational Annotation and Visualization of Nested Transposable Elements

Brent A. Kronmiller and Roger P. Wise

Abstract

Grass genomes harbor a diverse and complex content of repeated sequences. Most of these repeats occur as abundant transposable elements (TEs), which present unique challenges to sequence, assemble, and annotate genomes. Multiple copies of Long Terminal Repeat (LTR) retrotransposons can hinder sequence assembly and also cause problems with gene annotation. TEs can also contain protein-encoding genes, the ancient remnants of which can mislead gene identification software if not correctly masked. Hence, accurate assembly is crucial for gene annotation. We present TEnest v2.0. TEnest computationally annotates and chronologically displays nested transposable elements. Utilizing organism-specific TE databases as a reference for reconstructing degraded TEs to their ancestral state, annotation of repeats is accomplished by iterative sequence alignment. Subsequently, an output consisting of a graphical display of the chronological nesting structure and coordinate positions of all TE insertions is the result. Both linux command line and Web versions of the TEnest software are available at www.wiselab.org and www.plantgdb.org/tool/, respectively.

Key words Transposable elements, Retroelements, Bioinformatics, Sequence analysis, Annotation, Molecular evolution

1 Introduction

Transposable elements (TEs) contribute to a significant fraction of plant genomes, with considerable diversity in content, even between closely related species. Grass genomes are highly repetitive (e.g., *Oryza sativa* (rice) is 35 % repetitive [1], *Zea mays* (maize) is 75 % [2, 3], and *Triticum aestivum* (wheat) is approximately 80 % [4]. Classes of TEs also vary between grass species; rice contains 13 % DNA transposons and 19.3 % retrotransposons [1], while maize is made up of only 1.3 % DNA transposons and 63.3 % retrotransposons [3].

The high degree of TEs makes grass genomes difficult to sequence, assemble, and annotate [5]. Multiple copies of TEs can cause regions of the genome to collapse during sequence assembly.

Thomas Peterson (ed.), *Plant Transposable Elements: Methods and Protocols*, Methods in Molecular Biology, vol. 1057, DOI 10.1007/978-1-62703-568-2_22, © Springer Science+Business Media New York 2013

In addition, terminal repeats of retrotransposons, especially the longer copies seen in some plants can also cause mis-assemblies. In addition to hindering assembly, TEs can directly cause problems with gene annotation. Clusters of TEs also contain protein-encoding genes, and if not correctly masked they will be identified by gene finding software, further complicating annotation.

Long Terminal Repeat (LTR) retrotransposons are highly abundant in most grass genomes. An LTR retrotransposon is composed of a left flanking LTR, an internal region containing the *gag* and *pol* genes, and a right flanking LTR. When the retrotransposon integrates into the sequence, the left and right LTR are identical. The process of LTR retrotransposon insertion can be used to provide an estimate of the time since insertion in Million years ago (Mya) [6, 7]. While identical at insertion, LTRs will accumulate independent mutations over time. Summation of the mutation difference between LTR pairs will provide an estimate of the age of the TE insertion [3, 8].

With such a high density of TEs in grass genomes, more often than not, nesting of insertions occurs (i.e., a new TE integrates by inserting within a previously existing TE) [8, 9]. Nesting of TEs causes difficulties in repeat identification; the first TE insertion has been broken into two by the later TE insertion. Many levels of nesting can be seen in maize [10], their identification only limited by the degradation of ancient TE sequences. But careful reconstruction of TEs fragmented by nesting can give a detailed view of genome structure over evolutionary time.

Here we present TEnest version 2.0. Several enhancements have been added since the original version [3]. In addition to general bug fixes, enhanced processing speed for longer sequence submissions has been implemented. The BLAST phase of the alignment process has been changed from WU-BLAST [11] to NCBI BLAST [12]. The TE sequence database has been revamped to make construction of a novel database more user friendly. A secondary LTR pairing process has been included to prevent incorrect left and right joins and size restriction options have also been included (*see* Subheading 2.4). LTR reconstruction has been revamped to prevent memory problems that could occur when long input sequences have a large number of the same TE. A new long sequence submission script is available, split_TEnest.pl, replacing the original cluster_TEnest.pl to further speed up analysis. Finally, the TEnest display legend has been updated to include TE classifications.

2 Materials

2.1 Overview

TEnest annotates TE insertions in a given input sequence. A repeat database is used to identify TEs and to provide the reference for reconstructing degraded TEs to their ancestral state. An overview

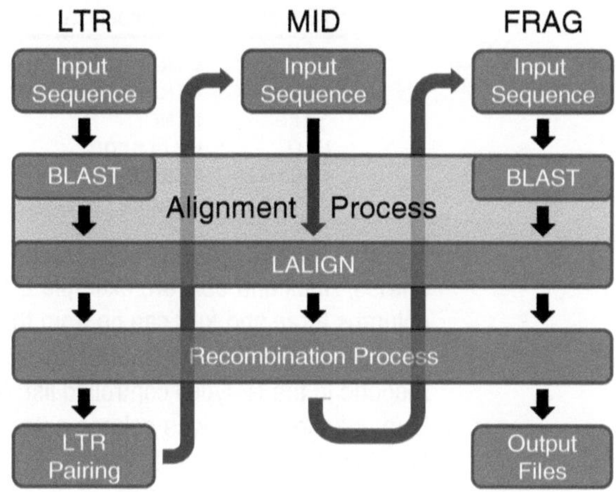

Fig. 1 General overview of TEnest processing. The input sequence cycles through three major phases: LTR, where LTRs are identified, reconstructed, and paired; MID, where LTR retrotransposon internal region are aligned and reconstructed; and FRAG, where regions that have not been previously identified as LTR retrotransposon are annotated as either fragmented TEs or non-LTR containing transposons

of the TEnest algorithm is shown in Fig. 1. The nesting structure of TEs is determined by analysis of TE insertion locations; age since insertion of LTR retrotransposons is calculated from LTR divergence. A graphical display of the annotated TEs is provided to visualize the chronological nesting structure. Both a Web version and a downloadable linux command line version of TEnest are available; in the following descriptions the command line version is discussed. Many, but not all of the explained options are also available on the Web version of TEnest.

2.2 Transposable Element Sequence Databases

TEnest utilizes organism-specific TE databases to annotate repeats by sequence alignment. Several preconstructed TE databases are supplied with TEnest and also available on the online PlantGDB [13] version of TEnest. Preconstructed TE databases include maize, rice, wheat, and *Hordeum vulgare* (barley). A TEnest database consists of two files per organism; full-length sequences for all TEs in multi-FASTA format [14], and a four column table (*see* Fig. 2) with an entry for each TE in the multi-FASTA file. The TEnest database sequence and table files are named first by organism and appended with "_TEnest.fasta" and "_TEnest.table", respectively. Users can switch between organism TE databases using the option "*--org organism*".

The TEnest database table has four tab-delimited columns (*see* Fig. 2). The first is the TE name, and corresponds to the header of the full-length TE sequence with the right chevron removed and trimmed at the first white space. Therefore, ">opie retrotransposon"

TE type	TE Classification	LTR start	LTR stop
huck	LTR_retrotransposon	12159	13896
opie	LTR_retrotransposon	1	1251
jare	LINE		
Hip1	Helitron		
CACTA16	TIR		
mutator	Ignore		

Fig. 2 A sample tab-delimited TEnest database table file with six hypothetical entries. *Huck* and *opie* are example LTR retrotransposons, their LTR positions in columns three and four can be from the left or right LTR. Non-LTR TEs *jare*, *Hip1*, *Helitron*, *CACTA16*, and *mutator* are also shown. The second column must correspond to the TE types controlled list. An easy way to omit a TE from an analysis is to change the TE type column entry to a non-approved value; the *mutator* TE type is entered here as "ignore"

becomes "opie". The second column is the classification of this TE. This is a one-word designator from a controlled list, and as such, only the following classification terms can be used: LTR_retrotransposon, Non-LTR_retrotransposon, DNA_transposon. However, this controlled list can be changed by using the options *--te_controlled_list_ltr* for LTR retrotransposons, *--te_controlled_list_other* for non-LTR retrotransposons and DNA transposons, and entering the lists of the altered classification names for each of the options in double quotes. The last two columns are reserved only for LTR retrotransposons. An LTR is defined here by the start and stop coordinates from the full-length TE. Either the right- or left-most LTR coordinate positions can be used.

Construction of a custom TE database is also possible. Users can add to one of the existing preconstructed databases, remove TEs from one of the available databases, or construct a fully novel TE database. To add TEs to an existing database, copy the original database files to a newly named file, "maize_TEnest.fasta" becomes "maize_custom_TEnest.fasta", and likewise for the table file. New full-length TE sequences can be appended to the new fasta file and new entries are added to the table file. If the new entry is an LTR retrotransposon, be sure to enter coordinates for the LTR location. Finally, point TEnest to the new sequence database by using the option *--org maize_custom*. The process is much easier to remove a TE from the sequence database. Simply changing the classification field in the sequence database table to a non-approved value will cause TEnest to exclude the TE from analysis. To construct a new TEnest sequence database, follow the above descriptions to create both the full-length multi-FASTA TE sequence file and the associated table, making sure the TE classifications agree with the controlled TE list.

2.3 Installing TEnest

TEnest comes packaged with the TEnest.pl perl program, the graphical display program of TEnest, svg_ltr.pl, the TEnest sequence databases, and a long-sequence submission script split_TEnest.pl explained in the special analysis section (*see* Subheading 2.4). TEnest utilizes two third-party software programs that must be obtained from their respective sources; NCBI BLAST [12] for the quick identification of potential TE alignment regions, and LALIGN [15] (available at http://faculty.virginia.edu/wrpearson/fasta/fasta2/) from the FASTA2 software suite [16] for local alignments of TEs to the reference sequence (*see* **Note 1**).

Once BLAST and LALIGN are installed, download the zipped TEnest.tar.gz file and the TE database files. Move them to the installation directory and unzip with the command "*tar –xzf *.tar.gz*". Several parameters in the TEnest program must be updated to integrate each of the BLAST and LALIGN program capabilities. Open the TEnest.pl file for editing and change each of the following location variables to point to the correct files or directories:

Database location: Point this to the directory where your TEnest database files are found. For example, if the maize_TEnest.fasta and maize_TEnest.table are found.

```
$db_directory='/home/user/TEnest/DB';
```

Output location: The default location for TEnest output is your current working directory. However, you can alter this to always write to the same location.

```
$output_directory='/home/user/TEnest';
```

NCBI-BLAST Executables: The path needed to run BLAST. If blastall is in your path this can be left blank.

```
$blast_directory='/usr/local/bin';
```

FASTA LALIGN Executable: The path needed to run LALIGN. If LALIGN is in your path this can be left blank.

```
$lal_directory='/usr/local/bin/FASTA2';
```

Once complete, save and close the TEnest.pl program. Make the file executable with the command "*chmod+x TEnest.pl*". Add the TEnest directory to your executable path.

2.4 Running TEnest

The general command for a TEnest analysis is "*TEnest.pl my_input.fasta [options]*". Many options are available for fine tuning and customizing TEnest runs. Each of these options can be entered via the command line; they are explained below by section corresponding to the different processes of TEnest.

2.5 LTR Identification

In the first stage of TEnest, LTR sequences are extracted from the TE database and aligned to the input sequence. A two-phase alignment process designed to provide quick but accurate alignments over long sequence regions is accomplished using BLAST coupled with a pairwise alignment. Nucleotide BLAST gives a TEnest a very general overview of where TEs may reside on the input sequence. Due to their sequence similarity multiple TEs may be identified at the same location on the input sequence by the BLAST alignment. Each of the regions identified by BLAST are then put into a pairwise local alignment process using LALIGN. Here a decision is made to identify the best-matching TE also finding the exact start and stop positions of the match.

Once alignments are complete, a set of fragmented LTR annotations have been created. Progressing through each TE, the fragments are joined to recreate full LTR sequences. There are multiple reasons LTRs might be fragmented: there may be differences between the input sequence version and the database version; nested insertions of TEs may have caused a sequence split and distance separation within the LTR; or sequence gaps or misassemblies have created an artificial split. In any case, recombination attempts to identify LTR fragments and create full-length groups. All sequential LTR fragment combinations are created and their LTR-based coordinates are considered for recombination possibilities.

Once full-length LTR sequences have been recombined, TEnest enters the LTR pairing phase where the left and right LTRs of a retrotransposon are identified. Due to the insertion process of LTR retrotransposons, the left and right LTR are sequence twins at the time of insertion. Over evolutionary time the two LTRs will accumulate mutations independently [17]. TEnest uses these mutations to identify a paired set of LTRs. The left and right LTRs of a retrotransposon will have more similar sequence to each other than to other LTRs of the same TE type. The base substitution rate (BSR) between all LTRs is calculated via LTR alignments, the LTR with most similar sequences are paired together. BSR is determined by the amount of single mutations between the two LTRs divided by the LTR alignment length. A maximum distance between left and right LTRs can be used to prevent unlikely long range TE insertions. Also, a filtering process for LTR pairing can be used to speed up the process and prevent incorrect combinations. When the option "*--ltr_find_LR*" is used, TEnest will attempt to designate each LTR as either a left or right copy prior to the pairing process by identifying a sequence junction between the LTR and internal region. Not all LTRs will be able to be classified as left or right if the junction sequence is not present. Left, right and unknown designated LTRs are then sent to the LTR pairing process where left-left, right-right and right-left (accounting for reverse-complementation) combinations are excluded. Finally, age since insertion in Mya is calculated: the BSR is divided by two times the substitution rate in repetitive regions in grasses (1.3×10^{-8}) [6, 7]. LTRs that are not identified as paired will be classified as solo-LTRs.

1. Alignment options [defaults in brackets].
   (a) --ltr_blast_cutoff Expectation value of the LTR BLASTN [1e-01].
   (b) --ltr_gap_open LTR alignment gap open (LALIGN –f parameter) [30].
   (c) --ltr_gap_ext LTR alignment gap extension (LALIGN –g parameter) [15].
   (d) --ltr_gap_rep LTR alignments to report within sub-region (LALIGN –k parameter) [7].
   (e) --ltr_lal_score Parsed expectation value of the LTR local alignment, where 20 equals 1e-20 [20].
2. LTR reconstruction options.
   (a) --ltr_smallest Shortest length in bp of LTR fragment allowed to attempt LTR reconstruction. Very small fragments may be non-TE hits and will exponentially inflate the powerset function [25].
   (b) --ltr_overlap Length in bp of LTR-based coordinates to allow when combining LTR fragments in the LTR alignment phase. A size of 0 will allow no overlap [25].
   (c) --ltr_full_shortest Shortest length in bp of LTRs after reconstruction. Any LTRs less than this value will be ignored [50].
   (d) --ltr_pwr_max Longest length in bp that the gap between LTR fragments can span. If the length between two fragments is greater the sections cannot be joined [100000].
   (e) --ltr_bad_set LTRs are reconstructed using their LTR-based coordinates to identify the entire LTR sequence. Two regions with alike positions cannot belong to the same LTR insertion and therefore do not need to attempt reconstruction. This parameter allows wobble between calling two LTR fragments alike, for example a value of 5 will identify LTRs with coordinates 100–200 and 102–197 as the same position [5].
3. LTR pairing options.
   (a) --ltr_find_LR Invoke the process to identify left and right LTRs prior to LTR pairing. T/F [T].
   (b) --ltr_bsr_length Greatest distance in bp to allow between paired left and right LTRs. A value of 0 will not invoke distance restriction [250000].
   (c) --bsr_gap_open Paired LTR alignment gap open (LALIGN –f parameter) [12].
   (d) --bsr_gap_ext Paired LTR alignment gap extension (LALIGN –g parameter) [4].
   (e) --bsr_max Max BSR value to allow between LTR pairs [0.6].

2.6 Identification of Internal Regions of LTR Retrotransposons

TEnest attempts to identify the internal regions for LTRs paired in the previous processes. Starting with the smallest distanced set first, TEnest cycles through all paired LTRs and aligns the internal sequence to the expected TE. Matching sequence again goes through the recombination process described above to create full-length joined internal regions. The annotated LTRs and internal region sequences are masked (changed to "N" but retaining the sequence length), and the next longest spanning LTR set is analyzed. Masking the sequence prior to the next alignment run prevents a nested TE from matching to multiple LTR pairs, especially a problem when the same TE types are nested within one another.

1. Local alignment options [defaults in brackets].
   (a) --mid_gap_open MID alignment gap open (LALIGN –f parameter) [75].
   (b) --mid_gap_ext MID alignment gap extension (LALIGN –g parameter) [75].
   (c) --mid_gap_rep MID alignments to report within sub-region (LALIGN –k parameter) [5].
   (d) --mid_lal_score Parsed expectation value of the MID local alignment, where 20 equals 1e-20 [20].
2. Internal region reconstruction options.
   (a) --mid_smallest Shortest length in bp of fragment allowed to enter reconstruction. Very small fragments may be non-TE hits and will exponentially inflate the powerset function [25].
   (b) --mid_overlap Length in bp of TE-based coordinates to allow when combining MID fragments in the alignment phase. A size of 0 will allow no overlap [25].
   (c) --mid_bad_set Basepair wobble range to consider MID fragments as identical and prevent both segments from entering powerset reconstruction (*see* --ltr_bad_set for more information) [5].

2.7 Fragmented Retrotransposons and Non-LTR Transposable Elements

All un-annotated regions make one final pass through TE identification. In this round the goal is to identify both fragmented LTR retrotransposons that did not have a paired set of LTRs identified and non-LTR containing TEs. The full set of TEs from the sequence database; LTR retrotransposons and other Class I retrotransposons such as LINEs and SINEs, along with all Class II TEs go through the previously described two phase alignment process. Again, split annotated TE regions are joined through the recombination process as previously described. After reconstruction, TEs are classified as fragmented. Their full annotation length is compared to their expected full TE length; if 80 % of the TE was identified the annotation is promoted to a non-LTR full-length classification.

Finally, a round of conflict checks takes place. All combinations of joined or paired TE sections are checked against all other joined groups to identify any disagreeing recombined sets. TEnest requires all TE insertions to be annotated in such a way as to show the chronological order of TE nesting. TEnest will not group TE sections if a later recombination has rearranged the sequence order. If a TE has nested within an older TE insertion, the full annotated sequence of the newer TE must be found within the older TE. If sequence order conflicts are seen, the TE with less sequence identity to the TE database entry is split into non-conflicting fragmented sections.

1. Local alignment options [defaults in brackets].
   (a) --fn_gap_open Fragment alignment gap open (LALIGN –f parameter) [75].
   (b) --fn_gap_ext Fragment alignment gap extension (LALIGN –g parameter) [75].
   (c) --fn_gap_rep Fragment alignments to report within subregion (LALIGN –k parameter) [5].
   (d) --fn_lal_score Parsed expectation value of the local alignment, where 20 equals 1e-20 [20].
2. Internal region reconstruction options.
   (a) --fn_smallest Shortest length in bp of fragment allowed to enter reconstruction. Very small fragments may be non-TE hits and will exponentially inflate the powerset function [25].
   (b) --fn_overlap Length in bp of TE-based coordinates to allow when combining fragments in the alignment phase. A size of 0 will allow no overlap [25].
   (c) --fn_bad_set Basepair wobble range to consider fragments as identical and prevent both segments from entering powerset reconstruction (*see* --ltr_bad_set for more information) [5].

2.8 TEnest Output Files

Upon completion TEnest.pl provides three files. The first is a repeat-masked FASTA format sequence file similar to the output of RepeatMasker [18]. The second is a generic feature format version 3 (GFF3) output of the nested TE annotations for use in genome viewers such as GBrowse [19]. The third is a coordinate file of the annotated TEs identified in the input sequence (*see* **Note 2**). This file is named with "LTR" as the file extension, and is used as the input for the TEnest graphical display program. As illustrated in Fig. 3, four sections in the LTR output file correspond to the following annotation classifications: (a) solo-LTRs identified as LTRs without pairs (SOLO), (b) Full LTR retrotransposon composed of paired LTRs and internal regions (PAIR), (c) Fragmented

Input length	194000
Input file	my_input.fasta
Solo 's0' header	SOLO s0 gyma R
Solo 's0' positions	s0 178958 182990 1 4198
Pair 'p0' header	PAIR p0 huck F 0.002
Pair 'p0' left LTR	p0 L 19850 21588 1 1713
Pair 'p0' right LTR	p0 R 32062 33743 1 1713
Pair 'p0' internal	p0 M 21589 28316 1606 9212 28423 28520 8366 8463 30736 32062 9689 11015
Fragment 'f0' header	FRAG f0 puck 0
Fragment 'f0' positions	f0 107339 108661 10462 11805
Fragment 'f1' header	FRAG f1 ruda 1
Fragment 'f1' positions	f1 61578 64119 182 2832
non-LTR 'n0' header	NLTR n0 cacta 0
non-LTR 'n0' positions	n0 75101 86111 1 11010

Fig. 3 A sample TEnest output annotation file. On the *left* a legend shows descriptions for each line. The *first two lines* show the input sequence length and the input sequence name. Sample SOLO, PAIR, FRAG, and NLTR annotation classifications are shown as header and coordinate position lines. Annotation coordinates are shown in sets of four numbers that correspond to input sequence based start and stop positions and TE based start and stop positions. Reconstructed fragmented TEs are shown on the same line as a concatenated set of four coordinates. See *line 8* pair "p0" internal, the *huck* PAIR LTR retrotransposon middle region, for an example of reconstructed coordinates

LTR retrotransposons (FRAG), and (d) full length non-LTR TEs (NLTR). The SOLO, FRAG, and NLTR entries each have a header line that contains the classification, a unique TE identifier, the TE name, and the direction of the TE relative to the TE database reference sequence. The following line contains the coordinates of the TE annotation, first the start and stop positions in the input sequence, secondly the start and stop positions in the reference TE sequence. The PAIR entries are similar to the other sections with two differences. The PAIR header line has a BSR value following the direction entry, and there are three coordinate lines corresponding to the left (L), right (R), and internal regions (M) (*see* Fig. 3). Output files will be written to an output directory named in the format "TEnest_date-time_my_input.fasta".

2.9 TEnest Graphical Display

Due to the nested structure of TE insertions, even with TEnest annotations the arrangement of the repetitive sequence can be difficult to understand. Therefore, TEnest provides an auxiliary graphical display program to visualize the coordinate annotation file. As shown in Fig. 4, svg_ltr.pl reads the TEnest output file "my_input. LTR" and provides a vector graphic based picture of nested TEs in the input sequence.

The svg_ltr.pl picture displays the original input sequence as a black line spanning the length of the display; however, the annotated TEs have been removed from the sequence length. TE insertions are shown as triangles, their flat top represents their sequence length, and the bottom point represents the point of TE insertion. Therefore, a summation of all horizontal lines including the black input sequence and all triangle tops would equal the length of the

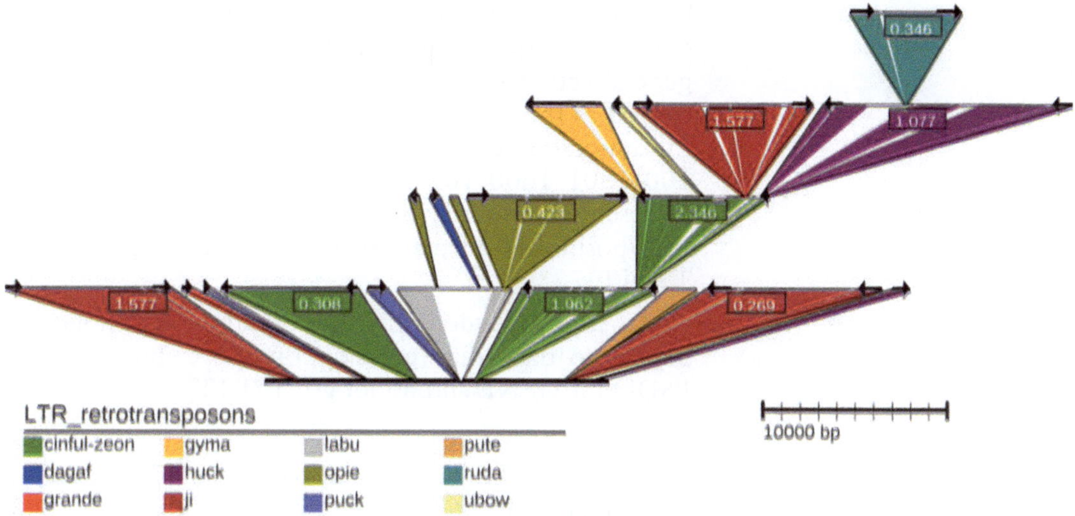

Fig. 4 Graphical output of TEnest; positions 21,811,324–21,924,324 of the maize AR182 contig on chromosome 4 of inbred B73. Annotated TE insertions are displayed as *triangles*. Older insertions are seen as integrated directly into the *black* DNA line while more recent insertions are *nested* within other TEs. Age since insertion is seen as a number in Mya within each LTR retrotransposon triangle and LTR are shown as *arrows*. White regions within TE triangles represent non-TE annotated sequence within a reconstructed TE. This sequence could non-TE or non-identified TE sequence that has insertion within the annotated TE, or TE sequence that has since degraded beyond alignment recognition. A legend at the bottom of the display matches TE types to triangle colors and gives a scale bar in basepairs

input sequence. LTR retrotransposons triangles have the LTRs shown with arrows at the triangle top and have age since insertion shown in an internal box. A legend at the bottom of the picture relates background triangle color to each displayed TE name.

The basic command to run svg_ltr.pl is "*svg_ltr.pl my_input.LTR [options]*". This will produce the scalable vector graphics (SVG) format picture my_input.svg. Many options are available for manipulating the TEnest graphic including subset views, displaying specific annotation types, and adding third-party annotations to the picture.

1. Mapping options [defaults in brackets].
   (a) --map_pair Include pair classified LTR retrotransposons. T/F [T].
   (b) --map_solo Include solo classified LTRs. T/F [T].
   (c) --map_nltr Include non-LTR classified TEs. T/F [T].
   (d) --map_frag Include fragmented TE annotations. T/F [T].
   (e) --map_gene Include third-party gene annotations. Gene annotations can be appended to the TEnest output file prior to running svg_ltr.pl. Similar to the description of the TEnest output file above, each gene entry has a first

line in the form "GENE g0 F the gene name", where "g0" is a unique count identifier for this gene, "F" is the gene direction and any additional entries in the line will be treated as the gene name. The second gene line contains the annotation coordinates in the form "g0 1001 1100 1 100 1501 1600 101 200", groups of four coordinates representing exons with input sequence based start and stop positions and gene based start and stop positions. T/F [T].

   (f) --map_psdo Include third-party pseudogene annotations. Similar to the --map_gene option above, but substitute PSDO for GENE and u0 for g0. T/F [T].

2. Coordinate options.
   (a) --print_coords Print coordinates on displayed annotations. Coordinates can be input sequence or TE based. A "F" entry overrides the individual coordinate options below, a "SEQ" or "TE" entry then checks the individual coordinate option. F/SEQ/TE [F].
   (b) --pair_coords Display pair classified coordinates. T/F [T].
   (c) --solo_coords Display solo classified LTRs. T/F [T].
   (d) --nltr_coords Display non-LTR classified TEs. T/F [T].
   (e) --frag_coords Display fragmented TE coordinates. T/F [T].
   (f) --gene_coords Display third-party gene coordinates. T/F [T].
   (g) --psdo_coords Display third-party pseudogene coordinates. T/F [T].

3. Display options.
   (a) --white_out Draw white triangles within TE triangles for un-identified regions. T/F [T].
   (b) --age Display MYA or BSR inside LTR retrotransposons. mya/bsr [mya].
   (c) --start Trim the display to this starting position. Any annotations prior to this position will be ignored. [].
   (d) --end Trim the display to this end position. Any annotations found after this position will be ignored. [].
   (e) --split If –start or –end are used the coordinates may sever a TE annotation. T will trim the TE annotation and display the viewable fragment. F will not display any portion of the TE. T/F [T].

SVG format pictures can be displayed directly in Mozilla Firefox version 2 or later (http://www.mozilla.com/firefox). The linux program rsvg (http://librsvg.sourceforge.net) is a convenient way to convert SVG files to png or jpeg format pictures. The command "*rsvg –h 1000 my_input.svg my_input.png*" will produce a publication quality image.

2.10 Additional TEnest Features

TEnest was built with multiprocessor support to facilitate high-throughput analysis. Several sections of the TEnest algorithm take advantage of perl multiprocessing by analyzing multiple instances of a routine at the same time. The LTR alignments, LTR reconstruction and LTR pair identification are each run on multiple processors per TE type. The internal region and TE fragment identification are run on multiple processors per sequence subsection. To invoke multiprocessing, change the *--processors* option from its default value of one to an appropriate value.

To further address annotation of long sequences, a wrapper script split_TEnest.pl is included with TEnest (*see* Subheading 3.2). Split_TEnest.pl has one additional parameter, *--node_length*, and splits the input into sequences of this size. Each split sequence is run individually in TEnest while only annotating the paired LTR retrotransposons. split_TEnest.pl can also utilize multiprocessing to analyze individual split input sequences simultaneously, using the option *--processors*. split_TEnest.pl will run concurrent sequence sub-sections equal to the *--processors* option, each of these TEnest runs will themselves be run on a single CPU. Once each split sequence TE annotation has been returned from TEnest, the masked sequence and TE annotations are joined and a final TEnest analysis is submitted to analyze solo LTRs, non-LTR TEs, fragmented TEs and any LTR retrotransposons not previously identified due to the sequence splits. This analysis will utilize multiple processors in the usual TEnest method.

3 Methods

3.1 Maize AR182

The maize accelerated region 182 (AR182) is a high quality assembly of a 22-Mb region on chromosome 4 of inbred B73 [20]. This assembly is not yet finished; the maizesequence.org (http://www.maizesequence.org) version of the contig contains 1,167 gaps, but it is nonetheless a good example to demonstrate the long-sequence capabilities of TEnest.

3.2 Maize TE Sequence Database

With the completion of the maize genome sequence [21], a comprehensive analysis identified maize genome TEs including more than 300 previously unidentified repeats [2]. This data set was configured into a TEnest sequence database following the descriptions outlined above. The resulting TEnest database contains 382 LTR retrotransposons, 35 non-LTR retrotransposons, and 897 DNA transposons. This version of the Baucom and colleagues TE set [2] is available with TEnest.

3.3 TEnest Run

The long-sequence TEnest submission script, split_TEnest.pl, was used to analyze the full 22Mb AR182 contig. The script was set to split the sequence into 223,100-kb sections (*--node_length 100000*), each was individually sent to TEnest. Using ten processors (*--processors 10*), the full TEnest analysis took approximately 12 h.

3.4 TEnest Results

TEnest identified 1,054 LTR retrotransposons with paired LTRs. Eight-hundred and sixty one of these were full-length after reconstruction from ancestral nested states. The 1,054 LTR retrotransposons identified across 22 Mb gives an average rate of 48 insertions per Mb. This rate is somewhat less than 58 LTR retrotransposons per Mb calculated by TEnest in the finished 1.5-Mb sequence from the *rf1*-associated region near the centromere of chromosome 3 [10]. While TE density differences will be observed, much of this drop in identification rate can also be explained by loss of TE sequences in the gapped regions of AR182.

The time since insertion of the LTR retrotransposons ranged from 0.04 to 7.70 Mya. TEnest identified 557 solo LTRs. This number is likely inflated and includes many LTRs that were unable to pair correctly due to their pair occurring in one of the sequence gaps. Approximately 64 non-LTR transposons including DNA transposons also were identified. Figure 4 illustrates a subset of the AR182 graphical output (the TEnest display of 113-Kb region from 21,811,324- to 21,924,324-bp)

In large grass genomes, TEs often occur as multi-tiered clusters with non-TE associated genes assembled in islands [10]. Hence, the graphical display (and its associated coordinate annotation file) is often useful to focus on "gene" regions for positional-based cloning projects, or identification of primer sets as markers for phenotypes that are influenced by incomplete penetrance or epistasis [22]. A full view of the maize 22-Mb AR182 contig can be seen at www.wiselab.org under the software and TEnest examples link.

4 Notes

1. TEnest is available in two forms. The linux command-line version is available from www.wiselab.org under the software section. The online interactive version is available at http://www.plantgdb.org/tool/ [13]. The Web version has many but not all configurable options found in the command-line version.
2. The highest quality output of a TEnest analysis will be from a finished sequence region. Gaps in the input sequence may be indications of missing TE sequence. If one LTR from a LTR retrotransposon is not found the LTR retrotransposon will be classified as fragmented instead of a full-length paired retrotransposon and age since insertion cannot be determined.

That said, TEnest can help order, orient, and visualize the repeat structure surrounding gaps for help with sequence assembly [10].

Acknowledgments

The authors thank Regina Baucom, James Estill, and Jeff Bennetzen for access to uncharacterized maize LTR retrotransposons prior to publication. This research was supported in part by USDA-National Research Initiative (NRI) grant no. 2002-35301-12064. Mention of trade names or commercial products in this publication is solely for the purpose of providing specific information and does not imply recommendation or endorsement by the US Department of Agriculture.

References

1. IRGSP (2005) The map-based sequence of the rice genome. Nature 436:793–800
2. Baucom RS et al (2009) Exceptional diversity, non-random distribution, and rapid evolution of retroelements in the B73 maize genome. PLoS Genet 5:e1000732
3. Kronmiller BA, Wise RP (2008) TEnest: automated chronological annotation and visualization of nested plant transposable elements. Plant Physiol 146:45–59
4. Sabot F et al (2005) Updating of transposable element annotations from large wheat genomic sequences reveals diverse activities and gene associations. Mol Genet Genomics 274:119–130
5. Rabinowicz PD, Bennetzen JL (2006) The maize genome as a model for efficient sequence analysis of large plant genomes. Curr Opin Plant Biol 9:149–156
6. Kimura M (1980) A simple method for estimating evolutionary rates of base substitutions through comparative studies of nucleotide sequences. J Mol Evol 16:111–120
7. Ma J, Bennetzen JL (2004) Rapid recent growth and divergence of rice nuclear genomes. Proc Natl Acad Sci USA 101:12404–12410
8. SanMiguel P et al (1998) The paleontology of intergene retrotransposons of maize. Nat Genet 20:43–45
9. SanMiguel P et al (1996) Nested retrotransposons in the intergenic regions of the maize genome. Science 274:765–768
10. Kronmiller BA, Wise RP (2009) Computational finishing of large sequence contigs reveals interspersed nested repeats and gene islands in the *rf1*-associated region of maize. Plant Physiol 151:483–495
11. Gish W (1996–2004) http://blast.wustl.edu
12. Altschul SF et al (1997) Gapped BLAST and PSI-BLAST: a new generation of protein database search programs. Nucleic Acids Res 25:3389–3402
13. Duvick J et al (2008) PlantGDB: a resource for comparative plant genomics. Nucleic Acids Res 36:D959–D965
14. Lipman DJ, Pearson WR (1985) Rapid and sensitive protein similarity searches. Science 227:1435–1441
15. Huang X, Miller W (1991) A time-efficient, linear-space local similarity algorithm. Adv Appl Math 12:337–357
16. Pearson WR, Lipman DJ (1988) Improved tools for biological sequence comparison. Proc Natl Acad Sci USA 85:2444–2448
17. Boeke JD, Corces VG (1989) Transcription and reverse transcription of retrotransposons. Annu Rev Microbiol 43:403–434
18. Smit AFA, Hubley R, Green P (1996–2004) RepeatMasker Open-3.0. http://www.repeatmasker.org
19. Stein LD et al (2002) The generic genome browser: a building block for a model organism system database. Genome Res 12:1599–1610
20. Wei F et al (2009) Detailed analysis of a contiguous 22-Mb region of the maize genome. PLoS Genet 5:e1000728
21. Schnable PS et al (2009) The B73 maize genome: complexity, diversity, and dynamics. Science 326:1112–1115
22. Meyer J, Pei D, Wise RP (2011) *Rf8*-Mediated T-*urf13* transcript accumulation coincides with a pentatricopeptide repeat cluster on maize chromosome 2L. Plant Genome 4(3):283–299

Index

Thomas Peterson (ed.), *Plant Transposable Elements: Methods and Protocols*, Methods in Molecular Biology, vol. 1057, DOI 10.1007/978-1-62703-568-2,

N

P

Q

R

S

T

U

V

W

Y

Z